Lecture Notes in Computer Science 16695

The series Lecture Notes in Computer Science (LNCS), including its subseries Lecture Notes in Artificial Intelligence (LNAI) and Lecture Notes in Bioinformatics (LNBI), has established itself as a medium for the publication of new developments in computer science and information technology research, teaching, and education.

LNCS enjoys close cooperation with the computer science R & D community, the series counts many renowned academics among its volume editors and paper authors, and collaborates with prestigious societies. Its mission is to serve this international community by providing an invaluable service, mainly focused on the publication of conference and workshop proceedings and postproceedings. LNCS commenced publication in 1973.

Taylor J. Smith
Editor

Implementation and Application of Automata

30th International Conference, CIAA 2026
Kingston, ON, Canada, August 5–8, 2026
Proceedings

Springer

Editor
Taylor J. Smith
St. Francis Xavier University
Antigonish, NS, Canada

ISSN 0302-9743 ISSN 1611-3349 (electronic)
Lecture Notes in Computer Science
ISBN 978-3-032-31176-4 ISBN 978-3-032-35967-4 (eBook)
https://doi.org/10.1007/978-3-032-31176-4

This Springer imprint is published by the registered company Springer Nature Switzerland AG
The registered company address is: Gewerbestrasse 11, 6330 Cham, Switzerland

Preface

The 30th International Conference on Implementation and Application of Automata (CIAA 2026) was held in Kingston, Canada on August 5–8, 2026, and was organized by the School of Computing at Queen's University.

CIAA is an annual international conference concerning research on all aspects of implementation and application of automata and related structures, including theoretical aspects. Topics of interest include but are not limited to: complexity of automata operations, compilers, computer-aided verification, concurrency, data structure design for automata, data and image compression, design and architecture of automata software, digital libraries, DNA/molecular/membrane computing, document engineering, editors and environments, experimental studies and practical experiences, industrial applications, natural language processing, networking, new algorithms for manipulating automata, object-oriented modeling, pattern-matching, quantum computing, speech and speaker recognition, structured and semi-structured documents, symbolic manipulation environments for automata, teaching, text processing, techniques for graphical display of automata, VLSI, and viruses and related phenomena.

CIAA originated in the mid-1990s as the International Workshop on Implementing Automata (WIA). This workshop was held in the following locations:

London, Canada (1996)
London, Canada (1997)
Rouen, France (1998)
Potsdam, Germany (1999)

In 2000, WIA became CIAA, and past editions of the conference were held in the following locations:

London, Canada (2000)
Pretoria, South Africa (2001)
Tours, France (2002)
Santa Barbara, USA (2003)
Kingston, Canada (2004)
Nice, France (2005)
Taipei, Taiwan (2006)
Prague, Czech Republic (2007)
San Francisco, USA (2008)
Sydney, Australia (2009)
Winnipeg, Canada (2010)
Blois, France (2011)
Porto, Portugal (2012)
Halifax, Canada (2013)
Gießen, Germany (2014)
Umeå, Sweden (2015)

Seoul, South Korea (2016)
Marne-la-Vallée, France (2017)
Charlottetown, Canada (2018)
Košice, Slovakia (2019)
Loughborough, UK (2020)
Bremen, Germany (2021)
Rouen, France (2022)
Famagusta, Cyprus (2023)
Akita, Japan (2024)
Palermo, Italy (2025)

The planned 2020 edition of CIAA was cancelled due to the pandemic. The in-person arrangements for the 2021 edition of CIAA were also cancelled due to the pandemic, though the conference went ahead as a virtual meeting.

This volume of *Lecture Notes in Computer Science* contains both the invited contributions and the accepted papers presented at CIAA 2026. We offer a special thanks to each of the invited speakers for accepting our invitation and for presenting their research:

- Anthony Bonato, Toronto Metropolitan University, Canada
- Nataša Jonoska, University of South Florida, USA
- Ian Mertz, Charles University, Czech Republic

34 papers were submitted to CIAA 2026 by a total of 77 authors from 15 countries: Belgium, Canada, China, Estonia, Finland, Germany, India, Japan, Portugal, Russia, Slovakia, South Africa, South Korea, the United Kingdom, and the United States. From these submissions, the program committee accepted 14 papers on the basis of three single-blind referee reports each. Additionally, 4 papers were selected for presentation in a special tool track session. We thank each of the members of the program committee for their excellent work in producing the program, and we thank the additional reviewers for their helpful contributions.

We also thank Salimur Choudhury and the members of the local organizing committee for all of their efforts to prepare the scientific sessions at Queen's University and the social events in Kingston.

Additionally, we thank the editorial staff at Springer for their collaboration and assistance in producing this volume in time for the conference, and we thank EasyChair for the use of their conference management system.

Finally, we thank the participants for attending CIAA 2026 and for making it another successful meeting of our research community. We hope that you found this year's conference as stimulating and as inspiring as previous years' conferences, and we look forward to seeing you again at CIAA 2027!

August 2026 Taylor J. Smith

Remembering Hsu-Chun Yen. The formal languages and automata theory research community lost our colleague and friend, Hsu-Chun Yen, in 2026. Hsu-Chun was a distinguished professor in the Department of Electrical Engineering at National Taiwan University. Among many other remarkable career contributions, Hsu-Chun was deeply involved with CIAA, serving on the program committees for each edition of the conference through 2003–2007 and 2009–2026, serving as the program committee co-chair for the 2006 edition in Taipei, delivering an invited talk at the 2008 edition in San Francisco, and sitting as a member of the CIAA steering committee since 2013. Hsu-Chun will be dearly missed.

Organization

Steering Committee

Markus Holzer (Chair)	Justus-Liebig-Universität Gießen, Germany
Oscar Ibarra	University of California, Santa Barbara, USA
Sylvain Lombardy	Université de Bordeaux, France
Nelma Moreira	Universidade do Porto, Portugal
Kai Salomaa (Co-chair)	Queen's University, Canada
Hsu-Chun Yen (Deceased)	National Taiwan University, Taiwan

Organizing Committee Chair

Salimur Choudhury	Queen's University, Canada

Program Committee Chair

Taylor J. Smith	St. Francis Xavier University, Canada

Program Committee

Frédérique Bassino	Université Sorbonne Paris Nord, France
Cezar Câmpeanu	University of Prince Edward Island, Canada
Pascal Caron	Université de Rouen Normandie, France
Giuseppa Castiglione	Università di Palermo, Italy
Szilárd Fazekas	Akita University, Japan
Yo-Sub Han	Yonsei University, South Korea
Markus Holzer	Justus-Liebig-Universität Gießen, Germany
Galina Jirásková	Slovak Academy of Sciences, Slovakia
Jarkko Kari	Turun yliopisto, Finland
Stavros Konstantinidis	Saint Mary's University, Canada
Martin Kutrib	Justus-Liebig-Universität Gießen, Germany
Andreas Malcher	Justus-Liebig-Universität Gießen, Germany
Florin Manea	Georg-August-Universität Göttingen, Germany
Sabrina Mantaci	Università di Palermo, Italy
Brink van der Merwe	Stellenbosch University, South Africa

Nelma Moreira	Universidade do Porto, Portugal
Cyril Nicaud	Université Gustave Eiffel, France
Giovanni Pighizzini	Università di Milano, Italy
Luca Prigioniero	Loughborough University, UK
Rogério Reis	Universidade do Porto, Portugal
Kai Salomaa	Queen's University, Canada
Shinnosuke Seki	University of Electro-Communications, Japan
Mikhail Volkov	Ural Federal University, Russia
Hsu-Chun Yen (Deceased)	National Taiwan University, Taiwan

Additional Reviewers

Duncan Adamson
Guilherme Duarte
Michal Hospodár
Annika Huch
Daihei Ise
Jarkko Kari
Tomas Masopust
Carlo Mereghetti
Priscilla Raucci
Christian Rauch
Pablo Rotondo
Aleksi Saarela
Wolfgang Steiner
Marek Szykuła
Gonçalo Teixeira
Stéphane Vialette
Matthias Wendlandt
Max Wiedenhöft

Sponsors

Invited Abstracts

Burning, Cooling, and Liminal Burning Graphs

Anthony Bonato

Department of Mathematics, Toronto Metropolitan University, Toronto, ON, M5B 2K3 Canada
abonato@torontomu.ca

Abstract. Burning processes on graphs may be viewed as finite automata whose states encode which vertices are burned and whose transitions describe the evolution of the process under prescribed choices of sources. We survey recent developments in graph burning, cooling, and liminal burning, with an emphasis on algorithmic and graph-theoretic aspects. Burning and cooling arise as opposite optimization problems, while liminal burning interpolates between them via a two-player game. We will discuss the burning number conjecture, complexity, graph products, and connections to isoperimetric problems.

Traversing Crystals by Automata

Nataša Jonoska

Department of Mathematics and Statistics, University of South Florida, Tampa, FL, 33620 USA
jonoska@usf.edu

Abstract. Recent advances in crystal nets and DNA crystal designs have highlighted the need for theoretical approaches that can distinguish and characterize these structures. One approach is to consider walks within the underlying structures. Labeled periodic graphs, where vertices correspond to molecular building blocks and labeled edges correspond to molecular bond types, are suitable crystal representations. The set of labels of all walks within a structure represents a language over the bond alphabet. The set of cyclic walks defines a language recognized by counter automata. These counter automata define a hierarchy of languages closely related to the hierarchy of finite intersections of deterministic context-free languages. We show a close relationship between crystallographic structures and the languages recognized by counter automata. Moreover, the counter automata can be seen as voltage graphs (graphs and digraphs whose edges are labeled with elements from a free abelian group), while the crystallographic structure represented by the voltage graphs are viewed as graphs generated by the counter automata. Using the notion of 'lifting voltages' in graphs, we show a polynomial-time algorithm for determining whether two finite bi-deterministic counter automata generate isomorphic periodic graphs. This algorithm aids in the classification of crystal structures and provides a method for determining whether two sets of building blocks can assemble isomorphic crystals.

Computing with a Full Hard Drive

Ian Mertz

Computer Science Institute, Faculty of Mathematics and Physics, Charles University, Praha 1, 118 00 Czech Republic
iwmertz@iuuk.mff.cuni.cz

Abstract. Space and time are the two most foundational resources in the study of computation. However, they behave in very different ways that can make our intuitions fail. One key difference is that time spent is gone forever, whereas space, once it has been used for one purpose, can be erased and re-used for another.

Just over a decade ago, Buhrman et al. discovered a shocking extension of this idea: even memory which is completely full and cannot be erased can be a powerful resource. Inspired by chemistry, they defined catalytic space as full memory which aids a computation without being altered by the process.

In this talk, we introduce the field of catalytic computation, from its inception to the recent explosion of results in the past few years. We will give an overview of the major techniques for using full memory as well as its greatest successes, including a novel relationship between space and time. Finally, we reflect on how re-using full memory forces us to reconsider our understanding of space.

Contents

Efficiently Testing Emptiness of Automata on Linear Orderings

Thomas Braipson(✉) and Tom Clara

Montefiore Institute, B28, University of Liège, Liège, Belgium
{Thomas.Braipson,Tom.Clara}@uliege.be

Abstract. Automata on linear orderings are a type of finite-state automata introduced by Bruyère and Carton, that unify and broadly extend the usual finite-state automata on finite words, (bi-)infinite words, and ordinal words. These automata read words indexed by linear orderings, defined as mappings from the elements of any linear ordering to a finite alphabet. Several authors have studied the emptiness problem for these automata, on several domains, mainly to obtain decision procedures for some logics. The resulting algorithms are polynomial. In this paper, we review them and provide new algorithms, whose complexity is improved by one power in time and space. In particular, their space complexity is linear.

Keywords: Finite Automata · Linear Orderings · Emptiness Test

1 Introduction

Finite-state automata are expressive objects that are easy to manipulate algorithmically. For these reasons, their various forms have been used for deciding several logical fragments. For instance, Büchi introduced automata on infinite words to establish the decidability of the second-order theory of one successor (S1S) [4]. The strategy is to transform a formula into an automaton whose accepted language corresponds to encodings of the models of the formula, and then test if this language is empty. Automata on linear orderings are a broad extension of automata on infinite words, that accept words indexed by linear orderings [1,3]. These automata have been used in decision procedures for linear-time temporal logic on any linear domain [6,8] and for the monadic first-order theory of order, on the domains $\mathbb{R}$ and $\mathbb{Q}$ [2]. In the literature, emptiness tests for automata on linear orderings on several domains are polynomial in time and have a quadratic space complexity. The contribution of this paper is to provide new algorithms, whose complexity is improved by one power in time and in space. This is an essential step towards the implementation of efficient decision procedures that use automata on linear orderings as data structures.

The paper is structured as follows. We start by recalling basic notions about words and automata on linear orderings in Sect. 2. Then in Sect. 3 we present an emptiness test for words indexed by scattered linear orderings due to [5] and a

T. J. Smith (Ed.): CIAA 2026, LNCS 16695, pp. 1–15, 2026.
https://doi.org/10.1007/978-3-032-31176-4_1

general one due to [7]. Finally, we introduce new tests of improved complexity in Sect. 4.

2 Basic Notions

2.1 Linear Orderings and Cuts

We recall elementary notions on linear orderings and refer to [9] for further details. A *linear ordering* is a set J equipped with a binary relation $<$ that is a *total order.* A *subordering* of J is a subset $I \subseteq J$ equipped with the same total order, restricted to I.

Two elements $j < k$ are said to be *consecutive* if there does not exist any element between them. In that case, j is the *predecessor* of k and k is the *successor* of j. If a linear ordering does not have any consecutive elements, it is *dense.* Conversely, a linear ordering without any infinite dense subordering is *scattered.*

A *cut* of a linear ordering J is a partition (K, L) of J such that for every $k \in K$ and $\ell \in L$, we have $k < \ell$. The set of cuts of the ordering J is written $\widehat{J}$. This set can be linearly ordered by the relation $(K_1, L_1) < (K_2, L_2)$ iff $K_1 \subsetneq K_2$ (or equivalently, $L_2 \subsetneq L_1$). For any linear ordering J, $\widehat{J}$ contains a least element $\widehat{J}_{min} = (\emptyset, J)$ called the *first cut* and a greatest element $\widehat{J}_{max} = (J, \emptyset)$ called the *last cut.* If (K, L) is a cut different from $\widehat{J}_{min}$ and $\widehat{J}_{max}$ such that K does not have a greatest element and L does not have a least element, this cut is a *gap.* An ordering for which there does not exist any gap is *complete.*

2.2 Words and Automata on Linear Orderings

The definitions of this section are borrowed from [1,3]. A *word indexed* by a linear ordering J is a mapping $w : J \to \Sigma$, where Σ is a finite set called the *alphabet.* We also say that J is the *length* of w and write it $|w| = J$.

Definition 1. *An* automaton on linear orderings *is a tuple* $\mathcal{A} = (Q, \Sigma, I, F, \Delta)$ *where:*

- Q *is a finite set of* states,
- Σ *is a finite* alphabet,
- $I \subseteq Q$ *is a set of* initial *states,*
- $F \subseteq Q$ *is a set of* final *states, and*
- $\Delta \subseteq (Q \times \Sigma \times Q) \cup (2^Q \times Q) \cup (Q \times 2^Q)$ *is a* transition relation.

The transition relation Δ is composed of three types of transitions. Those of the form $(p, a, q) \in Q \times \Sigma \times Q$ are called *successor* transitions and coincide with the usual notion of transition in automata. In a transition (p, a, q), p is called the *origin*, a the *label*, and q the *destination.* Transitions of the form $(P, q) \in 2^Q \times Q$ (alternatively noted $P \to q$) are called *left-limit* transitions. In that case, P is called the *limit set* and q is called the *destination.* Symmetrically, transitions of the form $(q, P) \in Q \times 2^Q$ (alternatively noted $q \to P$) are called *right-limit* transitions, have as origin q and as limit set P. Several limit transitions sharing the same limit set can be written in a compact form, by grouping them around the limit set, e.g. $P \to q_0$, $q_1 \to P$, and $q_2 \to P$ are written $q_1, q_2 \to P \to q_0$. The *cardinality* of a limit transition is the cardinality of its limit set.

Definition 2. *A* run *of an automaton on linear orderings $\mathcal{A} = (Q, \Sigma, I, F, \Delta)$ reading a word w indexed by J is a mapping $\rho : \widehat{J} \to Q$ that satisfies the following conditions:*

- *For every pair of cuts $c_1 = (K_1, L_1)$ and $c_2 = (K_2, L_2)$ such that c_2 is the successor of c_1 in $\widehat{J}$, $(\rho(c_1), w(j), \rho(c_2)) \in \Delta$, where $j \in J$ is such that $K_2 = K_1 \cup \{j\}$.*
- *For every cut $c \neq \widehat{J}_{min}$ which does not have any predecessor, $(\lim_{c^-} \rho, \rho(c)) \in \Delta$, where $\lim_{c^-} \rho$ is the* left-limit set *of ρ in c, defined as*

$$\lim_{c^-} \rho = \{q \in Q \mid \forall c_1 < c \, . \, \exists c_2 \, . \, c_1 < c_2 < c \wedge \rho(c_2) = q\}.$$

- *For every cut $c \neq \widehat{J}_{max}$ which does not have any successor, $(\rho(c), \lim_{c^+} \rho) \in \Delta$, where $\lim_{c^+} \rho$ is the* right-limit set *of ρ in c, defined as*

$$\lim_{c^+} \rho = \{q \in Q \mid \forall c_1 > c \, . \, \exists c_2 \, . \, c < c_2 < c_1 \wedge \rho(c_2) = q\}.$$

This run is accepting *if $\rho(\widehat{J}_{min}) \in I$ and $\rho(\widehat{J}_{max}) \in F$.*

A word w is *accepted* by an automaton $\mathcal{A}$ if there exists an accepting run of $\mathcal{A}$ that reads w. The class of the words accepted by $\mathcal{A}$ is its *accepted language.*

Intuitively, a run reads a word from a state to another and respects transitions locally. In the case of two successive cuts, the automaton must be able to read the letter that labels the element "between" these two cuts. If a cut is not the first one and does not have any predecessor, then it is the limit of an increasing sequence of cuts. In that case, this cut can be reached if a Muller-like condition is respected: a specific set of states must appear exactly infinitely often, arbitrarily close to that cut without predecessor. This is a left-limit transition. Right-limit transitions are symmetrical. If the run starts in an initial state and ends in a final state, the automaton accepts the word.

A left-limit transition $P \to q$ is *enabled* if there exists a run ρ reading a word w of length J and a cut $c \in \widehat{J}$ such that $\rho(c) = q$ and $\lim_{c^-} \rho = P$. A right-limit transition is enabled under symmetric conditions. We illustrate these definitions by means of some examples.

Example 1. The automaton $\mathcal{A}$ in Fig. 1.a accepts the words that can be described as ω copies of bi-infinite words on the alphabet $\{a, b\}$ that do not have two successive occurrences of the same letter[1]. Indeed, from q_0 to itself the automaton can read bi-infinite words, without occurrences of the factors aa or bb. If this is done infinitely often, the limit transition $\{q_0, q_1, q_2\} \to q_3$ can be followed and the resulting word is accepted.

[1] Actually, all these words are the same up to an isomorphism. We refer to [3] for a formal definition of isomorphism between words, a notion not necessary in this work.

Example 2. The automaton $\mathcal{B}$ in Fig. 1.b accepts all the non-empty words w on the alphabet $\{a\}$ such that

- the linear ordering $|w|$ does not have a least (resp. greatest) element, since the only transition leaving (resp. reaching) the initial (resp. final) state is a right-limit (resp. left-limit) transition,
- $|w|$ is dense, since it is impossible to read two consecutive letters,
- and for any two elements $i < j \in |w|$, there exists a gap (K, L) such that $i \in K$ and $j \in L$. Informally speaking, the elements of $|w|$ and its gaps need to be densely interleaved.

For instance, the word $w : \mathbb{Q} \to \{a\}$ is accepted by $\mathcal{B}$. Let us describe an accepting run ρ reading this word. It maps the first cut to q_0 and the last cut to q_4. The remaining cuts of $|w|$ form three groups: those immediately before a letter a, those immediately after a letter a, and the gaps. These three groups are mapped by ρ to q_1, q_2, and q_3, respectively. Notice that each irrational number x corresponds to a gap of $\mathbb{Q}$, since the sets $\{r \in \mathbb{Q} \mid r < x\}$ and $\{r \in \mathbb{Q} \mid r > x\}$ partition $\mathbb{Q}$ (here, $<$ denotes the ordering relation of $\mathbb{R}$). Therefore, it is clear that there exists a gap of $\mathbb{Q}$ between any two rational numbers. Hence in every right neighbourhood of a cut c without successor (different from the last one), there are cuts mapped to q_1, q_2 and q_3 by ρ. A right-limit transition involving $\{q_1, q_2, q_3\}$ can thus be followed to leave such a cut c. A similar analysis holds for a cut without predecessor. This shows that ρ is indeed a run reading w.

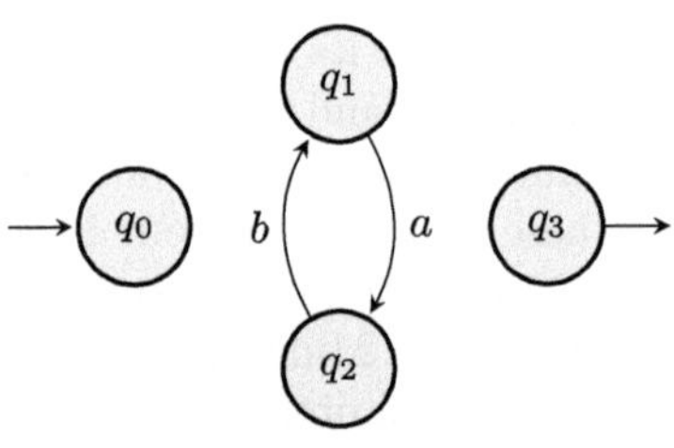

$$q_0 \to \{q_1, q_2\} \to q_0$$
$$\{q_0, q_1, q_2\} \to q_3$$

(a) The automaton $\mathcal{A}$.

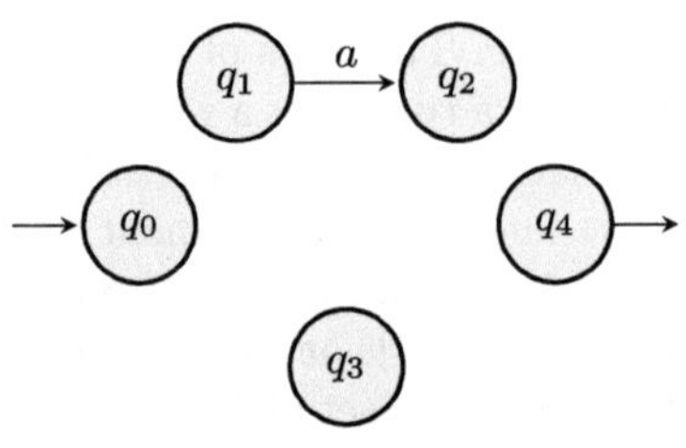

$$q_0, q_2, q_3 \to \{q_1, q_2, q_3\} \to q_1, q_3, q_4$$

(b) The automaton $\mathcal{B}$.

Fig. 1. Examples of automata on linear orderings.

3 State of the Art

In this section, we review several algorithms to check the emptiness of automata on linear orderings, for several classes of word lengths. We first recall the cases of finite and ω-words. Then, we present tests dedicated to any scattered linear ordering and any linear ordering. These tests are the basis of the efficient algorithms introduced in Sect. 4.

3.1 Finite and ω-Words

Automata on linear orderings without limit transitions read finite words, similarly to non-deterministic finite-state automata on finite words. Therefore, to test whether an automaton on linear orderings accepts a finite word, one can ignore its limit transitions and look for a final state reachable from an initial one, through a finite sequence of successor transitions.

Non-deterministic Muller automata can be seen as automata on linear orderings, where accepting sets of states play the role of left-limit transitions leading to a final state. The existence of a limit set that is reachable from an initial state and that forms a strongly connected component implies that such an automaton accepts an ω-word.

3.2 Words Indexed by Scattered Linear Orderings

An emptiness test for words indexed by scattered linear orderings is proposed in [5]. Given an automaton on linear orderings $\mathcal{A}$, it checks the existence of a final state reachable from an initial one, by following any kind of transitions, provided that they are compatible with words of scattered length. To do so, the algorithm progressively builds a *reachability automaton*. It is a finite-word automaton whose states are those of $\mathcal{A}$ and whose transitions encode the possibility of moving from one state to another by visiting some set of states in $\mathcal{A}$. Therefore, a transition in a reachability automaton is labelled by a subset of states of $\mathcal{A}$.

Concretely, the algorithm checks whether each limit transition of $\mathcal{A}$ is enabled. Whenever it is the case, fresh transitions are added to the reachability automaton to encode this information. Due to the Muller-like semantics of limit transitions in automata on linear orderings (i.e., states in the limit set must appear exactly infinitely often to enable the transition), a limit transition of limit set P can only be enabled by following successor transitions or limit transitions that involve limit sets included in P. Therefore, the test processes limit transitions by increasing cardinality (which strengthens the order given by inclusion). Once each limit transition has been processed, the emptiness of the resulting reachability automaton (which is a finite-word automaton) is tested.

We now formalise this algorithm. First, we define the notion of path in a reachability automaton.

Definition 3. *A* path $p \xrightarrow{P} q$ *in a reachability automaton is a finite sequence of states* $p_1, \ldots, p_n$*, with* $n > 1$ *such that there exists a transition* (p_i, P_i, p_{i+1}) *for all* $0 < i < n$*,* $p = p_1$*,* $q = p_n$*, and* $P = \bigcup_{0<i<n} P_i$*.*

The intuition of this definition is that a path collects the labels of the transitions that it follows by taking their union.

Formally, for an automaton on linear orderings $\mathcal{A} = (Q, \Sigma, I, F, \Delta)$, the algorithm computes the reachability-automaton sequence $\mathcal{A}_0, \mathcal{A}_1, \ldots, \mathcal{A}_M$, where M

is the cardinality of the largest limit set in $\mathcal{A}$. These automata are recursively defined as $\mathcal{A}_i = \left(Q, 2^Q, I, F, \Delta_i\right)$, with

$$\begin{aligned}
\Delta_0 &= \{(p, \{p, q\}, q) \mid (p, a, q) \in \Delta\} \\
\Delta_i &= \Delta_{i-1} \cup \Delta_i^L \cup \Delta_i^R \qquad \text{for } 0 < i \leq M \\
\Delta_i^L &= \{(p, P \cup \{q\}, q) \mid P \to q \in \Delta \text{ and } |P| = i \text{ and } C\} \\
\Delta_i^R &= \{(q, P \cup \{q\}, p) \mid q \to P \in \Delta \text{ and } |P| = i \text{ and } C\}
\end{aligned}$$

where the condition C imposes the existence of a path of the form $p \xrightarrow{P} p$ in $\mathcal{A}_{i-1}$.

The key property of the reachability automaton $\mathcal{A}_i$ is that the label of a path in between two states p and q is equal to a set of states that can be visited in $\mathcal{A}$, to move from p to q, while following limit transitions of cardinality at most i.

Example 3. Consider the automaton $\mathcal{A}$ in Fig. 1.a. First, $\mathcal{A}_0$ is built by replacing the labels of the successor transitions of $\mathcal{A}$ with their respective set of origin and destination states. Since there is not any limit transition of cardinality 1, $\mathcal{A}_1$ is equal to $\mathcal{A}_0$. Then, since the path $q_1 \xrightarrow{\{q_1,q_2\}} q_1$ exists in $\mathcal{A}_1$, the limit transitions $q_0 \to \{q_1, q_2\} \to q_0$ are transformed into the transitions $(q_0, \{q_0, q_1, q_2\}, q_1)$ and $(q_1, \{q_0, q_1, q_2\}, q_0)$. The same arguments yield the transitions $(q_0, \{q_0, q_1, q_2\}, q_2)$ and $(q_2, \{q_0, q_1, q_2\}, q_0)$. The result is $\mathcal{A}_2$, in Fig. 2.a, where we denote transitions of label $\{q_1, q_2\}$ by dashed arrows and those of label $\{q_0, q_1, q_2\}$ by dotted arrows. These transitions endow $\mathcal{A}_2$ with the path $q_0 \xrightarrow{\{q_0,q_1,q_2\}} q_0$. Together with the limit transition $\{q_0, q_1, q_2\} \to q_3$, this yields the transition $(q_0, \{q_0, q_1, q_2, q_3\}, q_3)$. The paths $q_1 \xrightarrow{\{q_0,q_1,q_2\}} q_1$ and $q_2 \xrightarrow{\{q_0,q_1,q_2\}} q_2$, together with $\{q_0, q_1, q_2\} \to q_3$, yield similar transitions. The result, $\mathcal{A}_3$, is depicted in Fig 2.b, where these transitions are written $\dashrightarrow$. In this reachability automaton, there exists a path from q_0 to q_3. This shows that $\mathcal{A}$ accepts at least one word of scattered length.

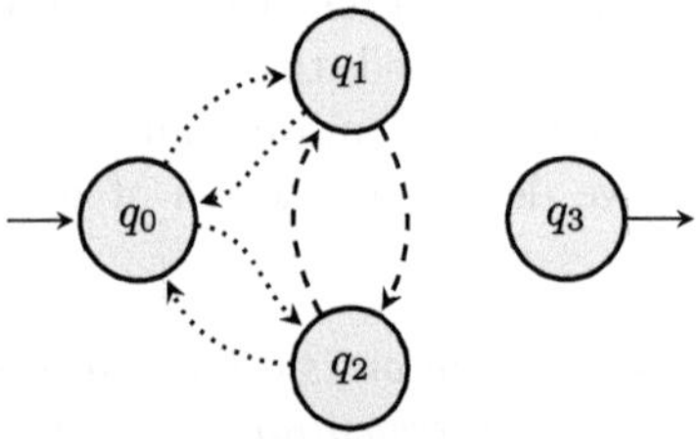

(a) The automaton $\mathcal{A}_2$.

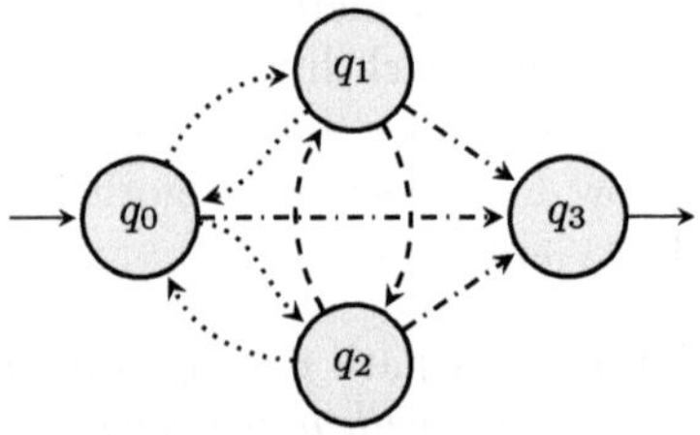

(b) The automaton $\mathcal{A}_3$.

Fig. 2. Illustration of the test for scattered orderings.

Lemma 1. *The emptiness for words indexed by scattered linear orderings can be tested in time* $\mathcal{O}\left(|\Delta|^2 \cdot |Q|^2\right)$ *and space* $\mathcal{O}\left(|\Delta| \cdot |Q|\right)$.

Proof. Each reachability automaton $\mathcal{A}_i$ has $\mathcal{O}(|\Delta| \cdot |Q|)$ transitions, since each enabled limit transition induces as many transitions as states in its limit set, and there are $\mathcal{O}(|\Delta|)$ limit transitions and $\mathcal{O}(|Q|)$ states in a limit set. Given that only the most recent reachability automaton needs to be stored in memory, the space complexity of the algorithm is $\mathcal{O}(|\Delta| \cdot |Q|)$.

For each limit transition $P \to q$ or $q \to P$, the algorithm needs to check the presence of a path $p \xrightarrow{P} p$ in $\mathcal{A}_{|P|}$, where p is an arbitrary state in P. This can be done by means of a depth-first search in $\mathcal{A}_{|P|}$ starting from p and following only transitions labelled by a subset of P. If this search reaches p after having followed transitions whose union of labels is exactly P, then the limit transition can be followed by some run. In the worst case, it is necessary to explore all the transitions of $\mathcal{A}_{|P|}$. Assuming that all the necessary manipulations of subsets of states (checking the inclusion of a subset in another, computing the union of two subsets or their difference) run in $\mathcal{O}(|Q|)$, the time complexity of the emptiness test is thus $\mathcal{O}\left(|\Delta| \cdot |\Delta_{|P|}| \cdot |Q|\right)$, that is, $\mathcal{O}\left(|\Delta|^2 \cdot |Q|^2\right)$. □

Note that the exploration of the final reachability automaton has a cost linear in its size, i.e., $\mathcal{O}(|\Delta| \cdot |Q|)$. Exploring this automaton thus has a negligible cost with respect to building it.

3.3 Words Indexed by Any Linear Ordering

An extension of the previous algorithm is proposed in [7] to deal with words indexed by any linear ordering. It is similar to the one presented in the scattered case, except for an additional rule that deals with density. It checks whether each limit set of cardinality i is strongly connected in $\mathcal{A}_{i-1}$, as previously, but it also checks whether it forms a *shuffle structure*, such as the automaton $\mathcal{B}$ in Fig. 1b. This is the mechanism that enables automata on linear orderings to read words of infinite dense length.

The transition relation of each automaton $\mathcal{A}_i$ is now defined as

$$\begin{aligned} \Delta_0 &= \{(p, \{p, q\}, q) \mid (p, a, q) \in \Delta\} \\ \Delta_i &= \Delta_{i-1} \cup \Delta_i^L \cup \Delta_i^R \cup \Delta_i^S \qquad \text{for } 0 < i \leq M \end{aligned}$$

where Δ_i^L and Δ_i^R are defined as previously and Δ_i^S is the following:

$$\Delta_i^S = \{(p, P \cup \{p, q\}, q) \mid p \to P, P \to q \in \Delta \text{ and } |P| = i \text{ and } C^S\}.$$

The condition C^S checks the presence of a shuffle structure, by imposing the existence of $n > 0$ paths $p_1 \xrightarrow{P_1} q_1, \ldots, p_n \xrightarrow{P_n} q_n$ in $\mathcal{A}_{i-1}$ and $m \geq 0$ states $r_1, \ldots, r_m$ such that:

$$\begin{aligned} &q_i \to P \to p_i \in \Delta \text{ for each } 0 < i \leq n \\ &r_i \to P \to r_i \in \Delta \text{ for each } 0 < i \leq m \\ &\bigcup_{0<i\leq n} P_i \cup \{r_1, \ldots, r_m\} = P \end{aligned}$$

Remark 1. This rule generates transitions witnessing shuffles of simpler languages (see [1] for a formal definition) that can be read in $\mathcal{A}$, from p_i to q_i, respectively. If these languages are the empty word, they form gaps in the shuffle and these gaps are mapped to states r_i. The existence of at least one path in $\mathcal{A}_{i-1}$ ensures that the shuffle is not composed of empty words only. If it were the case, there would only be one cut in the resulting ordering and the limit transition would not be enabled. For instance, adding the limit transitions $q_0, q_3 \to \{q_3\} \to q_3, q_4$ in the automaton $\mathcal{B}$ from Fig. 1.b would not change its accepted language. Indeed, assume that there exists a run ρ reading a word w of length J in this automaton, with $\rho(\widehat{J}_{min}) = q_0$, $\rho(\widehat{J}_{max}) = q_4$, and $\rho(c) = q_3$ for all the cuts c different from the first and the last one. Then it is clear that ρ cannot follow any successor transition. Hence w needs to be equal to the empty word. This is a contradiction, given that $q_0 \neq q_4$.

Example 4. We illustrate the algorithm on the automaton $\mathcal{B}$ in Fig. 1.b. First, the reachability automaton $\mathcal{B}_0$ is built by replacing the unique successor transition (q_1, a, q_2) of $\mathcal{B}$ by the transition $(q_1, \{q_1, q_2\}, q_2)$. Automata $\mathcal{B}_1$ and $\mathcal{B}_2$ are equal to $\mathcal{B}_0$ since the only limit-set cardinality is 3. In the automaton $\mathcal{B}_2$, the path $q_1 \xrightarrow{\{q_1,q_2\}} q_2$ exists and the set $\{q_1, q_2\} \cup \{q_3\}$ is exactly equal to the limit set $P = \{q_1, q_2, q_3\}$. Moreover, the limit transitions $q_2 \to P \to q_1$ and $q_3 \to P \to q_3$ belong to the transition relation of $\mathcal{B}$. Therefore, for every pair of limit transitions $p \to P \to q$, the transition $(p, P \cup \{p, q\}, q)$ is added to $\mathcal{B}_3$. The result is shown in Fig. 3, where dashed transitions are those added to $\mathcal{B}_3$, regardless of their label. There exists a path from q_0 to q_4 in the reachability automaton $\mathcal{B}_3$, hence $\mathcal{B}$ accepts at least one word.

Lemma 2. *The emptiness for words indexed by any linear ordering can be tested in time* $\mathcal{O}\left(|\Delta|^2 \cdot |Q|^2\right)$ *and space* $\mathcal{O}\left(|\Delta| \cdot |Q|\right)$.

Proof. The arguments are similar to those of Lemma 1. The only difference is the presence of additional transitions in the reachability automata, to take into account the potential shuffle structures. For each limit transition $q \to P$ or $P \to q$, at most $\mathcal{O}(|Q|)$ such transitions are added to $\Delta^S_{|P|}$, hence each reachability automaton still has $\mathcal{O}\left(|\Delta| \cdot |Q|\right)$ transitions.

Notice that checking that the limit transitions involving a limit set P form a shuffle structure (i.e., checking the condition C^S) can be done in a time $\mathcal{O}\left(|\Delta_{|P|}| \cdot |Q|\right)$, as for checking the presence of a path $p \xrightarrow{P} p$ in Lemma 1. Indeed, as already pointed out in [7] (in Lemma 26), checking the condition C^S amounts to

1. keeping only the transitions of $\mathcal{A}_{|P|}$ whose labels are included in P,
2. keeping only the states of $\mathcal{A}_{|P|}$ that are reachable from some state p such that $P \to p \in \Delta$, and co-reachable from some state q such that $q \to P \in \Delta$,
3. and checking that the union of the labels of the transitions that remain is exactly P.

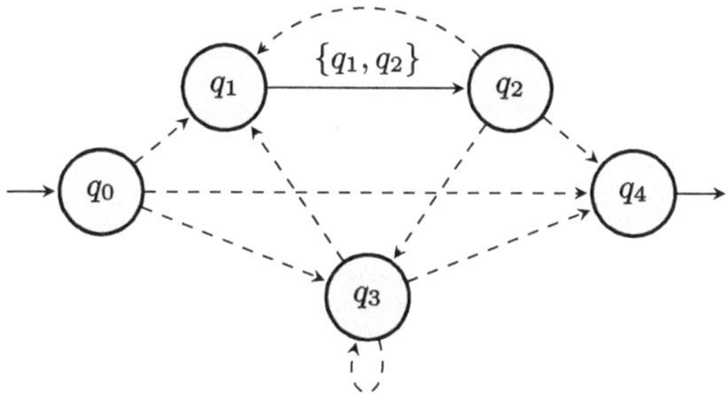

Fig. 3. Illustration of the test for any ordering.

Assuming, as in Lemma 1, that all the necessary manipulations of subsets of states run in $\mathcal{O}(|Q|)$, this yields an algorithm running in $\mathcal{O}\left(|\Delta_{|P|}| \cdot |Q|\right)$.

The overall time complexity of the emptiness test is thus $\mathcal{O}\left(|\Delta| \cdot |\Delta_{|P|}| \cdot |Q|\right)$, that is, $\mathcal{O}\left(|\Delta|^2 \cdot |Q|^2\right)$. □

4 Efficient Test of Emptiness

4.1 Preliminaries

The algorithms testing the emptiness for automata on linear orderings presented in the previous section convert each enabled limit transition into $\mathcal{O}(|Q|)$ transitions.

We introduce a new algorithm, inspired by the previous ones, that converts each enabled limit transition into a single transition. This is achieved by selecting a specific state in the limit set associated with the enabled transition, and using it as a representative. Intuitively, the limit set must form some loop to enable its limit transition. Therefore, choosing one member of this loop is enough to connect its content to the origin or destination state.

Importantly, there exist refined emptiness tests in the literature, such as those checking the existence of at least one accepted word indexed by $\mathbb{R}$ or $\mathbb{Q}$ [2]. These tests follow the same principles as those presented in Sect. 3, and they check additional conditions (closely related to specific properties of $\mathbb{R}$ and $\mathbb{Q}$). The algorithm introduced below can be adapted to these tests with the same gain in complexity.

4.2 Algorithm

In this section, we present an efficient algorithm testing the emptiness for words indexed by any linear ordering. We first introduce an arbitrary total order $<_Q$ on the set of states Q and define a function $min : 2^Q \setminus \{\emptyset\} \to Q$ that returns the least element of any non-empty subset of Q, according to this total order.

The transition relation Δ_i of each reachability automaton $\mathcal{A}_i$ is recursively defined as follows:

$$\begin{aligned}
\Delta_0 &= \{(p, \{p, q\}, q) \mid (p, a, q) \in \Delta\} \\
\Delta_i &= \Delta_{i-1} \cup \Delta_i^L \cup \Delta_i^R \\
\Delta_i^L &= \{(p, P \cup \{q\}, q) \mid P \to q \in \Delta \text{ and } |P| = i \text{ and } p = min\,(P) \text{ and } C\} \\
\Delta_i^R &= \{(q, P \cup \{q\}, p) \mid q \to P \in \Delta \text{ and } |P| = i \text{ and } p = min\,(P) \text{ and } C\}
\end{aligned}$$

The condition C is defined as:

- there exists a path $p \xrightarrow{P} p$ in $\mathcal{A}_{i-1}$, or
- there exist $n > 0$ paths $p_1 \xrightarrow{P_1} q_1, \ldots, p_n \xrightarrow{P_n} q_n$ in $\mathcal{A}_{i-1}$ and $m \geq 0$ states $r_1, \ldots, r_m$ such that:

$$\begin{aligned}
& q_i \to P \to p_i \in \Delta \text{ for each } 0 < i \leq n \\
& r_i \to P \to r_i \in \Delta \text{ for each } 0 < i \leq m \\
& \bigcup_{0<i\leq n} P_i \cup \{r_1, \ldots, r_m\} = P
\end{aligned}$$

Intuitively, for each enabled limit transition, the algorithm now chooses one specific state of the limit set to generate only one transition in the reachability automaton. When the condition to enable a limit transition is the existence of a cycle from that state to itself, there is no need to add more than one transition to obtain the desired reachability properties. When the condition is the existence of a shuffle structure, picking a single state remains sufficient, provided that each new transition is connected to it. This improves time and space complexities by one power, with respect to previously existing algorithms. Notice that the case of a shuffle structure is covered by the transition relations Δ_i^L and Δ_i^R. It is not necessary to handle this case separately. Therefore there is no need for a relation Δ_i^S, even though this algorithm checks emptiness for words indexed by any linear ordering.

Example 5. We apply the algorithm to the automaton $\mathcal{A}$ from Fig. 1.a. We decide to order its states according to their index. $\mathcal{A}_0$ is built by relabelling the successor transitions of $\mathcal{A}$ and is equal to $\mathcal{A}_1$ since there is no limit transition of cardinality 1. Then, the path $q_1 \xrightarrow{\{q_1,q_2\}} q_1$ exists in $\mathcal{A}_1$, hence the limit transitions $q_0 \to \{q_1, q_2\} \to q_0$ are converted into the transitions $(q_0, \{q_0, q_1, q_2\}, q_1)$ and $(q_1, \{q_0, q_1, q_2\}, q_0)$, since we have decided that $q_1 <_Q q_2$. This is depicted in Fig. 4.a. The transitions that were added at the previous step create the path $q_0 \xrightarrow{\{q_0,q_1,q_2\}} q_0$ in $\mathcal{A}_2$. The existence of this path shows that the limit transition $\{q_0, q_1, q_2\} \to q_3$ is enabled. It is thus converted into $(q_0, \{q_0, q_1, q_2, q_3\}, q_3)$. The result is shown in Fig. 4.b. There exists a path from q_0 to q_3, which shows that $\mathcal{A}$ accepts a non-empty language.

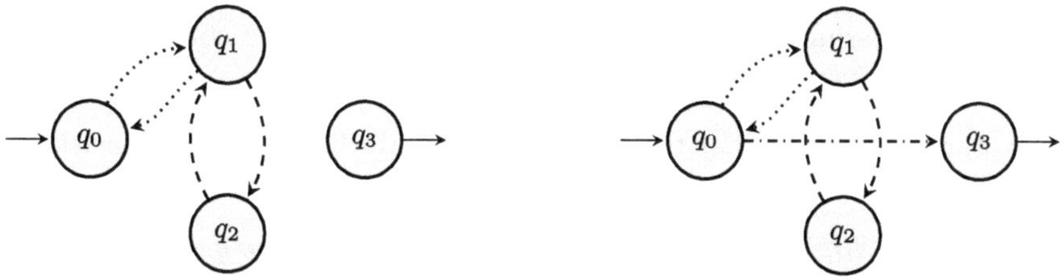

(a) The automaton $\mathcal{A}_2$. (b) The automaton $\mathcal{A}_3$.

Fig. 4. Illustration of the efficient test on the automaton $\mathcal{A}$.

Example 6. We apply the efficient emptiness test to the automaton $\mathcal{B}$ from Fig. 1.b. The automata $\mathcal{B}_0, \ldots, \mathcal{B}_2$ and the criteria to determine that every limit transition is enabled are the same as in Example 4. The difference is about the transitions added to $\mathcal{B}_3$ (Fig. 5). For this, we first choose q_1 as representative in the limit set $P = \{q_1, q_2, q_3\}$. Left-limit transitions $P \to q$ are converted into transitions that start in q_1 and end in q. Symmetrically, right-limit transitions $q \to P$ are converted into transitions that start in q and end in q_1. The whole limit set forms a strongly connected component in $\mathcal{B}_3$, because there exist transitions from q_1 to the origin of the path $q_1 \xrightarrow{\{q_1,q_2\}} q_2$, from q_1 to q_3, from the destination of the path $q_1 \xrightarrow{\{q_1,q_2\}} q_2$ to q_1 and from q_3 to q_1. The state q_0 is connected to q_1, which is itself connected to q_4. Hence there exists a path between q_0 and q_4, which shows that $\mathcal{B}$ accepts a non-empty language.

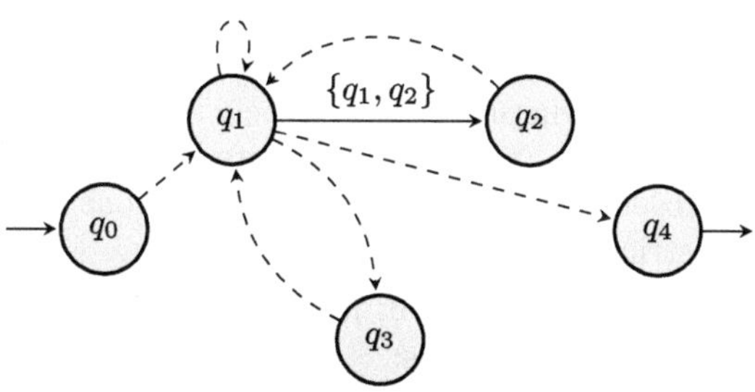

Fig. 5. Illustration of the efficient test on the automaton $\mathcal{B}$.

Lemma 3. *The emptiness for words indexed by (scattered) linear orderings can be tested in time* $\mathcal{O}\left(|Q| \cdot |\Delta|^2\right)$ *and space* $\mathcal{O}\left(|\Delta|\right)$.

Proof. Each enabled limit transition induces one transition in the reachability automaton, therefore $|\Delta_i|$ scales as $|\Delta|$. This is the only difference with respect to Lemmas 1 and 2, where $|\Delta_i|$ scales as $|\Delta| \cdot |Q|$. The remainder of the analysis is unchanged. Removing the second item in Condition C, restricts the test to scattered orderings without affecting its complexity. The correctness of the construction is given by Theorem 1. □

We start by showing that this algorithm adds all the necessary transitions in each reachability automaton.

Lemma 4. *If a limit transition $P \to q$ of cardinality i (resp. $q \to P$) is enabled in $\mathcal{A}$, the set of paths $\{p \xrightarrow{P \cup \{q\}} q \mid p \in P\}$ exists in $\mathcal{A}_i$ (resp. $\{q \xrightarrow{P \cup \{q\}} p \mid p \in P\}$).*

Proof. The proof is by induction on i. If $i = 0$, a limit transition with an empty limit set can never be enabled and the lemma is correct. The interesting case is the induction step. If a limit transition $P \to q$ of cardinality $i > 0$ can be followed in $\mathcal{A}$ ($q \to P$ is symmetric and not discussed here), there are two cases:

1. There exists a cycle in $\mathcal{A}$ that visits exactly P and involves limit transitions of cardinality at most equal to $i - 1$ and successor transitions that start in a state of P and end in a state of P. By induction hypothesis, any left-limit transition of cardinality at most $i - 1$ is encoded in $\mathcal{A}_{i-1}$ by the presence of paths that join each state of its limit set to its destination. Symmetrically, for any right-limit transition of cardinality at most $i - 1$, there exist paths in $\mathcal{A}_{i-1}$ from its origin to each state of its limit set. Moreover, every successor transition of $\mathcal{A}$ corresponds to a transition in $\mathcal{A}_i$, for every i. Thus in $\mathcal{A}_{i-1}$, there exists the path $p \xrightarrow{P} p$ for any $p \in P$. In particular, this path can be composed of the two paths $p_{min} \xrightarrow{P_1} p$ and $p \xrightarrow{P_2} p_{min}$, with $p_{min} = min\,(P)$ and $P_1 \cup P_2 = P$. Since the transition $(p_{min}, P \cup \{q\}, q)$ belongs to Δ_i, the path $p \xrightarrow{P \cup \{q\}} q$ can be built in $\mathcal{A}_i$ by the concatenation of the path $p \xrightarrow{P_2} p_{min}$ with the path $p_{min} \xrightarrow{P \cup \{q\}} q$, formed by the freshly added transition.
2. There exists a shuffle structure in P. Namely, there exist limit transitions $q_1, \ldots, q_n \to P \to p_1, \ldots, p_n$ such that the state q_j is reachable from the state p_j, by visiting a subset P_j of P via successor transitions and limit transitions whose limit set is strictly included in P.
   Now, we build the path $p \xrightarrow{P \cup \{q\}} q$, for any $p \in P$. It is composed of the concatenation of two or three paths:
   - $p \xrightarrow{P} p_{min}$ and $p_{min} \xrightarrow{P \cup \{q\}} q$, if $p \to P \in \Delta$,
   - $p \xrightarrow{P'_k} q_k$, $q_k \xrightarrow{P} p_{min}$, and $p_{min} \xrightarrow{P \cup \{q\}} q$ otherwise, for some $0 < k \leq n$ and where P'_k is a subset of P_k containing p. This is illustrated in Fig. 6. The existence of the path $p \xrightarrow{P'_k} q_k$ is guaranteed by induction hypothesis and Δ_0, since it is a portion of the path $p_k \xrightarrow{P_k} q_k$ that is necessary to enable the limit transition through a shuffle.

   The paths $p \xrightarrow{P} p_{min}$ and $p_{min} \xrightarrow{P \cup \{q\}} q$ exist, since they are both composed of a transition added to Δ_i at step i.

Cases 1 and 2 are exhaustive, as already shown in Lemma 24 of [7] . □

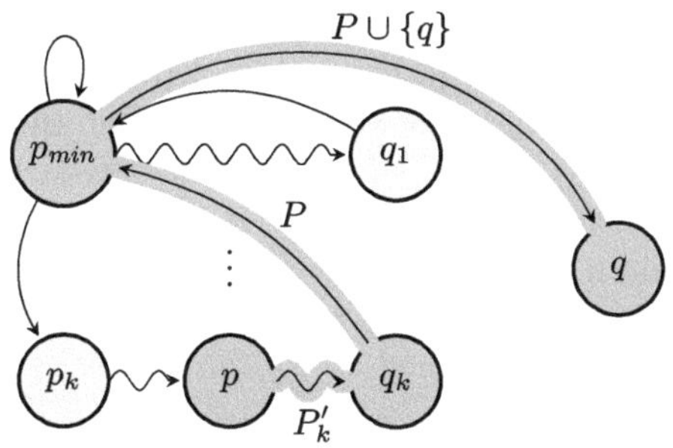

Fig. 6. Construction of the path $p \xrightarrow{P\cup\{q\}} q$.

Theorem 1. *In the reachability automaton $\mathcal{A}_i$, there exists a path $p \xrightarrow{P} q$ if and only if there exists a word w that can be read in $\mathcal{A}$, from p to q, by visiting exactly the set of states P, and by following limit transitions of cardinality at most equal to i.*

Proof. We first show that if there exists a path in $\mathcal{A}_i$, then w exists, by induction on i. When $i = 0$, only finite words can be read by $\mathcal{A}$, and $\mathcal{A}_0$ inherits the successor transition relation of $\mathcal{A}$. For the induction step, we only need to show that a word w can be associated with the path $p \xrightarrow{P\cup\{p,q\}} q$, composed of a single transition added at step i, resulting from the left-limit transition $P \to q$ (the case $q \to P$ is symmetric). There are two cases to consider:

1. This transition results from the existence of a path $p \xrightarrow{P} p$ in $\mathcal{A}_{i-1}$. By induction hypothesis, this path can be associated with a word u, that can be read in $\mathcal{A}$ from p to itself, by visiting exactly P. The word $w = u^\omega$ can be read in $\mathcal{A}$ from p to q, by visiting exactly $P \cup \{p, q\}$. Indeed, since u can be read from p to itself, by visiting exactly P and $P \to q$ is a left-limit transition, u can be read infinitely often to eventually reach q.
2. This transition results from the existence of $n > 0$ paths $p_1 \xrightarrow{P_1} q_1, \ldots,$ $p_n \xrightarrow{P_n} q_n$ in $\mathcal{A}_{i-1}$ and $m \geq 0$ states $r_1, \ldots, r_m$ such that:

$$\begin{aligned} &q_j \to P \to p_j \in \Delta && \text{for } 0 < j \leq n \\ &r_j \to P \to r_j \in \Delta && \text{for } 0 < j \leq m \\ &\bigcup_{0<j\leq n} P_j \cup \{r_1, \ldots, r_m\} = P \\ &p = min(P) \end{aligned}$$

By induction hypothesis, there exist words $u_1, \ldots, u_n$ such that each u_j can be read in $\mathcal{A}$ from p_j to q_j, by visiting exactly P_j. Moreover, the empty word can be read from each r_j to itself, by visiting $\{r_j\}$. Let $0 < k \leq n$ be such that $p \in P_k$. There are two cases: either there exists a path in $\mathcal{A}_{i-1}$ going from p to q_k, or $p = q_k$. In the first case, there exists a non-empty word u'_k

that can be read in $\mathcal{A}$ from p to q_k, by induction hypothesis. In the second one, the empty word $u'_k = \varepsilon$ can be read in $\mathcal{A}$ from p to $q_k = p$.
We are now ready to build the word w. Consider the partition of $\mathbb{R}$ into $\ell = n + m$ subsets $R_1, \ldots, R_\ell$, such that between any two elements of $\mathbb{R}$, there exists an element of R_j, for any $0 < j \leq \ell$ (i.e., these subsets are dense in $\mathbb{R}$). We first build a word v by replacing each element of the subset R_j by a copy of the word u_j, for $0 < j \leq n$, and we erase each element of R_j for $n < j \leq \ell$. The word w is obtained by concatenating u'_k and v, i.e., we define $w = u'_k \cdot v$.
Let us show that w can be read in $\mathcal{A}$ from p to q, while visiting exactly $P \cup \{p, q\}$. First, u'_k can be read from p to q_k, by visiting some set of states $P'_k \subseteq P_k$ that contains p. Then, v can be read from q_k to q, by visiting $P \cup \{q\}$. Indeed, consider a cut $c \in \widehat{|w|}$ that coincides with the first cut of some u_j. This cut does not have any predecessor and is not the first one in $\widehat{|w|}$, since $\mathbb{R}$ is dense and does not have a least element. Moreover, to the left of this cut, arbitrarily close to it, there are infinitely many occurrences of each state in P. Indeed, the subsets R_j defined above are dense in $\mathbb{R}$, and any state r_j can read the empty word (therefore, gaps can be mapped to the states r_j). Therefore, the transition $P \to p_j$ is enabled. Moreover u_j can be read from p_j to q_j by visiting exactly P_j, by induction hypothesis. The same arguments show that q_j can be left by the limit transition $q_j \to P$. This analysis holds for all the limit transitions $q_j \to P \to p_j$ and $r_j \to P \to r_j$. It can also be applied to show that the state q can be eventually reached by following the transition $P \to q$. The word v can thus be read in $\mathcal{A}$ from q_k to q.
Hence, one concludes that w can be read in $\mathcal{A}$, from p to q, by visiting exactly $P \cup \{p, q\}$, and following limit transitions of cardinality at most equal to i.

For the other direction, we consider a word w and a run ρ of $\mathcal{A}$ reading it from a state p to a state q, visiting exactly the set of states P, and following limit transitions of cardinality at most equal to i. We show that there exists a path $p \xrightarrow{P \cup \{p,q\}} q$, in $\mathcal{A}_i$. This is again done by induction on i. When $i = 0$, since $\mathcal{A}_0$ inherits the successor transition relation of $\mathcal{A}$, the property holds. Indeed, if the paths $p \xrightarrow{P_1} q'$ and $q' \xrightarrow{P_2} q$ exist, then the path $p \xrightarrow{P_1 \cup P_2} q$ exists as well. For the induction step, we use Lemma 4. By induction hypothesis, to runs reading words in $\mathcal{A}$ that follow limit transitions of cardinality less than i, correspond paths in $\mathcal{A}_{i-1}$ that share the same origin, destination, and states visited in between. Let $c \in \widehat{|w|}$ be a cut that is not the first one, that does not have a predecessor, and that ρ reaches by following a limit transition $P \to q$, with $|P| = i$. Since ρ uses this limit transition, it is enabled (either by an infinite repetition or by a shuffle structure). Hence, in $\mathcal{A}_i$, there exist paths from every state $p \in P$ to q, labelled by $P \cup \{q\}$. The same holds for right-limit transitions, and thus there exists the path $\rho(\widehat{|w|}_{min}) \xrightarrow{\{\rho(c) | c \in \widehat{|w|}\}} \rho(\widehat{|w|}_{max})$ in $\mathcal{A}_i$. $\square$

5 Conclusions and Perspectives

The contribution of this paper is to provide a family of efficient emptiness tests for automata on linear orderings. The gain in complexity with respect to previously existing algorithms comes from the choice of a specific representative state per limit transition.

Automata on linear orderings can serve as data structures to represent the models of formulas belonging to some logical fragments, for instance the monadic first-order theory of order on $\mathbb{R}$ or $\mathbb{Q}$ [2]. These decidable fragments might be of practical interest in verification or theorem proving, hence the need for decision procedures that are efficient in practice. In this context, checking that an automaton accepts at least one word is a crucial problem, since it amounts to checking the satisfiability of a logical formula. Therefore, efficient emptiness tests for automata on linear orderings are an important step towards the implementation of new practically usable decision procedures. The next task is to implement and benchmark these tests.

Acknowledgments. Thomas Braipson has been granted with a FRIA grant and Tom Clara is a Research Fellow of the F.R.S.-FNRS.

References

1. Bès, A., Carton, O.: A Kleene theorem for languages of words indexed by linear orderings. Int. J. Found. Comput. Sci. **17**(03), 519–541 (2006)
2. Boigelot, B., Fontaine, P., Vergain, B.: Non-emptiness test for automata over words indexed by the reals and rationals. In: Proc. 28th International Conference on Implementation and Application of Automata (CIAA). Lecture Notes in Computer Science, vol. 15015, pp. 94–108. Springer, Cham (2024)
3. Bruyère, V., Carton, O.: Automata on linear orderings. J. Comput. Syst. Sci. **73**(1), 1–24 (2007)
4. Büchi, J.R.: On a decision method in restricted second order arithmetic. In: Proceedings of International Congress on Logic, Methodology and Philosophy of Science, pp. 1–12. Stanford University Press (1962)
5. Carton, O.: Accessibility in automata on scattered linear orderings. In: Mathematical Foundations of Computer Science 2002, pp. 155–164. Springer Berlin Heidelberg (2002)
6. Cristau, J.: Automata and temporal logic over arbitrary linear time. In: Proceedings of IARCS 29th Annual Conference on Foundations of Software Technology and Theoretical Computer Science (FSTTCS). LIPIcs, vol. 4, pp. 133–144. (2009)
7. Cristau, J.: Jeux et automates sur les ordres, Université Paris-Diderot - Paris VII, Ph.D. thesis (2010)
8. Rabinovich, A.: Temporal logics over linear time domains are in PSPACE. Inf. Comput. **210**, 40–67 (2012)
9. Rosenstein, J.G.: Linear orderings, Academic Press (1982)

Boolean Products of Languages

Sabine Broda, António Machiavelo, Nelma Moreira(✉), and Rogério Reis

CMUP & DM-DCC, Faculdade de Ciências da Universidade do Porto, Rua do Campo Alegre, 4169-007 Porto, Portugal
{sabine.broda,antonio.machiavelo,nelma.moreira,rogerio.reis}@fc.up.pt

Abstract. We introduce Boolean product operations on languages that generalise several synchronised shuffle operators, which can be used to model synchronisation of concurrent systems. Boolean products can have arbitrary arities and allow to associate different synchronisation policies to different alphabetic symbols. Given an n-ary Boolean product operator $\|_\psi$ and n automata $\mathcal{A}_1, \ldots, \mathcal{A}_n$, we define the Boolean product automaton $\|_\psi(\mathcal{A}_1, \ldots, \mathcal{A}_n)$, which recognises $\|_\psi(\mathcal{L}(\mathcal{A}_1), \ldots, \mathcal{L}(\mathcal{A}_n))$. We relate right-invariant relations w.r.t. $\mathcal{A}_i$, for $1 \leq i \leq n$, with right-invariant relations w.r.t. $\|_\psi(\mathcal{A}_1, \ldots, \mathcal{A}_n)$. For regular expressions extended with Boolean products, we analyse several NFA constructions, such as the Glushkov automaton, and its quotients, the partial derivative automaton and the follow automaton. Finally, we discuss how to obtain smaller automata for Boolean product expressions.

1 Introduction

We introduce Boolean product operations on languages that generalise several synchronised shuffle operators, which can be used to model synchronisation of concurrent systems [18,24,25]. Boolean products can have arbitrary arities and allow to associate each alphabetic symbol with a Boolean function, which indicates if the operands must synchronise or interleave when that symbol occurs. For instance, the intersection of the operands (full synchronisation) corresponds to the case where for each alphabetic symbol the Boolean function is conjunction. Although defined independently, Boolean products are formal language variants of the Team automata framework for modelling collaboration between components of (groupware) systems [13,21–23]. Team automata are sets of labeled transition systems (components) that can communicate via external shared actions. Different types of synchronisation are obtained by the specification of predicates. In particular, several synchronised shuffles on languages were defined to characterise traces of some Team automata specifications [24,25]. Recently, for regular expressions with several synchronised shuffle operators, partial derivative and Glushkov (position) automata were defined and their average-state

This work was partially supported by CMUP, which is financed by national funds through FCT under the project with DOI 10.54499/UID/00144/2025; and by AVATAR through the FCT project 2023.12169.PEX.

T. J. Smith (Ed.): CIAA 2026, LNCS 16695, pp. 16–31, 2026.
https://doi.org/10.1007/978-3-032-31176-4_2

complexity was studied [6–9,20]. Boolean products allow to generalize some of those constructions and simplify some of the proofs given before. Given a Boolean product operator $\|_\psi$ and NFAs $\mathcal{A}_1, \ldots, \mathcal{A}_n$, we define the automaton $\|_\psi(\mathcal{A}_1, \ldots, \mathcal{A}_n)$, which recognises the Boolean product of $\mathcal{L}(\mathcal{A}_1), \ldots, \mathcal{L}(\mathcal{A}_n)$. We relate right-invariant relations w.r.t. $\mathcal{A}_i$, for $i \in [1, n]$, with right-invariant relations of $\|_\psi(\mathcal{A}_1, \ldots, \mathcal{A}_n)$. In particular, for regular expressions extended with Boolean products, we analise well known constructions such as the Glushkov automaton, and its quotients, the partial derivative automaton and the follow automaton. We also consider the bisimilarity of the Glushkov automaton. We show that the Glushkov automaton of a Boolean product is isomorphic to the Boolean product automaton of the Glushkov automaton of its operands. While the same holds for the partial derivative automaton, it is not true for the follow automaton, neither for the bisimilarity of the Glushkov automaton. Finally, we discuss how to obtain a smaller automaton for a Boolean product by combining different right-invariant relations for each of its operands.

2 Preliminaries

In this section we review some basic definitions about finite automata and regular languages and fix notation. Let Σ be an alphabet and $\Sigma^\star$ the set of words over Σ, where ε denotes the empty word. A language is any subset of $\Sigma^\star$. A *non-deterministic finite automaton* (NFA) is a five-tuple $\mathcal{A} = \langle Q, \Sigma, \delta, \iota, F\rangle$ where Q is a finite set of states, $\iota \in Q$ is the initial state, $F \subseteq Q$ is the set of final states, and $\delta : Q \times \Sigma \to 2^Q$ is the transition function. We consider the size of an NFA to be its number of states. The transition function can be extended to words and to sets of states in the natural way. The language accepted by an NFA $\mathcal{A}$ is $\mathcal{L}(\mathcal{A}) = \{\, w \in \Sigma^\star \mid \delta(\iota, w) \cap F \neq \emptyset \,\}$. Two automata are *equivalent* if they accept the same language. An NFA is *initially connected* if each state is reachable from the initial state. Given an NFA $\mathcal{A} = \langle Q, \Sigma, \delta, \iota, F\rangle$ and $p \in Q$, we consider the NFA $\mathcal{A}^p = \langle Q^p, \Sigma, \delta, p, F^p\rangle$ where $Q^p = \{\, q \in Q \mid \exists w \in \Sigma^\star : q \in \delta(p, w) \,\}$, and $F^p = F \cap Q^p$. Note that for $q \in Q^p$, if $q' \in \delta(q, \sigma)$ then $q' \in Q^p$. Hence, we can use the same name for the transition functions of $\mathcal{A}$ and $\mathcal{A}^p$. In this paper, whenever not explicitly stated otherwise, we consider initially connected NFAs, i.e., $\mathcal{A} = \mathcal{A}^\iota$. Two automata $\mathcal{A}_1 = \langle Q_1, \Sigma, \delta_1, \iota_1, F_1\rangle$ and $\mathcal{A}_2 = \langle Q_2, \Sigma, \delta_2, \iota_2, F_2\rangle$ are *isomorphic*, $\mathcal{A}_1 \simeq \mathcal{A}_2$, if there is a bijection $\varphi : Q_1 \longrightarrow Q_2$ such that $\varphi(\iota_1) = \iota_2$, $\varphi(F_1) = F_2$, and $\varphi(\delta_1(q_1, \sigma)) = \delta_2(\varphi(q_1), \sigma)$, for all $q_1 \in Q_1$, $\sigma \in \Sigma$.

Lemma 1. *Given $\mathcal{A}$ and $\mathcal{B}$ two isomorphic NFAs, i.e., $\mathcal{A} \simeq \mathcal{B}$, with isomorphism $\varphi : Q_\mathcal{A} \longrightarrow Q_\mathcal{B}$, and a state $p \in Q_\mathcal{A}$, we have $\mathcal{A}^p \simeq \mathcal{B}^{\varphi(p)}$.*

An equivalence relation $\equiv$ on Q is *right-invariant* w.r.t. an NFA $\mathcal{A}$ (or a *bisimulation*) if $\equiv$ is a refinement of $(Q-F)^2 \cup F^2$, i.e., $\equiv\ \subseteq (Q-F)^2 \cup F^2$, and for every $p, q \in Q$ with $p \equiv q$, one has $\forall \sigma \in \Sigma, \forall p' \in \delta(p, \sigma)\ \exists q' \in \delta(q, \sigma) : p' \equiv q'$.

Let $Q/_\equiv\ = \{\, [p]_\equiv \mid p \in Q \,\}$, where $[p]_\equiv\ = \{q \in Q \mid p \equiv q\}$. If $\equiv$ is a right-invariant relation on Q, the *quotient automaton* $\mathcal{A}/_\equiv$ is given by

$$\mathcal{A}/_\equiv = \langle Q/_\equiv, \Sigma, \delta/_\equiv, [\iota], F/_\equiv \rangle,$$

where $\delta/_\equiv([p],\sigma) = \{[q] \mid q \in \delta(p,\sigma)\} = \delta(p,\sigma)/_\equiv$. It is easy to see that $\mathcal{L}(\mathcal{A}/_\equiv) = \mathcal{L}(\mathcal{A})$.

Lemma 2. *Consider $\mathcal{A} = \langle Q, \Sigma, \delta, \iota, F\rangle$ and a right-invariant equivalence relation $\equiv$ on Q. Then, for any $p \in Q$, the restriction $\equiv^p$ of $\equiv$ to Q^p is right-invariant w.r.t. $\mathcal{A}^p$. Furthermore, $\mathcal{A}^p/_{\equiv^p} \simeq (\mathcal{A}/_\equiv)^{[p]}$.*

2.1 Synchronised Shuffle Operators

In this section we briefly review several (synchronised) shuffle operators that were studied by Beek et al. [25] and by Manea et al. [18]. We treat those operators in a unified and generalised framework in the forthcoming sections.

The *shuffle* of two words $u, v \in \Sigma^\star$ is the set of words $u \sqcup\!\sqcup v$ obtained by arbitrarily interleaving the letters in u and v. Formally, we have

$$u \sqcup\!\sqcup \varepsilon = \varepsilon \sqcup\!\sqcup u = \{u\},$$

and for any $\sigma \in \Sigma$,

$$\sigma u \sqcup\!\sqcup \tau v = \{\sigma w \mid w \in u \sqcup\!\sqcup \tau v\} \cup \{\tau w \mid w \in \sigma u \sqcup\!\sqcup v\}.$$

Given $\Gamma \subseteq \Sigma$ and $u, v \in \Sigma^\star$, the *strongly synchronised shuffle* $u \,{}^{\mathsf{s}}\|_\Gamma\, v$ imposes synchronisation on all letters of Γ, while in the *arbitrary synchronised shuffle* $u \,{}^{\mathsf{a}}\|_\Gamma\, v$ the synchronisation of the symbols in Γ is not mandatory . Given two integers i, j with $i \leq j$, let $[i,j] = \{m \mid i \leq m \leq j\}$. Formally, for non-empty words $u, v \in \Sigma^+$, the finite sets of words $u \,{}^{\mathsf{s}}\|_\Gamma\, v$ and $u \,{}^{\mathsf{a}}\|_\Gamma\, v$ are defined, respectively, as follows [25],

$$\begin{aligned} u \,{}^{\mathsf{s}}\|_\Gamma\, v = \{x \mid\ & (\exists n \geq 1)(\forall i \in [1,n])\ \sigma_i \in \Gamma \wedge u_i, v_i \in (\Sigma \setminus \Gamma)^\star \\ & \wedge\ u = u_1\sigma_1u_2\cdots u_{n-1}\sigma_{n-1}u_n\ \wedge\ v = v_1\sigma_1v_2\cdots v_{n-1}\sigma_{n-1}v_n \\ & \wedge\ x \in (u_1 \sqcup\!\sqcup v_1)\sigma_1(u_2 \sqcup\!\sqcup v_2)\cdots(u_{n-1} \sqcup\!\sqcup v_{n-1})\sigma_{n-1}(u_n \sqcup\!\sqcup v_n)\}, \\ u \,{}^{\mathsf{a}}\|_\Gamma\, v = \{x \mid\ & (\exists n \geq 1)(\forall i \in [1,n])\ (\sigma_i \in \Gamma \wedge u_i, v_i \in \Sigma^\star \\ & \wedge\ u = u_1\sigma_1u_2\cdots u_{n-1}\sigma_{n-1}u_n\ \wedge\ v = v_1\sigma_1v_2\cdots v_{n-1}\sigma_{n-1}v_n \\ & \wedge\ x \in (u_1 \sqcup\!\sqcup v_1)\sigma_1(u_2 \sqcup\!\sqcup v_2)\cdots(u_{n-1} \sqcup\!\sqcup v_{n-1})\sigma_{n-1}(u_n \sqcup\!\sqcup v_n)\}. \end{aligned}$$

Manea et al. [18] introduced yet another kind of shuflle operator, where two words synchronise on a predefined set of words, called the backbones. Informally, for $u, v, t \in \Sigma^\star$ the *synchronised shuffle of u and v on the backbone t*, $u \,|||_t\, v$, requires that t is a subword of both u and v, and they have to synchronise on the positions forming t in both words. Formally, $u \,|||_t\, v$ is defined as follows:

$$\begin{aligned} u \,|||_t\, v = \{x \mid\ & t = \sigma_1\cdots\sigma_{n-1}\ \wedge\ u_i, v_i \in \Sigma^\star\ \wedge u = u_1\sigma_1u_2\cdots\sigma_{n-1}u_n\ \wedge \\ & v = v_1\sigma_1v_2\cdots\sigma_{n-1}v_n\ \wedge x \in (u_1 \sqcup\!\sqcup v_1)\sigma_1(u_2 \sqcup\!\sqcup v_2)\cdots\sigma_{n-1}(u_n \sqcup\!\sqcup v_n)\}. \end{aligned}$$

Moreover, given a language $L \subseteq \Sigma^\star$, we define

$$x \, |||_L \, y = \bigcup_{t \in L} x \, |||_t \, y.$$

Given $L_1, L_2, L \subseteq \Sigma^\star$, $\Gamma \subseteq \Sigma$, and $\circ \in \{{}^{\mathsf{s}}\|_\Gamma \, , {}^{\mathsf{a}}\|_\Gamma \, , |||_L\}$ one has, as usual,

$$L_1 \circ L_2 = \bigcup_{u \in L_1, v \in L_2} u \circ v.$$

3 Boolean Product Operators

In this section we define a family of operators that generalise the (synchronised) shuffle operators considered before. Let $\mathbb{B} = \{\top, \bot\}$ be a set of *Boolean values*, where $\top$ is true and $\bot$ is false. We start by motivating this approach with an example. For instance, given words $u, v, \sigma w \in \Sigma^\star$ such that $\sigma w \in u \, {}^{\mathsf{s}}\|_\Gamma \, v$, we have that either

1. $\sigma \in \Gamma$, $u = \sigma u'$, $v = \sigma v'$, and $w \in u' \, {}^{\mathsf{s}}\|_\Gamma \, v'$, or
2. $\sigma \notin \Gamma$, $u = \sigma u'$, and $w \in u' \, {}^{\mathsf{s}}\|_\Gamma \, v$, or
3. $\sigma \notin \Gamma$, $v = \sigma v'$, and $w \in u \, {}^{\mathsf{s}}\|_\Gamma \, v'$.

To express this, for Case 1. we associate to every letter $\sigma \in \Gamma$ the pair of Boolean values $(\top, \top)$, meaning that σ is the first letter of **both** u and v, and w belongs to the result of applying ${}^{\mathsf{s}}\|_\Gamma$ to the remaining parts of u and v. To every $\sigma \notin \Gamma$ we associate two pairs $(\top, \bot)$ and $(\bot, \top)$. The former corresponds to Case 2. and means that σ is the first letter of u, and w belongs to the result of applying ${}^{\mathsf{s}}\|_\Gamma$ to the remaining part of u and to v. The latter corresponds to Case 3. and means that σ is the first letter of v, and w belongs to the result of applying ${}^{\mathsf{s}}\|_\Gamma$ to u and to the remaining part of v. As a result, it makes sense to associate to each letter $\sigma \in \Gamma$ a binary Boolean function f_1, such that $f_1(\mathsf{b}_1, \mathsf{b}_2) = \top$ if and only if $\mathsf{b}_1 \wedge \mathsf{b}_2 = \top$, and to each letter $\sigma \notin \Gamma$ a binary Boolean function f_2, such that $f_2(\mathsf{b}_1, \mathsf{b}_2) = \top$ if and only if $\mathsf{b}_1 \oplus \mathsf{b}_2 = \top$, where $\oplus$ represents the Boolean function *exclusive or*. We can generalise this behaviour to n-ary, $n \geq 2$, operators in the following way.

Consider the set $\mathcal{F}_n$ of *Boolean functions* $f : \mathbb{B}^n \longrightarrow \mathbb{B}$. For $f \in \mathcal{F}_n$, let

$$\mathsf{sat}(f) = \{ (\mathsf{b}_1, \ldots, \mathsf{b}_n) \in \mathbb{B}^n \mid f(\mathsf{b}_1, \ldots, \mathsf{b}_n) = \top \} \setminus \{(\bot, \ldots, \bot)\}.$$

Example 1. We have $\mathsf{sat}(\wedge) = \{(\top, \top)\}$ and $\mathsf{sat}(\oplus) = \{(\top, \bot), (\bot, \top)\}$.

Remark 1. Whenever convenient we shall denote a tuple $(x_1, \ldots, x_n)$ by x. Moreover, we shall use x_i without explicitly stating the range of i.

Given $\psi : \Sigma \longrightarrow \mathcal{F}_n$ we define an n-ary operator $\|_\psi$ on words $u_1, \ldots, u_n \in \Sigma^\star$, which we call the *Boolean product under* ψ, as follows:

$$\begin{aligned} \|_\psi(\varepsilon, \ldots, \varepsilon) &= \{\varepsilon\}, \qquad (1) \\ \|_\psi(u_1, \ldots, u_n) &= \{\sigma w \mid (\exists \mathsf{b} \in \mathsf{sat}(\psi(\sigma))) \\ &\qquad [u_i = \sigma v_i \text{ if } \mathsf{b}_i; u_i = v_i \text{ otherwise}] \wedge w \in \|_\psi \psi(v_1, \ldots, v_n)\}. \end{aligned}$$

Example 2. Considering $\psi(a) = \wedge$ and $\psi(b) = \oplus$, we have

$$\|_{\psi}(ab, ba) = \{bab\} = ab\,{}^{\mathsf{s}}\|_{\{a\}}\, ba.$$

The operator $\|_{\psi}$ is naturally extended to languages by

$$\|_{\psi}(L_1, \ldots, L_n) = \bigcup_{u \in L_1 \times \cdots \times L_n} \|_{\psi}(u_1, \ldots, u_n),$$

where $u = (u_1, \ldots, u_n)$. For some operation op, if one has $\psi(\sigma) = \mathsf{op}$ for all $\sigma \in \Sigma$, we often write $\|_{\mathsf{op}}(L_1, \ldots, L_n)$ instead of $\|_{\psi}(L_1, \ldots, L_n)$.

The following lemmas are direct consequences of the definitions of the operations.

Lemma 3. *Let $L_1, L_2, L_3 \subseteq \Sigma^\star$, then*

$$\begin{aligned}\|_{\wedge}(L_1, L_2) &= L_1 \cap L_2,\\ \|_{\vee}(L_1, L_2) &= L_1\, {}^{\mathsf{a}}\|_{\Sigma}\, L_2,\\ \|_{\oplus}(L_1, L_2) &= L_1 \sqcup L_2,\end{aligned}$$

where $\oplus$ denotes the Boolean binary function exclusive or. *Moreover,*

$$\|_{\mathsf{op}}(L_1, L_2, L_3) = L_1 \,|||_{L_2}\, L_3,$$

where $\mathsf{op}(\mathsf{b}_1, \mathsf{b}_2, \mathsf{b}_3) = (\mathsf{b}_2 \leftrightarrow (\mathsf{b}_1 \wedge \mathsf{b}_3))$*, with*

$$\mathsf{sat}(\mathsf{op}) = \{(\top, \top, \top), (\top, \bot, \bot), (\bot, \bot, \top)\}.$$

Lemma 4. *Let $\Gamma \subseteq \Sigma$ and $L_1, L_2 \subseteq \Sigma^\star$, then*

$$\begin{aligned}\|_{\psi_s}(L_1, L_2) &= L_1\, {}^{\mathsf{s}}\|_{\Gamma}\, L_2,\\ \|_{\psi_a}(L_1, L_2) &= L_1\, {}^{\mathsf{a}}\|_{\Gamma}\, L_2,\end{aligned}$$

where

$$\psi_s(\sigma) = \begin{cases} \wedge, & \textit{if } \sigma \in \Gamma,\\ \oplus, & \textit{otherwise,}\end{cases} \quad \textit{and} \quad \psi_a(\sigma) = \begin{cases} \vee, & \textit{if } \sigma \in \Gamma,\\ \oplus, & \textit{otherwise.}\end{cases}$$

Example 3. The operators above can be generalised to arities other than two. For instance, we have $\|_{\wedge}(L_1, \ldots, L_n) = L_1 \cap \cdots \cap L_n$ and $\|_{\oplus}(L_1, \ldots, L_n) = L_1 \sqcup \cdots \sqcup L_n$, where $\wedge$ and $\oplus$ are n-ary Boolean operators for $n \geq 2$.

3.1 Boolean Product Automata

In this section we define operations on finite automata whose languages correspond to Boolean products of the languages of their operands. In the rest of the paper we fix a function $\psi : \Sigma \longrightarrow \mathcal{F}_n$.

Given NFAs $\mathcal{A}_1, \ldots, \mathcal{A}_n$, with $\mathcal{A}_i = \langle Q_i, \Sigma, \delta_i, \iota_i, F_i \rangle$ $(i \in [1, n])$, the *Boolean product automaton* is

$$\|_\psi(\mathcal{A}_1, \ldots, \mathcal{A}_n) = \langle Q_1 \times \cdots \times Q_n, \Sigma, \delta_\psi, \iota, F_1 \times \cdots \times F_n \rangle^\iota, \tag{2}$$

where $\iota = (\iota_1, \ldots, \iota_n)$, and

$$\delta_\psi(s, \sigma) = \bigcup_{\mathsf{b} \in \mathsf{sat}(\psi(\sigma))} \{ t \in Q_1 \times \cdots \times Q_n \mid t_i \in \delta_i(s_i, \sigma) \text{ if } \mathsf{b}_i; t_i = s_i \text{ otherwise} \}.$$

It follows from the definitions on page 2 that for $\|_\psi(\mathcal{A}_1, \ldots, \mathcal{A}_n) = \langle Q, \Sigma, \delta, \iota, F \rangle$ we have $Q = (Q_1 \times \cdots \times Q_n)^\iota$, $F = (F_1 \times \cdots \times F_n)^\iota$, and $\delta = (\delta_\psi)_{|Q}$.

Proposition 1. $\mathcal{L}(\|_\psi(\mathcal{A}_1, \ldots, \mathcal{A}_n)) = \|_\psi (\mathcal{L}(\mathcal{A}_1), \ldots, \mathcal{L}(\mathcal{A}_n))$.

Corollary 1. *If $L_1, \ldots, L_n$ are regular languages, then the Boolean product $\|_\psi (L_1, \ldots, L_n)$ is also a regular language.*

Corollary 2. *Consider $n \geq 2$ deterministic finite automata $\mathcal{A}_1, \ldots, \mathcal{A}_n$. Then, the Boolean product automaton $\|_\psi(\mathcal{A}_1, \ldots, \mathcal{A}_n)$ is deterministic if and only if*

$$\forall \sigma \in \Sigma, \ (|\,\mathsf{sat}(\psi(\sigma)| = 1).$$

3.2 Right-Invariant Relations and Quotient Automata

In this section we show that, given NFAs $\mathcal{A}_1, \ldots, \mathcal{A}_n$, $n \geq 2$, and n right-invariant equivalence relations $\equiv_1, \ldots, \equiv_n$ defined on their sets of states, one can obtain from the Boolean product of the quotient automata $\mathcal{A}_1/_{\equiv_1}, \ldots, \mathcal{A}_n/_{\equiv_n}$ a quotient automaton of the Boolean product of $\mathcal{A}_1, \ldots, \mathcal{A}_n$.

Given n equivalence relations $\equiv_1, \ldots, \equiv_n$, with $n \geq 2$, defined on sets X_1, …, X_n respectively, the equivalence relation $\equiv_\pi$ on $X = X_1 \times \cdots \times X_n$, defined by

$$(x_1, \ldots, x_n) \equiv_\pi (y_1, \ldots, y_n) \quad \text{if and only if} \quad x_1 \equiv_1 y_1, \ldots, x_n \equiv_n y_n,$$

is the *natural extension* of $\equiv_1, \ldots, \equiv_n$ to set X. In particular, for every $(x_1, \ldots, x_n) \in X$, we have

$$[(x_1, \ldots, x_n)]_{\equiv_\pi} = [x_1]_{\equiv_1} \times \cdots \times [x_n]_{\equiv_n}.$$

Furthermore, whenever convenient, we represent the natural extension $\equiv_\pi$ by $(\equiv_1, \ldots, \equiv_n)$.

Lemma 5. *Let $\mathcal{A}_i = \langle Q_i, \Sigma, \delta_i, \iota_i, F_i \rangle$, for $i \in [1, n]$, with $n \geq 2$, be NFAs over Σ. For each $i \in [1, n]$, let $\equiv_i$ be a right-invariant equivalence relation on Q_i w.r.t. $\mathcal{A}_i$. Furthermore, consider the natural extension $\equiv_\pi$ of the relations $\equiv_1, \ldots, \equiv_n$ to $Q_1 \times \cdots \times Q_n$. Then, the restriction $\equiv_\pi^\iota$ of $\equiv_\pi$ to $Q = (Q_1 \times \cdots \times Q_n)^\iota$ where $\iota = (\iota_1, \ldots, \iota_n)$, is right-invariant w.r.t. the automaton $\|_\psi(\mathcal{A}_1, \ldots, \mathcal{A}_n)$.*

Proposition 2. *Let $\mathcal{A}_i = \langle Q_i, \Sigma, \delta_i, \iota_i, F_i \rangle$, for $i \in [1, n]$, with $n \geq 2$, be NFAs over Σ. For each $i \in [1, n]$, let $\equiv_i$ be an equivalence relation on Q_i, which is right-invariant w.r.t. $\mathcal{A}_i$. Then,*

$$\|_\psi(\mathcal{A}_1/_{\equiv_1}, \ldots, \mathcal{A}_n/_{\equiv_n}) \simeq \|_\psi(\mathcal{A}_1, \ldots, \mathcal{A}_n)/_{\equiv_\pi^\iota}.$$

The following result is straightforward and will be used in Sect. 4.

Lemma 6. *Consider two families of automata $\mathcal{A}_1, \ldots, \mathcal{A}_n$ and $\mathcal{B}_1, \ldots, \mathcal{B}_n$, such that $\mathcal{A}_i \simeq \mathcal{B}_i$ for $i \in [1, n]$. Then, one has $\|_\psi(\mathcal{A}_1, \ldots, \mathcal{A}_n) \simeq \|_\psi(\mathcal{B}_1, \ldots, \mathcal{B}_n)$.*

3.3 Regular Expressions with Boolean Products

In this section we extend the set of standard regular expressions with Boolean product operators. Given $\psi : \Sigma \longrightarrow \mathcal{F}_n$, with $n \geq 2$, we use $\|_\psi$ to denote the corresponding regular operator. Let RE($\|_\psi$) denote the set of expressions containing $\emptyset$ and all terms generated by the following grammar:

$$\alpha := \varepsilon \mid \sigma \mid (\alpha + \alpha) \mid (\alpha \cdot \alpha) \mid (\alpha^\star) \mid \|_\psi(\alpha, \ldots, \alpha), \tag{3}$$

where $\sigma \in \Sigma$, and $\psi \in \mathcal{F}_n^\Sigma$. The *language* associated to α is denoted by $\mathcal{L}(\alpha)$, and defined as usual for standard operators, and

$$\mathcal{L}(\|_\psi(\alpha_1, \ldots, \alpha_n)) = \|_\psi(\mathcal{L}(\alpha_1), \ldots, \mathcal{L}(\alpha_n)).$$

We define $\varepsilon(\alpha)$ by $\varepsilon(\alpha) = \varepsilon$ if $\varepsilon \in \mathcal{L}(\alpha)$, and $\varepsilon(\alpha) = \emptyset$, otherwise. We also consider $\varepsilon S = S\varepsilon = S$ and $\emptyset S = S\emptyset = \emptyset$, for any set S of expressions.

4 From Boolean Product Expressions to Automata

Several constructions converting regular expressions into automata have previously been extended to accommodate different synchronised shuffle operators [5–7,20]. In this section, we revisit a number of standard constructions and extend them to expressions in RE($\|_\psi$), generalising results obtained previously for several synchronised operators.

4.1 Glushkov Automata for RE($\|_\psi$)

Glushkov automata for expressions in RE($\|_\psi$) are an extension of the classic position automata for regular expressions with shuffle operators [5–7, 20], therein also called location automata. Given a standard expression α, one can mark each occurrence of an alphabet symbol σ with its position i in α, reading it from left to right, i.e., representing this occurrence by σ^i. The resulting regular expression is a *marked* regular expression $\overline{\alpha}$ with all alphabet symbols occurring only once (linear) and belonging to $\Sigma_{\overline{\alpha}}$, which contains the (marked) letters occurring in $\overline{\alpha}$. The *position* or *Glushkov automaton*, $\mathcal{A}_{\mathcal{G}\ell}(\alpha)$, is computed from sets of positions, specifying which occurrences of letters can appear at the beginning, which can follow each other, or appear as the last letter in a word generated by α. Let $\mathsf{Pos}(\overline{\alpha}) = [1, \Sigma_{\overline{\alpha}}]$, and $\mathsf{Pos}_0(\overline{\alpha}) = \mathsf{Pos}(\overline{\alpha}) \cup \{0\}$. The sets that are used to define the position automaton are

$$\mathsf{First}(\overline{\alpha}) = \{\, i \mid (\exists w \in \Sigma_{\overline{\alpha}}^{\star})\, (\sigma^i w \in \mathcal{L}(\overline{\alpha}))\,\},$$
$$\mathsf{Last}(\overline{\alpha}) = \{\, i \mid (\exists w \in \Sigma_{\overline{\alpha}}^{\star})\, (w\sigma^i \in \mathcal{L}(\overline{\alpha}))\,\},$$

and, given $i \in \mathsf{Pos}(\overline{\alpha})$,

$$\mathsf{Follow}(\overline{\alpha}, i) = \{\, j \mid (\exists u, v \in \Sigma_{\overline{\alpha}}^{\star})\, (u\sigma^i\sigma^j v \in \mathcal{L}(\overline{\alpha}))\,\}.$$

For expressions containing the shuffle operator $\sqcup\!\sqcup$, Broda et al. [4, 7] considered more complex structures, called *locations*. Locations were defined in such a way that, given an expression with nested shuffles, they allow to specify how far a word has advanced in each of the shuffle components of this expression. Locations were also used in [6] to define a Glushkov automaton for several synchronised shuffles. Now, we define locations for expressions in RE($\|_\psi$). In this case, given $\alpha \in$ RE($\|_\psi$), each location $p \in \mathsf{Loc}(\overline{\alpha})$ is an n-ary tree, whose leaves are labelled either with 0, or with a position $i \in \mathsf{Pos}(\overline{\alpha})$. Additionally, instead of defining the set Last explicitly, we annotate each location p with a Boolean b. Then, $p \in \mathsf{Last}(\overline{\alpha})$ if and only if the annotation of p is $\top$, i.e. $\langle p : \top \rangle \in \mathsf{Loc}(\overline{\alpha})$. Furthermore, we define $\mathsf{Last}_0(\overline{\alpha}) = \mathsf{Last}(\overline{\alpha}) \cup \{0\}$ if $\varepsilon(\alpha)$ is ε, and $\mathsf{Last}_0(\overline{\alpha}) = \mathsf{Last}(\overline{\alpha})$ otherwise. Note that one has $\varepsilon(\alpha) = \varepsilon(\overline{\alpha})$ for any expression α. In the definition below, the term $\varepsilon(\alpha) = \varepsilon$ denotes the Boolean value $\top$ if the equation is true, and $\bot$ otherwise. The set of *annotated locations* $\mathsf{Loc}(\overline{\alpha})$ is inductively defined as follows:

$$\mathsf{Loc}(\varepsilon) = \emptyset, \quad \mathsf{Loc}(\sigma^i) = \{\langle i : \top \rangle\}, \quad \mathsf{Loc}(\overline{\alpha}^{\star}) = \mathsf{Loc}(\overline{\alpha}),$$
$$\mathsf{Loc}(\overline{\alpha} + \overline{\beta}) = \mathsf{Loc}(\overline{\alpha}) \cup \mathsf{Loc}(\overline{\beta}),$$
$$\mathsf{Loc}(\overline{\alpha}\overline{\beta}) = \begin{cases} \mathsf{Loc}(\overline{\alpha}) \cup \mathsf{Loc}(\overline{\beta}), & \text{if } \varepsilon(\beta) = \varepsilon, \\ \{\, \langle p : \bot \rangle \mid \langle p : \mathsf{b} \rangle \in \mathsf{Loc}(\overline{\alpha}) \,\} \cup \mathsf{Loc}(\overline{\beta}), & \text{otherwise,} \end{cases}$$
$$\mathsf{Loc}(\|_\psi(\overline{\alpha_1}, \ldots, \overline{\alpha_n})) = \{\langle (p_1, \ldots, p_n) : \mathsf{b}_1 \wedge \cdots \wedge \mathsf{b}_n \rangle \mid \forall i\, [\langle p_i : \mathsf{b}_i \rangle \in \mathsf{Loc}(\overline{\alpha_i}) \vee (p_i = 0 \wedge \mathsf{b}_i = (\varepsilon(\alpha_i) = \varepsilon))]\} \setminus \{\langle (0, \ldots, 0) : \mathsf{b} \rangle\}$$

Furthermore, let $\mathsf{Loc}_0(\overline{\alpha}) = \mathsf{Loc}(\overline{\alpha}) \cup \{\langle 0 : \varepsilon(\alpha) = \varepsilon \rangle\}$. It follows from the definition that $\langle p : \mathsf{b}\rangle$, $\langle p : \mathsf{b}'\rangle \in \mathsf{Loc}(\overline{\alpha})$ implies $\mathsf{b} = \mathsf{b}'$. As such, we shall omit the Boolean of an annotated location whenever convenient. Moreover, we shall use $\|_\psi(\alpha)$ (resp. $\|_\psi(\overline{\alpha})$) as an abbreviation for $\|_\psi(\alpha_1, \ldots, \alpha_n)$ (resp. $\|_\psi(\overline{\alpha_1}, \ldots, \overline{\alpha_n})$). Now, we extend the usual definitions of $\mathsf{First}(\overline{\alpha}) \subseteq \Sigma \times \mathsf{Loc}(\overline{\alpha})$ and $\mathsf{Follow}(\overline{\alpha}) \subseteq \mathsf{Loc}(\overline{\alpha}) \times \Sigma \times \mathsf{Loc}(\overline{\alpha})$ to include the operator $\|_\psi$:

$$\mathsf{First}(\|_\psi(\overline{\alpha})) = \bigcup_{\sigma \in \Sigma, \mathsf{b} \in \mathsf{sat}(\psi(\sigma))} \{(\sigma, t) \mid \forall i \, [(\sigma, t_i) \in \mathsf{First}(\overline{\alpha_i}) \text{ if } \mathsf{b}_i; t_i = 0 \text{ otherwise}]\},$$

$$\mathsf{Follow}(\|_\psi(\overline{\alpha}), s) = \bigcup_{\sigma \in \Sigma, \mathsf{b} \in \mathsf{sat}(\psi(\sigma))} \{(\sigma, t) \mid \forall i \, [(\sigma, t_i) \in \mathsf{Follow}(\overline{\alpha_i}, s_i) \text{ if } \mathsf{b}_i; t_i = s_i \text{ otherwise}]\}.$$

Example 4. Let $\alpha = \|_\wedge (a(ab^\star + bb^\star)^\star, (ab^\star)^\star) = a(ab^\star + bb^\star)^\star \cap (ab^\star)^\star$, with $\overline{\alpha} = \|_\wedge (a^1(a^2(b^3)^\star + b^4(b^5)^\star)^\star, (a^6(b^7)^\star)^\star)$. We have,

$$\mathsf{Loc}_0(\overline{\alpha}) = \{\langle 0, \bot \rangle\} \cup \{\langle (n, 6), \top \rangle, \langle (n, 7), \top \rangle \mid n \in [1, 5]\}.$$

Furthermore, $\mathsf{First}(\overline{\alpha}) = \{(1, 6)\}$, and for instance

$$\mathsf{Follow}(\overline{\alpha}, (2, 6)) = \{(a, (2, 6)), (b, (3, 7)), (b, (4, 7))\},$$

and $\mathsf{Follow}(\overline{\alpha}, (1, 7)) = \{(a, (2, 6)), (b, (4, 7))\}$.

Then, the *Glushkov automaton* for $\alpha \in \mathrm{RE}(\|_\psi)$ is

$$\mathcal{A}_{\mathcal{G}\ell}(\alpha) = \langle \mathsf{Loc}_0(\overline{\alpha}), \Sigma, \delta_{\mathcal{G}\ell}, 0, \mathsf{Last}_0(\overline{\alpha}) \rangle^0,$$

where $\mathsf{Follow}(\overline{\alpha}, 0) = \mathsf{First}(\overline{\alpha})$, and

$$\delta_{\mathcal{G}\ell}(s, \sigma) = \{\, t \mid (\sigma, t) \in \mathsf{Follow}(\overline{\alpha}, s) \,\},$$

for $s \in \mathsf{Loc}_0(\overline{\alpha})$ and $\sigma \in \Sigma$.

Example 5. The Glushkov automaton for α from Example 4 is depicted below. Although $|\mathsf{Loc}_0(\overline{\alpha})| = 11$, the automaton has only six states, since the remaining ones are not reachable from the initial state.

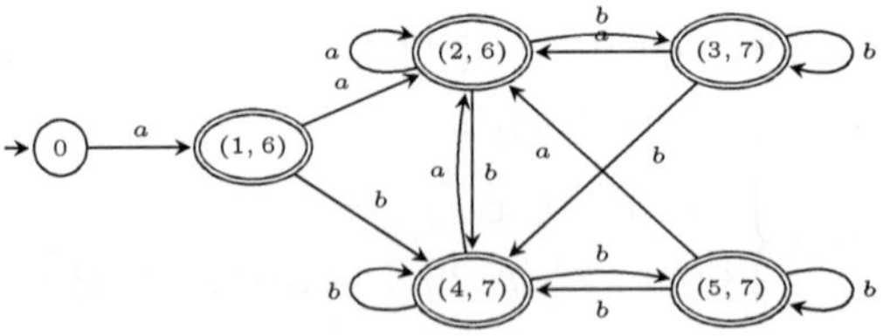

Remark 2. Note, that for expressions without the $\|_\psi$ operator, the Glushkov automaton is always initially connected, i.e.,

$$\mathcal{A}_{\mathscr{G}\ell}(\alpha) = \langle \mathsf{Loc}_0(\overline{\alpha}), \Sigma, \delta_{\mathscr{G}\ell}, 0, \mathsf{Last}_0(\overline{\alpha})\rangle^0 = \langle \mathsf{Loc}_0(\overline{\alpha}), \Sigma, \delta_{\mathscr{G}\ell}, 0, \mathsf{Last}_0(\overline{\alpha})\rangle.$$

The following results shows that the Glushkov automaton of a Boolean product expression could have been defined alternatively as the Boolean product of the Glushkov automata of its operands.

Proposition 3. *For* $\alpha_1, \ldots, \alpha_n \in \mathrm{RE}(\|_\psi)$ *we have,*

$$\|_\psi(\mathcal{A}_{\mathscr{G}\ell}(\alpha_1), \ldots, \mathcal{A}_{\mathscr{G}\ell}(\alpha_n)) \simeq \mathcal{A}_{\mathscr{G}\ell}(\|_\psi(\alpha_1, \ldots, \alpha_n)).$$

Proposition 4. *If* $\alpha \in \mathrm{RE}(\|_\psi)$*, then* $\mathcal{L}(\mathcal{A}_{\mathscr{G}\ell}(\alpha)) = \mathcal{L}(\alpha)$.

Since the synchronised operators ${}^{\mathsf{s}}\|_\Gamma$, ${}^{\mathsf{a}}\|_\Gamma$, $\sqcup\!\sqcup$, and $|||$ can be defined as Boolean product operators, cf. Lemma 3 and Lemma 4, the Glushkov automata for expressions with these operators can be obtained as the Boolean product of the Glushkov automata of their operands.

4.2 Partial Derivative Automata for RE($\|_\psi$)

Partial derivative automata for regular expressions is another well known automata construction [1]. Their extension for Boolean product automata was studied by Broda et al. [3]. In this section we revisit this construction and show that, again, the partial derivative automaton of a Boolean product expression $\|_\psi(\alpha_1, \ldots, \alpha_n)$ coincides with the Boolean product automaton of the partial derivative automata of its operands $\alpha_1, \ldots, \alpha_n$. Consider the set of partial derivatives of a standard regular expression by a symbol $\sigma \in \Sigma$, $\partial_\sigma(\alpha)$, to be defined as usual. The set of partial derivatives of an expression $\|_\psi(\alpha_1, \ldots, \alpha_n)$, by a symbol $\sigma \in \Sigma$, is then defined by,

$$\partial_\sigma(\|_\psi(\alpha)) \quad = \quad \bigcup_{\mathsf{b} \in \mathsf{sat}(\psi(\sigma))} \{\|_\psi(\beta) \mid \beta_i \in \partial_\sigma(\alpha_i) \text{ if } \mathsf{b}_i; \beta_i = \alpha_i \text{ otherwise}\}.$$

The following Lemma is a direct consequence of the definition above.

Lemma 7. $\mathcal{L}(\partial_\sigma(\|_\psi(\alpha_1, \ldots, \alpha_n))) = \sigma^{-1}\mathcal{L}(\|_\psi(\alpha_1, \ldots, \alpha_n))$.

The set of partial derivatives of an expression α w.r.t. a word $w \in \Sigma^\star$ is inductively defined by $\partial_\varepsilon(\alpha) = \{\alpha\}$ and $\partial_{w\sigma}(\alpha) = \partial_\sigma(\partial_w(\alpha))$.

Let $\partial(\alpha) = \bigcup_{w \in \Sigma^\star} \partial_w(\alpha)$ and $\partial^+(\alpha) = \bigcup_{w \in \Sigma^+} \partial_w(\alpha)$. The partial derivative automaton of $\alpha \in \mathrm{RE}(\|_\psi)$ is

$$\mathcal{A}_{\mathrm{PD}}(\alpha) = \langle \partial(\alpha), \Sigma, \alpha, \delta_{\mathrm{PD}}, F_{\mathrm{PD}}\rangle,$$

with $F_{\mathrm{PD}} = \{\beta \in \partial(\alpha) \mid \varepsilon(\beta) = \varepsilon\}$ and $\delta_{\mathrm{PD}}(\beta, \sigma) = \partial_\sigma(\beta)$, for $\beta \in \partial(\alpha)$, $\sigma \in \Sigma$.

Example 6. The partial derivative automaton for $\alpha = a(ab^\star + bb^\star)^\star \cap (ab^\star)^\star$ from Example 4 is depicted below, where we use the abbreviations $\beta = (ab^\star + bb^\star)^\star$ and $\gamma = (ab^\star)^\star$.

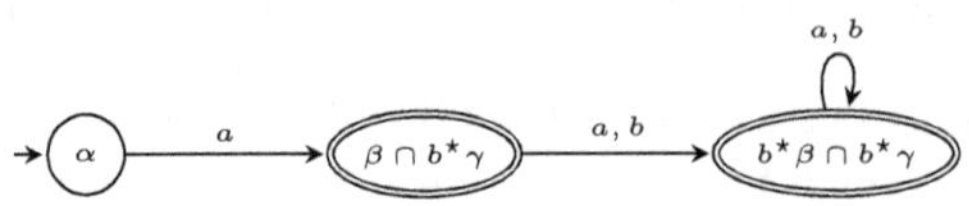

The following result shows that the automaton $\mathcal{A}_{\mathrm{PD}}(\|_\psi(\alpha_1, \ldots, \alpha_n))$ coincides with $\|_\psi(\mathcal{A}_{\mathrm{PD}}(\alpha_1), \ldots, \mathcal{A}_{\mathrm{PD}}(\alpha_n))$.

Proposition 5. $\|_\psi(\mathcal{A}_{\mathrm{PD}}(\alpha_1), \ldots, \mathcal{A}_{\mathrm{PD}}(\alpha_n)) \simeq \mathcal{A}_{\mathrm{PD}}(\|_\psi(\alpha_1, \ldots, \alpha_n))$.

For standard regular expressions extended with several regular operators, it is the case that the partial derivative automaton is a quotient of the Glushkov automaton [6,7,11]. We show that the same is true for expressions in $\mathrm{RE}(\|_\psi)$. To this end, we define the set of unmarked c-continuations, as defined in [11], which we denote by $\mathsf{d}(\overline{\alpha}, p)$ for $\alpha \in \mathrm{RE}(\|_\psi)$ and $p \in \mathsf{Loc}_0(\overline{\alpha})$. One has $\mathsf{d}(\overline{\alpha}, 0) = \alpha$. For $p, p_1, \ldots, p_n \in \mathsf{Loc}(\overline{\alpha})$ the definition is as follows:

$$\mathsf{d}(\sigma^i, i) = \varepsilon, \ \mathsf{d}(\overline{\alpha^\star}, p) = \mathsf{d}(\overline{\alpha}, p)\alpha^\star,$$

$$\mathsf{d}(\overline{\alpha_1 + \alpha_2}, p) = \begin{cases} \mathsf{d}(\overline{\alpha_1}, p), & \text{if } p \in \mathsf{Loc}(\alpha_1), \\ \mathsf{d}(\overline{\alpha_2}, p), & \text{if } p \in \mathsf{Loc}(\alpha_2), \end{cases}$$

$$\mathsf{d}(\overline{\alpha_1\alpha_2}, p) = \begin{cases} \mathsf{d}(\overline{\alpha_1}, p)\alpha_2, & \text{if } p \in \mathsf{Loc}(\alpha_1), \\ \mathsf{d}(\overline{\alpha_2}, p), & \text{if } p \in \mathsf{Loc}(\alpha_2), \end{cases}$$

$$\mathsf{d}(\overline{\|_\psi(\alpha_1, \ldots, \alpha_n)}, (p_1, \ldots, p_n)) = \|_\psi(\mathsf{d}(\overline{\alpha_1}, p_1), \ldots, \mathsf{d}(\overline{\alpha_n}, p_n)).$$

For standard regular expressions α it has been shown that $\mathcal{A}_{\mathrm{PD}}(\alpha)$ is a quotient of $\mathcal{A}_{\mathcal{G}\ell}(\alpha)$ by the right-invariant relation $\equiv_c$, defined by

$$p \equiv_c q \quad \text{iff} \quad \mathsf{d}(\overline{\alpha}, p) = \mathsf{d}(\overline{\alpha}, q),$$

for $p, q \in \mathsf{Loc}_0(\overline{\alpha})$. We extend $\equiv_c$ to expressions of the form $\|_\psi(\alpha_1, \ldots, \alpha_n)$. However, as we show below, for those expressions the relation $\equiv_c$ coincides with the natural extension of the equivalence relations $\equiv_c^1, \ldots, \equiv_c^n$ to the set $\mathsf{Loc}_0(\overline{\alpha_1})^2 \times \cdots \times \mathsf{Loc}_0(\overline{\alpha_n})^2$.

Lemma 8. *Given* $\|_\psi(\alpha) \in \mathrm{RE}(\|_\psi)$*, the natural extension* $\equiv_\pi \subseteq \mathsf{Loc}_0(\|_\psi(\alpha))^2$ *of the relations* $\equiv_c^1 \subseteq \mathsf{Loc}_0(\overline{\alpha_1})^2, \ldots, \equiv_c^n \subseteq \mathsf{Loc}_0(\overline{\alpha_n})^2$ *coincides with* $\equiv_c$*, i.e., given* $p, q \in \mathsf{Loc}_0(\|_\psi(\overline{\alpha}))$*, we have* $p \equiv_\pi q \iff \mathsf{d}(\|_\psi(\overline{\alpha}), p) = \mathsf{d}(\|_\psi(\overline{\alpha}), q)$.

Proposition 6. *Given* $\alpha \in \mathrm{RE}(\|_\psi)$*, one has* $\mathcal{A}_{\mathrm{PD}}(\alpha) \simeq \mathcal{A}_{\mathcal{G}\ell}(\alpha)/_{\equiv_c}$.

Proof. The proof proceeds by induction on the structure of α. The result is true for the standard regular operators. It remains to consider expressions of the form $\|_\psi(\alpha_1,\ldots,\alpha_n)$. In this case, suppose that $\mathcal{A}_{\rm PD}(\alpha_i) \simeq \mathcal{A}_{\mathscr{G}\!\ell}(\alpha_i)/_{\equiv_c}$, for $i \in [1,n]$. Then, it follows by Proposition 5, Lemma 6, the induction hypothesis, Lemma 8, Proposition 2, and finally by Proposition 3, that,

$$\begin{aligned}
\mathcal{A}_{\rm PD}(\|_\psi(\alpha_1,\ldots,\alpha_n)) &\simeq (\|_\psi(\mathcal{A}_{\rm PD}(\alpha_1),\ldots,\mathcal{A}_{\rm PD}(\alpha_n)))\\
&\simeq (\|_\psi(\mathcal{A}_{\mathscr{G}\!\ell}(\alpha_1)/_{\equiv_c},\ldots,\mathcal{A}_{\mathscr{G}\!\ell}(\alpha_n)/_{\equiv_c}))\\
&\simeq (\|_\psi(\mathcal{A}_{\mathscr{G}\!\ell}(\alpha_1),\ldots,\mathcal{A}_{\mathscr{G}\!\ell}(\alpha_n))/_{\equiv_c})\\
&\simeq (\mathcal{A}_{\mathscr{G}\!\ell}(\|_\psi(\alpha_1,\ldots,\alpha_n))/_{\equiv_c}).
\end{aligned}$$ □

Example 7. The partial derivative automaton in Example 6 is obtained from the Glushkov automaton in Example 5, by merging the states in $\mathcal{A}_{\mathscr{G}\!\ell}(\alpha)$ labelled with $(2,6)$, $(3,7)$, $(4,7)$, and $(5,7)$. In fact, $\mathsf{d}(\overline{\alpha},(2,6)) = \cdots = \mathsf{d}(\overline{\alpha},(5,7)) = b^\star\beta \cap b^\star\gamma$.

4.3 Follow Automata for RE($\|_\psi$)

Another well known quotient of the Glushkov automaton is the follow automaton [2,15]. Given $\alpha \in \mathrm{RE}(\|_\psi)$ let

$$\mathsf{F}(\alpha) = \{\, (\mathsf{Follow}(\overline{\alpha},p),\varepsilon(p)) \mid p \in \mathsf{Loc}_0(\overline{\alpha}) \,\} \subseteq 2^{\Sigma\times\mathsf{Loc}(\overline{\alpha})} \times \{\varepsilon,\emptyset\},$$

where $\varepsilon(p) = \varepsilon$ if $p \in \mathsf{Last}_0(\overline{\alpha})$, and $\varepsilon(p) = \emptyset$ otherwise. Then, the *Follow automaton* is defined by,

$$\mathcal{A}_{\rm F}(\alpha) = \langle \mathsf{F}(\alpha),\Sigma,\delta_{\rm F},\iota_{\rm F},F_{\rm F}\rangle^{\iota_{\rm F}},$$

where $\iota_{\rm F} = (\mathsf{Follow}(\overline{\alpha},0),\varepsilon(0))$, $F_{\rm F} = \{\,(S,c)\in\mathsf{F}(\alpha) \mid c=\varepsilon\,\}$, and for $(S,c) \in \mathsf{F}(\alpha)$ and $\sigma\in\Sigma$, $\delta_{\rm F}((S,c),\sigma) = \{\,(\mathsf{Follow}(\overline{\alpha},p),\varepsilon(p)) \mid (\sigma,p)\in S\,\}$. As usual, we consider the equivalence relation $\equiv_{\rm F}\subseteq \mathsf{Loc}_0(\overline{\alpha})\times\mathsf{Loc}_0(\overline{\alpha})$ defined by

$$p \equiv_{\rm F} q \;\Leftrightarrow\; (\mathsf{Follow}(\overline{\alpha},p),\varepsilon(p)) = (\mathsf{Follow}(\overline{\alpha},q),\varepsilon(q)).$$

Lemma 9. *The relation $\equiv_{\rm F}$ is right-invariant with respect to the automaton*

$$\mathcal{A}(\alpha) = \langle \mathsf{Loc}_0(\overline{\alpha}),\Sigma,\delta_{\mathscr{G}\!\ell},0,\mathsf{Last}_0(\overline{\alpha})\rangle.$$

Lemma 10. $\langle \mathsf{Loc}_0(\overline{\alpha}),\Sigma,\delta_{\mathscr{G}\!\ell},0,\mathsf{Last}_0(\overline{\alpha})\rangle/_{\equiv_{\rm F}} \simeq \langle \mathsf{F}(\alpha),\Sigma,\delta_{\rm F},\iota_{\rm F},F_{\rm F}\rangle$.

Proposition 7. *For $\alpha\in\mathrm{RE}(\|_\psi)$, we have $\mathcal{A}_{\mathscr{G}\!\ell}(\alpha)/_{\equiv_{\rm F}} \simeq \mathcal{A}_{\rm F}(\alpha)$.*

In contrast to what happens with the Glushkov and the partial derivative automaton, the follow automaton $\mathcal{A}_{\rm F}(\|_\psi(\alpha))$ is not necessarily isomorphic to the Boolean automaton $\|_\psi(\mathcal{A}_{\rm F}(\alpha_1),\ldots,\mathcal{A}_{\rm F}(\alpha_n))$, since in this case, the natural extension of $\equiv_{\rm F}^1,\ldots,\equiv_{\rm F}^n$ is not necessarily equal to $\equiv_{\rm F}$ defined in $\mathsf{Loc}_0(\overline{\|_\psi(\alpha)})$.

Example 8. Consider two expressions $\alpha = (a^\star + b^\star)^\star bb^\star$, with $\overline{\alpha} = ((a^1)^\star + (b^2)^\star)^\star b^3(b^4)^\star$, and $\beta = a(ab^\star + bb^\star)^\star$, with $\overline{\beta} = a^1((a^2)(b^3)^\star + (b^4)(b^5)^\star)^\star$. The follow automata are, respectively:

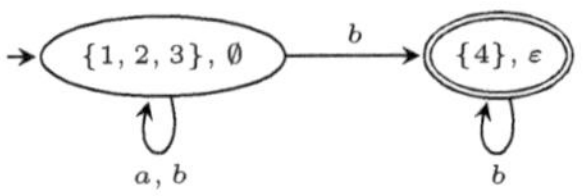

$\mathcal{A}_{\mathrm{F}}(\alpha)$

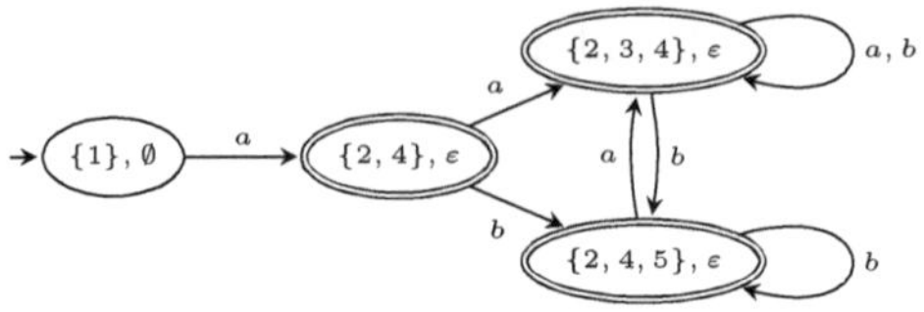

$\mathcal{A}_{\mathrm{F}}(\beta)$

Now, consider $\alpha\,{}^{\mathsf{s}}\|_{\{b\}}\,\beta$. The Glushkov automaton $\mathcal{A}_{\mathscr{G}\!\ell}(\alpha\,{}^{\mathsf{s}}\|_{\{b\}}\,\beta)$ has 21 states and 89 transitions; its quotient $\mathcal{A}_{\mathscr{G}\!\ell}(\alpha\,{}^{\mathsf{s}}\|_{\{b\}}\,\beta)/_{\equiv_{\mathrm{F}}} = \mathcal{A}_{\mathrm{F}}(\alpha\,{}^{\mathsf{s}}\|_{\{b\}}\,\beta)$ has 17 states and 71 transitions. Nevertheless, the Boolean product automaton $\mathcal{A}_{\mathrm{F}}(\alpha)\,{}^{\mathsf{s}}\|_{\{b\}}\,\mathcal{A}_{\mathrm{F}}(\beta) \simeq \mathcal{A}_{\mathscr{G}\!\ell}(\alpha\,{}^{\mathsf{s}}\|_{\{b\}}\,\beta)/_{(\equiv_{\mathrm{F}},\equiv_{\mathrm{F}})}$ is much smaller. Its diagram is

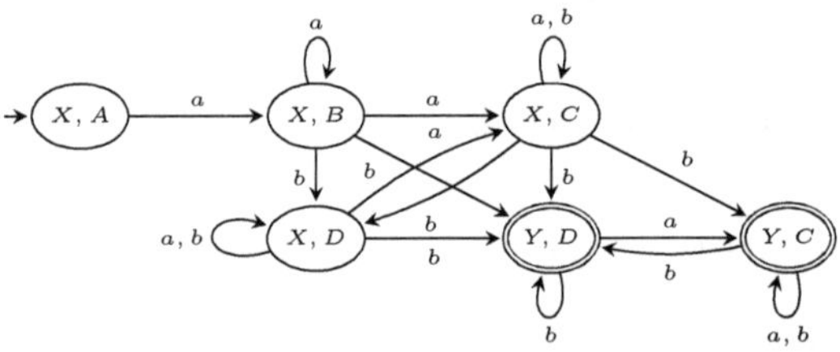

where $X = (\{1,2,3\},\varepsilon)$, $Y = (\{4\},\varepsilon)$, $A = (\{1\},\varepsilon)$, $B = (\{2,4\},\varepsilon)$, $C = (\{2,3,4\},\varepsilon)$, and $D = (\{2,4,5\},\varepsilon)$.

4.4 Bisimilarity of the Glushkov Automaton

Given an automaton $\mathcal{A}$, the set of right-invariant equivalent relations w.r.t. $\mathcal{A}$ is closed under the union operation. The coarsest right-invariant relation, i.e., the union of all right-invariant relations w.r.t. $\mathcal{A}$, is called the *bisimilarity* relation of $\mathcal{A}$, and is denoted by $\equiv_b$. As such, the *bisimilarity automaton* $\mathcal{A}/_{\equiv_b}$ is the smallest quotient automaton of $\mathcal{A}$ by a right-invariant relation. In particular, for $\alpha \in \mathrm{RE}(\|_\psi)$ the bisimilarity automaton of the Glushkov automaton, $\mathcal{A}_{\mathscr{G}\!\ell}(\alpha)/_{\equiv_b}$, is also a quotient automaton of $\mathcal{A}_{\mathrm{PD}}(\alpha)$, as well as of $\mathcal{A}_{\mathrm{F}}(\alpha)$. One can compute $\mathcal{A}_{\mathscr{G}\!\ell}(\alpha)/_{\equiv_b}$ in almost linear time by the Paige–Tarjan Algorithm [16,19]. However, $\mathcal{A}_{\mathscr{G}\!\ell}(\|_\psi(\alpha))/_{\equiv_b}$ is not necessarily isomorphic to the Boolean automaton $\|_\psi(\mathcal{A}_{\mathscr{G}\!\ell}(\alpha_1)/_{\equiv_b^1},\dots,\mathcal{A}_{\mathscr{G}\!\ell}(\alpha_1)/_{\equiv_b^n})$. Again, the natural extension of $\equiv_b^1,\dots,\equiv_b^n$ is not necessarily equal to $\equiv_b$ defined in $\mathsf{Loc}_0(\overline{\|_\psi(\alpha)})$.

Example 9. Consider the expressions $\alpha = (ab)^\star a^\star$ and $\beta = (ab^\star)^\star(\varepsilon + a)$. The bisimilarity automata of their Glushkov automata are the following.

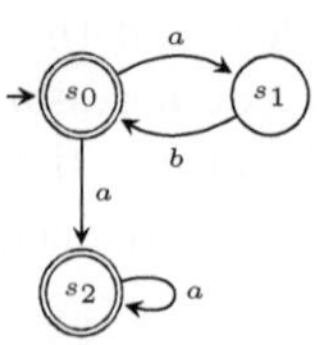

$\mathcal{A}_{\mathscr{G}\!\ell}(\alpha)/_{\equiv_b^1}$

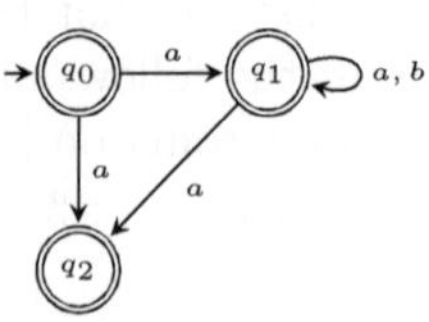

$\mathcal{A}_{\mathscr{G}\!\ell}(\beta)/_{\equiv_b^2}$

Now consider the expression $\|_\wedge(\alpha,\beta)$. The Boolean product automaton $\|_\wedge(\mathcal{A}_{\mathscr{G}}(\alpha)/_{\equiv_b^1}, \mathcal{A}_{\mathscr{G}}(\beta)/_{\equiv_b^2}) \simeq \mathcal{A}_{\mathscr{G}}(\|_\wedge(\alpha,\beta))/_{(\equiv_b^1,\equiv_b^2)}$ is depicted below on the left. On the other hand, the bisimilarity automaton of $\mathcal{A}_{\mathscr{G}}(\|_\wedge(\alpha,\beta))$, i.e. $\mathcal{A}_{\mathscr{G}}(\|_\wedge(\alpha,\beta))/_{\equiv_b}$, is smaller and depicted on the right.

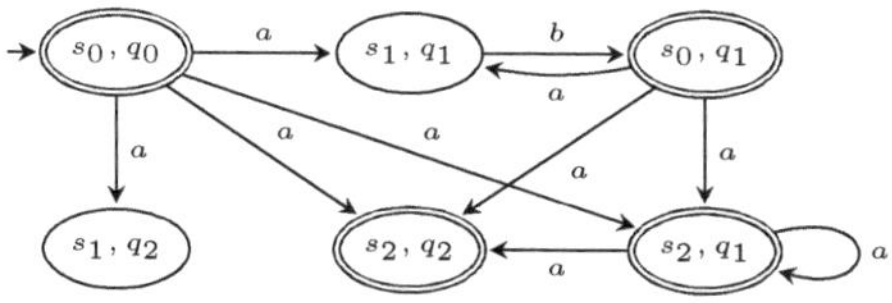

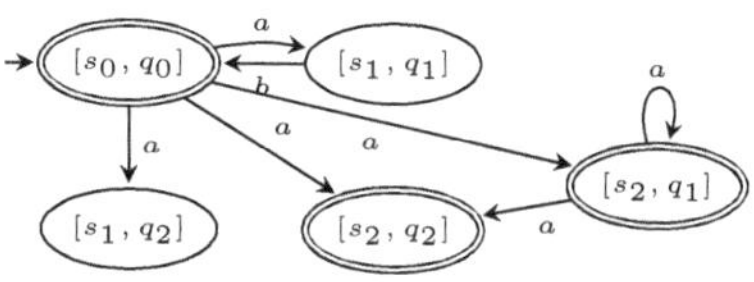

As a matter of fact, for every operator $\|_\psi$, since the bisimilarity relations $\equiv_b^1,\ldots,\equiv_b^n$ on automata $\mathcal{A}_1,\ldots,\mathcal{A}_n$ are right-invariant, it follows from Lemma 5 and Proposition 2 that $\|_\psi(\mathcal{A}_1,\ldots,\mathcal{A}_n)/_{\equiv_b}$ is always a quotient of $\|_\psi(\mathcal{A}_1,\ldots,\mathcal{A}_n)/_{(\equiv_b^1,\ldots,\equiv_b^n)} \simeq \|_\psi(\mathcal{A}_1/_{\equiv_b^1},\ldots,\mathcal{A}_n/_{\equiv_b^n})$. In this case, $[s_0,q_0]_{\equiv_b} = \{(s_0,q_0),(s_0,q_1)\}$ and $\mathcal{A}_{\mathscr{G}}(\|_\wedge(\alpha,\beta))/_{\equiv_b}$ is obtained from $\|_\wedge(\mathcal{A}_{\mathscr{G}}(\alpha)/_{\equiv_b^1}, \mathcal{A}_{\mathscr{G}}(\beta)/_{\equiv_b^2})$ by merging the states (s_0,q_0) and (s_0,q_1).

4.5 Combining Automata Constructions

One can consider the Boolean product of arbitrary automata $\mathcal{A}_1,\ldots,\mathcal{A}_n$, and, for each one, a right-invariant relation $\equiv_i$, for $i \in [1,n]$. Even if all $\mathcal{A}_i$ are Glushkov automata, one might benefit by considering different right-invariant relations $\equiv_i$ w.r.t. to $\mathcal{A}_{\mathscr{G}_i}$, for different $i \in [1,n]$. This will be illustrated in the following example.

Example 10. Consider again the expression $\alpha\,{}^{\mathsf{s}}\|_{\{b\}}\,\beta$, where $\alpha = (a^\star + b^\star)^\star bb^\star$ and $\beta = a(ab^\star + bb^\star)^\star$, from Example 8, as well as $\mathcal{A}_{\mathrm{F}}(\alpha)$ and $\mathcal{A}_{\mathrm{F}}(\beta)$. The corresponding partial derivative automata $\mathcal{A}_{\mathrm{PD}}(\alpha)$ and $\mathcal{A}_{\mathrm{PD}}(\beta)$, are respectively,

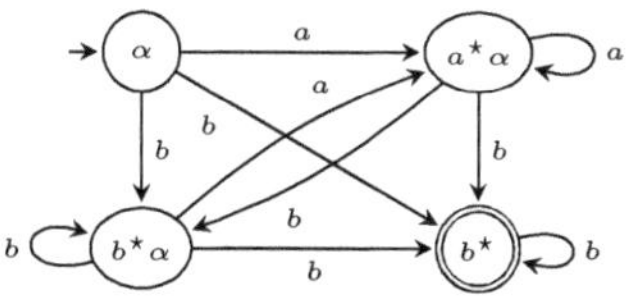

$\mathcal{A}_{\mathrm{PD}}(\alpha)$

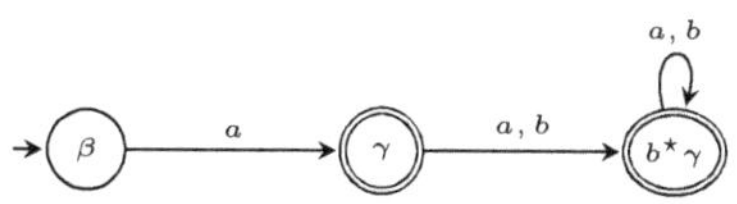

$\mathcal{A}_{\mathrm{PD}}(\beta)$

where $\gamma = (ab^\star + bb^\star)^\star$. The automaton

$$\mathcal{A}_{\mathscr{G}}(\alpha\,{}^{\mathsf{s}}\|_{\{b\}}\,\beta)/_{\equiv_c} \simeq \mathcal{A}_{\mathscr{G}}(\alpha\,{}^{\mathsf{s}}\|_{\{b\}}\,\beta)/_{(\equiv_c,\equiv_c)} \simeq$$
$$\simeq \mathcal{A}_{\mathrm{PD}}(\alpha)\,{}^{\mathsf{s}}\|_{\{b\}}\,\mathcal{A}_{\mathrm{PD}}(\beta) \simeq \mathcal{A}_{\mathrm{PD}}(\alpha\,{}^{\mathsf{s}}\|_{\{b\}}\,\beta)$$

has 8 states and 25 transitions, and is thus bigger than $\mathcal{A}_{\mathrm{F}}(\alpha\,{}^{\mathsf{s}}\|_{\{b\}}\,\beta)$, computed in Example 8. A smaller automaton is obtained when we combine the smallest automaton for each subexpression, namely

$$\mathcal{A}_{\mathrm{F}}(\alpha)\,{}^{\mathsf{s}}\|_{\{b\}}\,\mathcal{A}_{\mathrm{PD}}(\beta) \simeq \mathcal{A}_{\mathscr{G}}(\alpha\,{}^{\mathsf{s}}\|_{\{b\}}\,\beta)/_{(\equiv_{\mathrm{F}},\equiv_c)} \simeq \mathcal{A}_{\mathrm{F}}(\alpha)\,{}^{\mathsf{s}}\|_{\{b\}}\,\mathcal{A}_{\mathrm{PD}}(\beta),$$

that is depicted below.

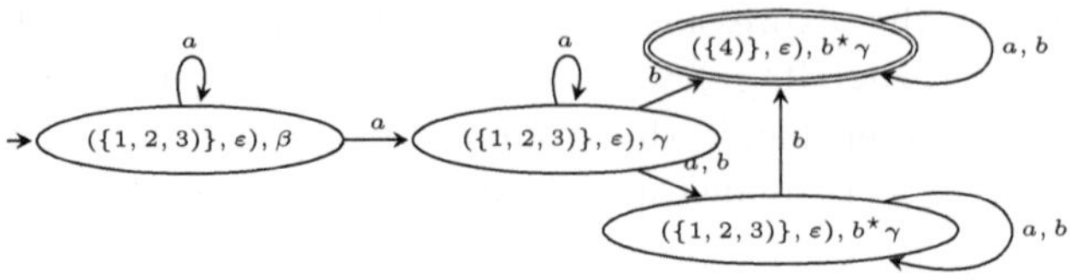

For standard regular expressions, there are conditions that ensure that the partial derivatives automaton is a quotient of the follow automaton, in particular if the expressions are in strong star-normal form [10,12,14,17]. In this case, β is in star-normal form and $\mathcal{A}_{\mathrm{PD}}(\beta)$ is smaller than $\mathcal{A}_{\mathrm{F}}(\beta)$.

References

1. Antimirov, V.M.: Partial derivatives of regular expressions and finite automaton constructions. Theoret. Comput. Sci. **155**(2), 291–319 (1996)
2. Broda, S., Holzer, M., Maia, E., Moreira, N., Reis, R.: A mesh of automata. Inf. Comput. **265**, 94–111 (2019)
3. Broda, S., Machiavelo, A., Moreira, N., Reis, R.: Partial derivatives for Boolean products and average complexity of multiple intersections. Submitted
4. Broda, S., Machiavelo, A., Moreira, N., Reis, R.: Location based automata for expressions with shuffle. In: Leporati, A., Martín-Vide, C., Shapira, D., Zandron, C. (eds.) LATA 2021. LNCS, vol. 12638, pp. 43–54. Springer, Cham (2021). https://doi.org/10.1007/978-3-030-68195-1_4
5. Broda, S., Machiavelo, A., Moreira, N., Reis, R.: Average complexity of partial derivatives for synchronised shuffle expressions. In: Nagy, B., (ed.), 27th CIAA, vol. 14151 of LNCS, pp. 103–115. Springer (2023)
6. Broda, S., Machiavelo, A., Moreira, N., Reis, R.: Location automata for synchronised shuffle expressions. J. Log. Algebr. Program. **132**, 100847 (2023)
7. Broda, S., Machiavelo, A., Moreira, N., Reis, R.: Location based automata for expressions with shuffle and intersection. Inf. Comput. **295**, 104917 (2023)
8. Broda, S., Machiavelo, A., Moreira, N., Reis, R.: Average complexity of partial derivatives for synchronised shuffle expressions. In: Prigioniero, L., Merças, R., (eds.), 26th DCFS, LNCS. Springer (2025)
9. Broda, S., Machiavelo, A., Moreira, N., Reis, R.: Average state complexity of partial derivative automata for synchronised shuffles. Int. J. Found. Comput. Sci. 1–21 (2026)
10. Brüggemann-Klein, A.: Regular expressions into finite automata. Theoret. Comput. Sci. **48**, 197–213 (1993)
11. Champarnaud, J.M., Ziadi, D.: Canonical derivatives, partial derivatives and finite automaton constructions. Theoret. Comput. Sci. **289**, 137–163 (2002)
12. Champarnaud, J.-M., Ouardi, F., Ziadi, D.: Follow automaton versus equation automaton. In: Ilie, L., Wotschke, D. (eds.) Proc 6th DCFS 2004, volume Report No. 619, pp. 145–153. Department of Computer Science, The University of Western Ontario, Canada (2004)
13. Ellis, C.A.: Team automata for groupware systems. In: Proceedings of GROUP'97, International Conference on Supporting Group Work: The Integration Challenge, pp. 415–424. ACM, USA (1997)

14. Gruber, H., Gulan, S.: Simplifying regular expressions. In: Dediu, A.-H., Fernau, H., Martín-Vide, C. (eds.) LATA 2010. LNCS, vol. 6031, pp. 285–296. Springer, Heidelberg (2010). https://doi.org/10.1007/978-3-642-13089-2_24
15. Ilie, L., Yu, S.: Follow automata. Inf. Comput. **186**(1), 140–162 (2003)
16. Ilie, L., Yu, S.: Reducing NFAs by invariant equivalences. Theoret. Comput. Sci. **306**(1–3), 373–390 (2003)
17. Ping, L., Chen, H.: Relations between equation automata and follow automata. Inf. Comput. **306**, 105333 (2025)
18. Manea, F., Mitrana, V., Voinescu, D.-C.: Synchronized shuffle on backbones. Fundam. Inform. **73**(1–2), 191–202 (2006)
19. Paige, R., Tarjan, R.E.: Three partition refinement algorithms. SIAM J. Comput. **16**(6), 973–989 (1987)
20. Sulzmann, M., Thiemann, P.: Derivatives and partial derivatives for regular shuffle expressions. J. Comput. Syst. Sci. **104**, 323–341 (2019)
21. ter Beek, M.H., Ellis, C.A., Kleijn, J., Rozenberg, G.: Synchronizations in team automata for groupware systems. Comput. Support. Cooper. Work. **12**(1), 21–69 (2003)
22. Beek, M.H., Hennicker, R., Proença, J.: Team automata: overview and roadmap. In: Castellani, I., Tiezzi, F., (eds.), 26th IFIP WG 6.1 COORDINATION 2024, volume 14676 of LNCS, pp. 161–198. Springer (2024)
23. ter Beek, M.H., Kleijn, J.: Team automata satisfying compositionality. In: Araki, K., Gnesi, S., Mandrioli, D. (eds.) FME 2003. LNCS, vol. 2805, pp. 381–400. Springer, Heidelberg (2003). https://doi.org/10.1007/978-3-540-45236-2_22
24. ter Beek, M.H., Kleijn, J.: Shuffles and synchronized shuffles: a survey. In: Paun, G., Rozenberg, G., Salomaa, A., (eds.), Discrete Mathematics and Computer Science. In Memoriam Alexandru Mateescu (1952-2005), pp. 37–50. The Publishing House of the Romanian Academy (2014)
25. ter Beek, M.H., Martín-Vide, C., Mitrana, V.: Synchronized shuffles. Theoret. Comput. Sci. **341**, 263–275 (2005)

Bandwidth of Nondeterministic Finite Automata

Da-Jung Cho[1(✉)], Szilárd Zsolt Fazekas[2], Daihei Ise[3], Shinnosuke Seki[3], Wataru Tamehira[3], and Max Wiedenhöft[4]

[1] Department of Software and Computer Engineering, Ajou University, Suwon, Republic of Korea
dajungcho@ajou.ac.kr
[2] Graduate School of Engineering Science, Akita University, Akita, Japan
szilard.fazekas@ie.akita-u.ac.jp
[3] University of Electro-Communications, Tokyo, Japan
{daihei.ise,s.seki}@uec.ac.jp, t2210411@gl.cc.uec.ac.jp
[4] Department of Computer Science, Kiel University, Kiel, Germany
maw@informatik.uni-kiel.de

Abstract. Co-transcriptional splicing generates RNA sequences from a DNA template by deleting subsequences nondeterministically . Recent work showed how to encode an NFA into such a template, but the construction requires deleting subsequences whose length grows with the distance between states, which makes such deletions unlikely under the local nature of co-transcriptional splicing. We introduce k-bandwidth NFAs, in which transitions span at most k states. These automata form a strict hierarchy of language classes. For finite languages, bandwidth 2 suffices, and bandwidth 1 can be decided in polynomial-time when the language is presented as a list of words. Minimizing the bandwidth is NP-hard even for fixed $k \geq 2$.

1 Introduction

Co-transcriptional splicing (Fig. 1) is a process in which an RNA sequence is edited by letting its subsequences fold into a hairpin and cut from its root while being sequentially synthesized from its DNA template, or *transcribed* [6]. Co-transcriptional *folding*, by which hairpins to be spliced out form therein, has been proven programmable *in vitro* for self-assembly of a 2-dimensional tile-like structure [5] and *in papyro* for universal computation [4]. The former architecture (*RNA origami*) demonstrated how to program a chain of folding of constitutional hairpins and their interactions into an RNA sequence in such a way that the sequence co-transcriptionally folds according to the event chain. With rapid progress in experiments in mind, it is not unrealistic to envision applications of co-transcriptional splicing.

For instance, artificial molecular systems generate RNA sequences from DNA templates during transcription to reduce cost, since DNA is cheaper to synthesize. As of now, many templates have to be ordered as target RNA sequences, but what if multiple targets could be accommodated on a single template?

T. J. Smith (Ed.): CIAA 2026, LNCS 16695, pp. 32–46, 2026.
https://doi.org/10.1007/978-3-032-31176-4_3

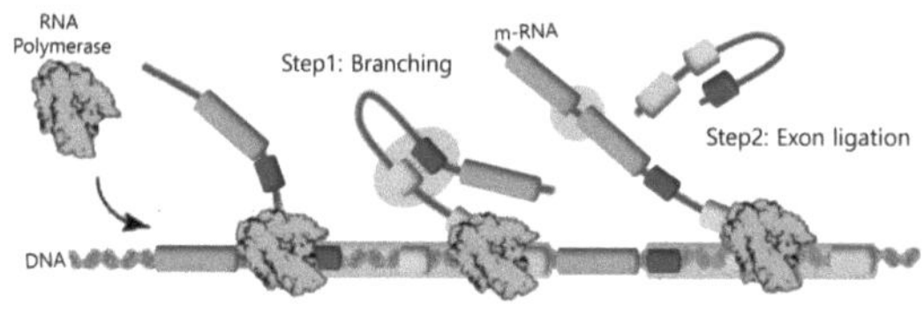

Fig. 1. Co-transcriptional splicing. While an RNA transcript is synthesized (*transcribed*), a hairpin forms via hybridization of its complementary subsequences (colored in red and yellow) and spliced out. (Color figure online)

Co-transcriptional splicing can be modeled as a greedy deletion of hairpins one after another from a sequence [2]. A hairpin is represented, using two sequences x, ℓ and an antimorphic involution θ, as $x\ell\theta(x)$, where x hybridizes with its "Watson-Crick complement" $\theta(x)$ into a stem, leaving ℓ as a loop. Using this model, we have recently demonstrated how to encode a nondeterministic finite automaton into a circular template w in such a way that all words accepted by the NFA can be co-transcriptional spliced out of w^* [3]. The states $q_1, q_2, \cdots, q_n$ of a target NFA are encoded as sequences over a 4-letter alphabet and concatenated together with a special factor s_e for termination into the template as $w = s_1 s_2 \cdots s_n s_e$ in such a way that

1. $s_i \cdots s_{j-1}$ can be decomposed as $h_1 a h_2$ for a letter a and two hairpins h_1, h_2 as long as the NFA can transition from state q_i to state q_j by reading a;
2. h_1 is a prefix of s_i;
3. h_2 does not share its start point with any other hairpin.

Starting at the beginning of s_1, one of the outgoing transitions from q_1, or particularly, their prefix h_1's is chosen nondeterministically and deleted, guiding the input head to the letter to be read, and then h_2 is deleted deterministically, guiding the head to the beginning of the encoding of destination state; the transition from q_i to q_j by a is thus implemented. A drawback of this architecture is that h_2 can be arbitrarily long. The most serious threat is self-loop. In addition, the more outgoing transitions a state q_i has, the longer the corresponding encoding s_i becomes. From a biological perspective, such long-range interactions are unlikely to occur. Co-transcriptional splicing operates in a local and greedy manner and acts on splice sites as soon as they are transcribed. This limits interactions between distant regions of the transcript and motivates restricting transitions to nearby states along the template. This leads to a linearization problem: given an NFA, can its states be arranged along the template so that every transition spans only a bounded number of segments?

The goal of this paper is to introduce the notion of *k-bandwidth NFAs*, whose states can be ordered in such a way that all transitions skip at most $k-1$ states in-between, and investigate their properties. We first show an infinite hierarchy among the class of such NFAs with respect to the bandwidth k (Fig. 2). The class $\mathcal{F}$ of finite languages is of practical interest as no molecular system involves

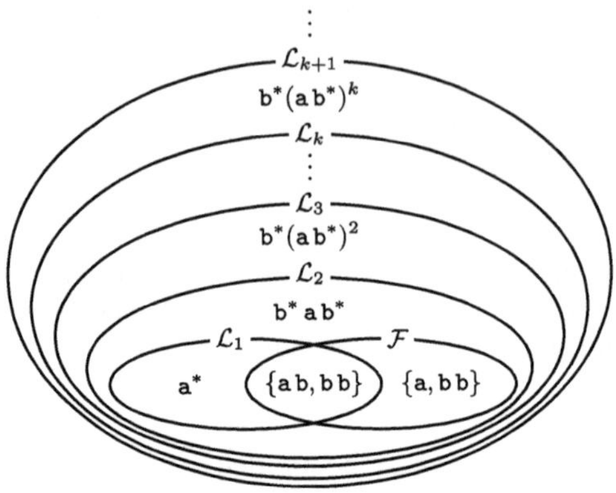

Fig. 2. A relationship among $\mathcal{L}_1, \mathcal{L}_2, \ldots$ and $\mathcal{F}$, established as Proposition 5, Lemma 6, and Lemma 7.

infinitely many kinds of RNA sequences. In this line, we will show how to convert an arbitrary finite language into a bandwidth-2 NFA that accepts it (Lemma 6). This result turns out to be optimal in the sense that there exists a finite language that cannot be accepted by any bandwidth-1 NFA (Lemma 7), but we provide a polynomial-time bandwidth-1 test for finite languages (Theorem 9). Minimizing the bandwidth turns out to be NP-hard even for fixed $k \geq 2$ (Theorem 10).

2 Preliminaries

Let $\mathbb{N}$ denote the set of positive integers. For some $m \in \mathbb{N}$, denote by $[m]$ the set $\{1, 2, ..., m\}$. With Σ, we denote a finite set of symbols, called an *alphabet*. The elements of Σ are called *letters*. A *word* over Σ is a finite sequence of letters from Σ. With ε, we denote the *empty word*. The i-th element of a word w is denoted by $w[i]$. The set of all words over Σ is denoted by Σ^*. Additionally, set $\Sigma^+ = \Sigma^* \setminus \{\varepsilon\}$. The *length* of a word $w \in \Sigma^*$, i.e., the number letters in w, is denoted by $|w|$; hence, $|\varepsilon| = 0$. For some $w \in \Sigma^*$, if we have $w = xyz$ for some $x, y, z \in \Sigma^*$, then we say that x is a *prefix*, y is a *factor*, and z is a *suffix* of w. A nondeterministic finite automaton (NFA) is a tuple $A = (\Sigma, Q, q_0, \delta, F)$ where Σ is the input alphabet, Q is the finite set of states, $\delta\colon Q \times \Sigma \to 2^Q$ is the multivalued transition function, $q_0 \in Q$ is the initial state, and $F \subseteq Q$ is the set of final states. In the usual way, δ is extended as a function $Q \times \Sigma^* \to 2^Q$ and the language accepted by A is $L(A) = \{w \in \Sigma^* \mid \delta(q_0, w) \cap F \neq \emptyset\}$.

Definition 1 (k-bandwidth NFA). *Given $k \in \mathbb{N}$, an NFA $A = (Q, \Sigma, \delta, q_0, F)$ is* k-bandwidth *(k-BW-NFA) if the states can be indexed as $Q = \{q_0, ..., q_{n-1}\}$ such that for all $q_i, q_j \in Q$ and $\mathtt{a} \in \Sigma \cup \{\varepsilon\}$, $q_j \in \delta(q_i, \mathtt{a})$ implies $0 < (j - i) \bmod n \leq k$.*

This definition corresponds to the notion of distance bounded NFA from [3], where they just have been briefly considered and not formally investigated in a lot of detail. As bandwidth is an established name for the underlying graph property [1], we have chosen this name in this paper. Intuitively, the bandwidth k defines a maximum distance that can be traversed by a single transition between states. This results in either a *line of states* where transitions may *jump* a certain number of states ahead, or it results in a *circle of states* that can be traversed in a similar manner, but repeatingly.

2.1 Constantly-Bounded Co-Transcriptional Splicing as a Programming Platform

Motivated by the practical problems discussed in the introduction, we introduce the following problems. First, the most general one asks whether a given regular language represented by some NFA can be obtained with a given bandwidth k.

Problem 2 (Linearization of NFA). Given an NFA A and a number $k \geq 1$, does there exist a k-BW-NFA B such that $L(A) = L(B)$?

In many cases, especially in the context of synthesizing a certain set of RNA sequences, we do not need the full power of NFA to express target/input languages. Hence, we consider the following subproblem as highly practically relevant.

Problem 3 (Linerization of Finite Languages). Given some finite language $L_f \subset \Sigma^*$ and a number $k \geq 1$, does there exist a k-BW-NFA A such that $L_f = L(A)$? Does there exist a constant bound for k such that all finite languages lie in the class of languages that can be expressed with k-BW-NFA?

Up to this point, we ask basic existence questions but do not specify any constraints with respect to the size of the obtained automata. Taking the embedding in co-transcriptional splicing into account, minimizing the size of the automata and, thus, minimizing the necessary size of the encoding as a DNA sequence also remains of highly practical relevance, as shorter DNA strands promise more stability and reduced cost. Deciding the existence of some n state k-BW-NFA A for a target finite language L_f has been shown to be NP-hard if in addition to n and L_f, k is also left as part of the input. Hence, we wondered whether this problem can be efficiently decided if k is bound by a constant, effectively resulting in an FPT algorithm for the following problem.

Problem 4 (Minimization of k-BW-NFA). Let $k \geq 1$ be a constant. Given a finite language L_f and a number $n \geq 1$, does there exist a k-BW-NFA A with n states such that $L(A) = L_f$?

As mentioned in the introduction, for all of the highly relevant problems we obtained answers with respect to their computational complexity. The more general variants, as to be seen, remain open for future research. We continue with some general observations regarding the accepted languages by k-BW-NFA.

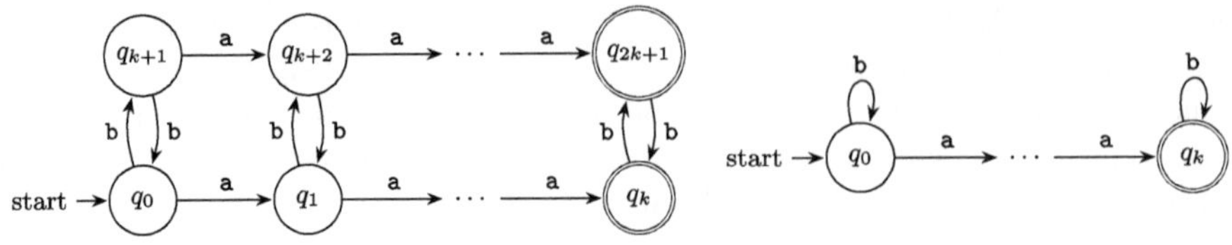

Fig. 3. (Left) The construction of the $(k+1)$-BW-NFA accepting $\mathtt{b}^*(\mathtt{a}\,\mathtt{b}^*)^k$. The double circles represents its accepting states. (Right) The NFA accepting $\mathtt{b}^*(\mathtt{a}\,\mathtt{b}^*)^k$.

3 The Hierarchy of k-Bandwidth NFAs

We begin with a central result that the classes recognizable by k-bandwidth NFAs form a strict hierarchy based on the selection of k. Consider the following definition. Let $\mathcal{L}_k$ be the class of languages recognizable by k-BW-NFAs.

Proposition 5. *The proper inclusion $\mathcal{L}_k \subsetneq \mathcal{L}_{k+1}$ holds, for any k.*

Proof. Consider the regular expression $\mathtt{b}^*(\mathtt{a}\,\mathtt{b}^*)^k$. We will prove $\mathtt{b}^*(\mathtt{a}\,\mathtt{b}^*)^k \in \mathcal{L}_{k+1} \setminus \mathcal{L}_k$. Firstly, we can see $\mathtt{b}^*(\mathtt{a}\,\mathtt{b}^*)^k \in \mathcal{L}_{k+1}$ by Fig. 3 (Left) which illustrates a $(k+1)$-BW-NFA accepting the language. Note that while the NFA illustrated in Fig. 3 (Right) accepts the language, it is not a $(k+1)$-BW-NFA. Secondly, we will prove $\mathtt{b}^*(\mathtt{a}\,\mathtt{b}^*)^k \notin \mathcal{L}_k$ by contradiction. Assume $\mathtt{b}^*(\mathtt{a}\,\mathtt{b}^*)^k \in \mathcal{L}_k$. Then, there exists a k-BW-NFA $A = (Q, \Sigma, \delta, q_0, F)$ with $Q = \{q_0, q_1, \ldots, q_{n-1}\}$ such that $L(A) = \mathtt{b}^*(\mathtt{a}\,\mathtt{b}^*)^k$. Consider the sequence $w = \beta(\mathtt{a}\beta)^k \in L(A)$ such that $\beta = \mathtt{b}^{n+1}$. Then, while reading each occurrence of β, the automaton visits at least one state more than once. For each ℓ with $1 \leq \ell \leq k+1$, let q'_ℓ be a state that the automaton visits at least twice while reading the ℓ-th occurrence of β. We define C_ℓ as the set of all states that the cycle (from q'_ℓ to q'_ℓ) includes.

Then, we will prove $C_i \cap C_j = \emptyset$ for any $i \neq j$ by contradiction. Assume $C_i \cap C_j \neq \emptyset$ for some $i \neq j$. Without loss of generality, we can assume $i < j$. Then, there exists a common state $q \in C_i \cap C_j$. Therefore, the automaton A includes a loop from q to q_i, then via an $\mathtt{a}$-transition to q_j, and finally back to q. This loop includes the symbol $\mathtt{a}$, which allows the automaton A to accept words with more than k occurrences of $\mathtt{a}$, which contradicts the definition of $\mathtt{b}^*(\mathtt{a}\,\mathtt{b}^*)^k$. Therefore, we have $C_i \cap C_j = \emptyset$ for any $i \neq j$.

Let $Q_{x,y}$ ($\subseteq Q$) be the set of y consecutive states starting from q_x, defined as $Q_{x,y} = \{q_{(x+i) \bmod n} \mid i = 0, \ldots, y-1\}$. Let $\ell \in \{1, \ldots, k+1\}$ be any integer. Then, since A is a k-BW-NFA, for all $q_i, q_j \in C_\ell$ and $\mathtt{a} \in \Sigma \cup \{\varepsilon\}$, $q_j \in \delta(q_i, \mathtt{a})$ implies $0 < (j-i) \bmod n \leq k$. Therefore, for any integer $m \in \{1, 2, \ldots\}$, the set $Q_{m,k}$ includes at least one element from C_ℓ. Since $C_i \cap C_j = \emptyset$ holds for any $i \neq j$, the set $Q_{m,k}$ includes at least $k+1$ distinct states, which contradicts the fact $|Q_{m,k}| = k$. Therefore, $\mathtt{b}^*(\mathtt{a}\,\mathtt{b}^*)^k \notin \mathcal{L}_k$ holds. □

Hence, we know that for each fixed k, the class $\mathcal{L}_k$ is indeed a sub-regular language and a distinct class of languages compared to all $\mathcal{L}_{k'}$ where either $k' < k$

or $k' > k$. Related to the practical motivation for this model this implies that maximizing the achievable size of a removed factor by co-transcriptional splicing also strictly increases the languages that can be expressed using this operation.

4 Finite Languages and *k*-Bandwidth NFAs

In this section, we consider the class of finite languages over k-BW-NFAs. From a practical viewpoint, this class of languages seems most interesting as a set of target RNA sequences encoded on DNA templates is typically finite. Let $\mathcal{F}$ be the set of all finite languages. First, we can show for finite languages, a sort of trivial concatenation of words can be done to obtain a k-BW-NFA that represents this language. Indeed, even for a very small constant, $k = 2$, we can encode any finite language. The construction uses an alternating structure between two paths, one for reading the current word and the other one to skip ahead to a point after the current word is read.

Lemma 6. *We have* $\mathcal{F} \subseteq \mathcal{L}_2$.

Proof. Let Σ be some alphabet. Let $L_f = \{w_1, w_2, ..., w_m\} \subseteq \Sigma^*$ for some $m \geq 1$. Let $k = \sum_{i=1}^{m} |w_i|$. We construct a 2-BW-NFA $A = (Q, \Sigma, \delta, q_0, F)$ s.t. $L(A) = L_f$. Set $Q = \{q_0, q_1, ..., q_{2k-|w_m|+1}\}$ and define δ and F by the following: For each $i \in [m]$, we distinguish between 2 cases.

First, if $m = 1$, we do the following: We set $q_1 \in \delta(q_0, w_1[1])$ as well as $q_2 \in \delta(q_0, \varepsilon)$. Then, for each $j \in [|w_1| - 1]$, we add the transition $q_{2(j+1)-1} \in \delta(q_{2j-1}, w_1[j + 1])$ and the transition $q_{2(j+1)} \in \delta(q_{2j}, \varepsilon)$. Finally, we define $q_{2|w_1|-1} \in F$. In total, we obtain two distinct outgoing paths from q_0, one reading w_1 and ending in a final state, the other reading ε and ending in a non final state (see Fig. 4).

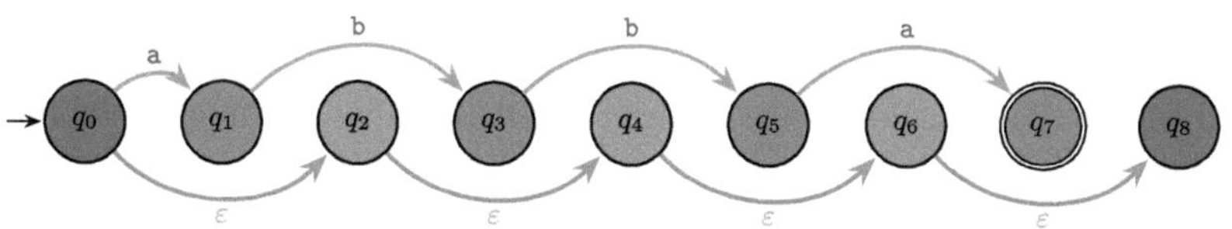

Fig. 4. The initial part of the automaton that accepts the first word of L_f. The teal path correspondings to w_1 (here, for example **abba**) while the orange path corresponds to the skipping ε-path that can be used to move to q_8.

Now, if $m \geq 2$, we inductively continue as follows. Let the largest state reached by the previous step be the state q_s, i.e., $s = \sum_{i'=0}^{i-1} 2|w_{i'+1}|$. We reach q_s only by the path that reads ε. Now we analogously construct two paths, one reading w_i and the other reading ε. First, set $q_{s+1} \in \delta(q_s, w_i[1])$ and $q_{s+2} \in \delta(q_s, \varepsilon)$. Now, for each $j \in [|w_i| - 1]$, we define $q_{s+2(j+1)-1} \in \delta(q_{s+2j-1}, w_i[j+1])$ and $q_{s+2(j+1)} \in \delta(q_{s+2j}, \varepsilon)$. Finally, set $q_{s+2|w_i|-1} \in F$. As before, we constructed

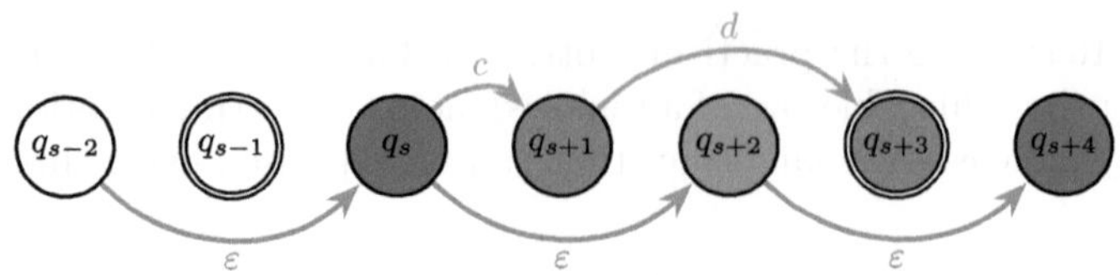

Fig. 5. The inductive step of adding additional words to the automaton by the exemplary word cd. Notice that q_s could correspond to q_8 from the automaton in Fig. 4.

two outgoing paths of the state q_s, one reading only ε and ending in state $q_{s+2|w_i|}$, the other reading w_i and ending in state $q_{s+2|w_i|-1}$ (see Fig. 5).

For the last word w_m, we only have to construct one of the paths as the skipping ε path is not necessary anymore, leading to $2k - |w_m| + 1$ instead of $2k+1$ states (The $+1$ coming from the initial state q_0). Finally, we have to verify that A actually has bandwidth 2, i.e., each outgoing transition moves at most two states ahead. As all cases are constructed analogously, we only show this for the base step. First, there are transitions from q_0 to q_1 and from q_0 to q_2, which fulfill the constant. Next, for each $j \in [|w_1| - 1]$, there is one transition from q_{2j-1} to $q_{2(j+1)-1}$ and one transition from q_{2j} to $q_{2(j+1)}$. For the first transition, we obtain a distance of $2(j+1) - 1 - (2j-1) = 2j + 2 - 1 - 2j + 1 = 2$. For the second transition, we obtain a distance of $2(j+1) - (2j) = 2j + 2 - 2j = 2$. Hence, for all transitions, $q_j \in \delta(q_i, a)$ implies $0 < (j - i) \bmod |Q| \leq 2$. By the inductive construction, each and only the words w_1 to w_m are accepted by A, resulting in $L_f = L(A)$. No transition skips over more than 2 states and no state has more than 2 outgoing transitions. □

4.1 The Special Case of Bandwidth 1

It is rather apparent that 1-BW-NFA do not have the same expressibility as 2-BW-NFA. Not all finite languages can be encoded in a 1-BW-NFA, even though ε-transitions are allowed. The main reasoning lies in the fact that we cannot skip ahead an arbitrary distance without implicitly allowing more combinations of letters to the read.

Lemma 7. *We have $\mathcal{F} \not\subseteq \mathcal{L}_1$.*

Proof. We will show $\{\mathtt{a}, \mathtt{b}\,\mathtt{b}\} \notin \mathcal{L}_1$. Assume that there exists a 1-BW-NFA that accepts the language $\{\mathtt{a}, \mathtt{b}\,\mathtt{b}\}$. Due to its linear structure, such an automaton contains exactly one transition labeled with a and two transitions labeled with b along a single path. Therefore, there exist words w_1, w_2, and w_3 such that the automaton accepts either $w_1\,\mathtt{a}\,w_2\,\mathtt{b}\,w_3$ or $w_1\,\mathtt{b}\,w_2\,\mathtt{a}\,w_3$. A contradiction. □

Nevertheless, they are of high interest as in the context of co-transcriptional splicing, a bandwidth of 1 in an encoded automaton could result in only very small factors having to be spliced out (see, e.g., [2]). As each 1-BW-NFA can be understood as a line of states in which some states might be set to final states, we can give a full characterization of all non-empty finite languages that can be expressed by some 1-BW-NFA.

Lemma 8. *Let L_f be some finite language. There exists a 1-BW-NFA A with $n \geq 1$ states and $L(A) = L_f$ if and only if there exist languages $S_1, ..., S_{n-1} \subseteq \Sigma \cup \{\varepsilon\}$, for some $n \geq 1$, and a set $I \subseteq [n]$ such that*

$$L_f = \bigcup_{i \in I} S_1 \cdot ... \cdot S_{i-1}.$$

In the case of $i = 1$, let the empty product be defined as $\{\varepsilon\}$.

Proof. First, assume there exists a 1-BW-NFA $A = (Q, \Sigma, \delta, q_1, F)$ with $n \geq 1$ states and $L(A) = L_f$. Assume w.l.o.g. that $Q = \{q_1, ..., q_n\}$ such that for each $i \in [n-1]$ and each $a \in \Sigma \cup \{\varepsilon\}$ we have $\delta(q_i, a) \subseteq \{q_{i+1}\}$. For each $i \in [n-1]$, we define $S_i = \{ a \in \Sigma \cup \{\varepsilon\} \mid q_{i+1} \in \delta(q_i, a) \}$. Now let $t \in [n]$. A word $w \in \Sigma^*$ labels a path from q_1 to q_t if and only if it can be written as $w = w[1]w[2] \cdots w[t-1]$ with $w[i] \in S_i$ for every $i \in [t-1]$ (recall that $w[i] = \varepsilon$ is allowed). Thus, the set of words on the path from q_1 to q_t is exactly $S_1 \cdot S_2 \cdot ... \cdot S_{t-1}$. By that,

$$L_f = L(A) = \bigcup_{q_t \in F} S_1 \cdot ... \cdot S_{t-1}.$$

So, setting $I = \{i \mid q_i \in F\}$, the claim holds.

For the other direction, assume there exist languages $S_1, ..., S_{n-1} \subseteq \Sigma \cup \{\varepsilon\}$, for some $n \geq 1$, and a set $I \subseteq [n]$ such that

$$L_f = \bigcup_{i \in I} S_1 \cdot ... \cdot S_{i-1}.$$

We construct an NFA $A = (Q, \Sigma, \delta, q_1, F)$ with states $Q = \{q_1, ..., q_n\}$, initial state q_1, final states $F = \{ q_i \mid i \in I\}$, and transitions $q_{j+1} \in \delta(q_j, a)$ iff $a \in S_j$, for $j \in [n-1]$. This automaton clearly has bandwidth 1. Following the construction, we observe the following, concluding this direction:

$$L(A) = \bigcup_{i \in I} S_1 \cdot ... \cdot S_{i-1} = L_f.$$

□

So, by now we know that not all finite languages can be expressed by a 1-BW-NFA but we know their general structure by this previous characterization. Given a set of words, however, it might not be immediately clear whether it can be structured with respect to the previous characterization. Hence, we consider the feasibility of Problem 3 for the case $k = 1$.

Theorem 9. *Let L_f be a finite language over an alphabet Σ, given explicitly as a list of words, and let $N = \max_{w \in L_f} |w|$. We can decide in $O(|L_f|^2 N^4)$ time whether there exists a bandwidth-1 NFA A such that $L(A) = L_f$.*

Proof. Let $L_{\max} = \{w \in L_f \mid |w| = N\}$. If $L_f = \emptyset$, the problem is trivial. Hence assume $L_f \neq \emptyset$. As in Lemma 8, a bandwidth-1 NFA accepting a finite language may be viewed as a path

$$q_0 \to q_1 \to \cdots \to q_N,$$

where the transition from q_{i-1} to q_i has a label set $S'_i \subseteq \Sigma \cup \{\varepsilon\}$. Every word of length N must use a non-ε label on every one of the N transitions. Hence the non-ε backbone is uniquely forced by $L_{\max}$. For $i \in [N]$, define

$$S_i = \{w[i] \mid w \in L_{\max}\}.$$

A necessary condition is $L_{\max} = S_1 S_2 \cdots S_N$. If this condition fails, we reject, so we assume from now on that $L_{\max} = S_1 S_2 \cdots S_N$.

Phase 1: Guessing Nonsingleton ε-assignments.). Let $J = \{i \in [N] \mid |S_i| \geq 2\}$. Since

$$|L_{\max}| = \prod_{i=1}^{N} |S_i|,$$

we have $2^{|J|} \leq |L_{\max}| \leq |L_f|$. Therefore $|J| \leq \log_2 |L_f|$. Thus all possible choices of which transitions $i \in J$ also receive an ε-label can be enumerated in polynomial time in $|L_f|$.

Fix one such choice $E_J \subseteq J$. We now decide whether this choice can be completed by assigning ε-labels to singleton transitions.

Phase 2: Right-End Normal Form for Singleton Blocks. We group the singleton transitions into maximal consecutive blocks with the same singleton label. A singleton block has the form a^m for some $a \in \Sigma$, and two consecutive singleton blocks have different letters. Non-singleton transitions separate singleton blocks.

We use the following normal form. In each maximal singleton block, if d transitions are assigned ε, then these d transitions are placed at the right end of the block. Consider one maximal singleton block a^m. Suppose d of its transitions are assigned ε. For any final state after the end of the block, the contribution of this block depends only on d, not on the exact positions of the d optional transitions: it is

$$\{a^{m-d}, a^{m-d+1}, \ldots, a^m\}.$$

Now consider a final state inside the block, after h positions of this block. If e of the first h positions are optional, then the block contribution is

$$\{a^{h-e}, a^{h-e+1}, \ldots, a^h\}.$$

After moving all d optional positions to the right end of the block, this same interval of exponents can be obtained by setting the corresponding states in the block final. By the way we defined the language accepted with safe final states, every state whose prefix language is contained in L_f will be final, and no unsafe state will be, in other words, moving the optional transitions to the right end of the block cannot introduce words outside L_f, in other words, every word previously accepted inside the block is still accepted at one of the corresponding safe endpoints. Hence it is enough to store, for each singleton block B_r, a number d_r of rightmost ε-transitions.

Let $D = (d_1, \ldots, d_t)$ denote the current vector of singleton ε-counts, together with the fixed non-singleton choice E_J, where t is the number of transition blocks consisting of identical labels. For $0 \leq i \leq N$, let $P_i(D)$ be the language accepted by setting q_i final under the current ε-assignment D.

A state q_i is safe if $P_i(D) \subseteq L_f$. Define $F_{\mathsf{safe}}(D) = \{q_i \mid P_i(D) \subseteq L_f\}$, and

$$L(D) = \bigcup_{q_i \in F_{\mathsf{safe}}(D)} P_i(D).$$

Thus $L(D)$ is the largest language accepted by the current transition assignment without accepting any word outside L_f.

Phase 3: Forced Inference on Singleton Blocks. Initialize $D = (0, \ldots, 0)$.

At every step compute $U = L_f \setminus L(D)$. If $U = \emptyset$, the current assignment realizes L_f, so we accept this guess E_J and the current D.

Otherwise let $\ell = \max\{|w| \mid w \in U\}$, and let $U_\ell = \{w \in U \mid |w| = \ell\}$.

For a singleton block B_r, let $D + e_r$ denote the assignment obtained from D by adding one more rightmost ε-transition to B_r. We say that r is a candidate for $w \in U_\ell$ if

$$w \in L(D + e_r)$$

and the final state q_N remains safe, that is,

$$P_N(D + e_r) \subseteq L_f.$$

Let $C_w = \{r \mid r \text{ is a candidate for } w\}$. If $C_w = \emptyset$ for some $w \in U_\ell$, then this guess E_J cannot be completed, and we reject this guess.

Otherwise, by the forcing claim below, if this guess admits a valid completion, then there exists some $w^\star \in U_\ell$ with $|C_{w^\star}| = 1$. Let $C_{w^\star} = \{r\}$. We increment d_r by one and continue.

Forcing Claim for Singleton Blocks. We prove the claim used above. Suppose that the current assignment D has a valid completion $D^\star \geq D$, still under the fixed nonsingleton choice E_J, such that

$$L(D^\star) = L_f.$$

We prove that there exists a word $w^\star \in U_\ell$ with $|C_{w^\star}| = 1$.

Since $U \neq \emptyset$ and $D^\star$ realizes L_f, some word of U_ℓ is accepted at a safe final state under $D^\star$. Consider an accepting computation of such a word, chosen so that the number of singleton transitions that are skipped using an ε-transition under $D^\star$ but not yet skippable under D is minimal. At least one such new ε-transition (singleton skip) is used, otherwise the word would already belong to $L(D)$. In particular, some singleton block contributes one new skip to the current highest uncovered layer.

We now show that one may choose such a contribution with a private witness. Fix one singleton block B_i whose next rightmost transition is skipped in the completion. Freeze the choices of all other transitions as follows. For singleton blocks, use the choices of the computation described above. For the nonsingleton transitions, choose non-ε letters, subject only to the following separator convention: whenever two singleton runs with the same letter a would become adjacent after deleting a non-singleton block that has some letter $b \neq a$ (such a letter exists, since the block is not a singleton one), keep b^k from the non-singleton block's accepted words, where k is the length of the block. If the singleton letters on the two sides are already different, no separator is needed.

With these choices fixed, the relevant part of the accepting path becomes an ordinary word. By the separator convention, distinct singleton runs do not merge. Thus its singleton-run skeleton has the form

$$a_1^{m_1} a_2^{m_2} \cdots a_s^{m_s}, \qquad m_j > 0, \quad a_j \neq a_{j+1}.$$

Let the chosen block B_i correspond to the run $a_i^{m_i}$. Form the word w_i by decreasing this run by one and leaving all other fixed choices unchanged.

Then $w_i \in L(D + e_i)$. Moreover,

$$P_N(D + e_i) \subseteq P_N(D^{\star}) \subseteq L_f,$$

so i is a candidate for w_i.

The word w_i cannot be obtained by adding one more ε-transition to any other singleton block. Indeed, if a different block B_j with $j < i$ produced w_i, then the run $a_i^{m_i}$ would have to start one position too early and consume the last letter before it. This is impossible because, by construction, that preceding letter is different from a_i. The case $j > i$ is symmetric: then the run $a_i^{m_i}$ would have to end one position too late and consume the first letter after it, again impossible because that letter is different from a_i. The boundary cases are one-sided versions of the same argument.

Hence no other singleton block is a candidate for w_i. Also $w_i \notin L(D)$, since producing it requires exactly the new skip in B_i, and no other block can simulate that skip. Since the final state used by the completion is safe, $w_i \in L_f$. Therefore $w_i \in U$. By construction it belongs to the current highest uncovered layer, so $w_i \in U_\ell$. Thus $C_{w_i} = \{i\}$. This proves that whenever a valid completion exists, the algorithm finds a word with a unique candidate singleton block. Therefore rejecting a branch only when no such word exists is sound.◀

Verification and Complexity. If the algorithm accepts, it has explicitly constructed a bandwidth-1 NFA whose safe final states accept exactly L_f. Conversely, suppose a valid bandwidth-1 NFA exists. Choose the branch E_J corresponding to its ε-choices on non-singleton transitions, and put its singleton part into right-end normal form. By the forcing claim, at every step of Phase 3 the algorithm chooses a singleton block that is forced in every valid completion of the current assignment. Therefore this branch never rejects and eventually reaches an assignment D with $L(D) = L_f$. Thus the algorithm accepts exactly the bandwidth-1 finite languages.

Phase 1 enumerates at most $|L_f|$ choices for the ε-assignments on non-singleton transitions.

Fix one such choice. The singleton-inference loop performs at most N iterations, since each iteration adds one new ε-transition to a singleton block. In one iteration, there are at most N singleton blocks to test. For each candidate increment $D+e_r$, we recompute the prefix languages $P_i(D+e_r)$, the safe states, and the language $L(D+e_r)$. Using the prefix tree of L_f, every product enumeration is stopped as soon as a word leaves the tree. If the product remains safe, it contains at most $|L_f|$ words. Therefore one such recomputation costs $O(|L_f|N^2)$. Thus one iteration costs $O(N \cdot |L_f|N^2) = O(|L_f|N^3)$, and one fixed non-singleton guess costs $O(N \cdot |L_f|N^3) = O(|L_f|N^4)$. Multiplying by the at most $|L_f|$ non-singleton guesses gives the total running time

$$O(|L_f|^2N^4).$$

□

This settles this part of the problem on a positive note with respect to practical feasibility, but since the running time depends polynomially on the number of words in the language, the algorithm does not work in polynomial time w.r.t. to an arbirary DFA (let alone NFA) input. This leaves open the question whether an efficient decision procedure exists when the input language is presented as a finite automaton rather than a list of words (as the latter can be exponentially larger than the former for the same language). For $k \geq 2$, we know that each finite language can be encoded and for $k = 1$ we now know that we can decide in polynomial time whether a finite language given as a list of words can be encoded in a k-BW-NFA. For $k = 1$, the size of the resulting automata is determined by the longest words in the given language F. For $k \geq 2$, the size of the resulting automaton is not apparent. Hence, asking whether such an automaton with n states can be obtained, remains of high practical interest.

4.2 Minimization of k-BW-NFAs Is Hard for Fixed $k \geq 2$

In this last part, we consider the computational complexity of the minimization problem for k-bandwidth NFAs. In [2], it was shown for an analogous model that if k is left as part of the input, the problem is generally NP-hard. As it turns out, even though its expressiveness is severely restricted if the parameter k is chosen to be fixed as a small number, the problem of deciding the existence of an n state k-bandwith NFA is also already NP-hard in the case of fixing $k = 2$ and setting the required language to be a finite set of words.

Theorem 10. *Let L_f be a finite language and $n \in \mathbb{N}$. Deciding the existence of a 2-BW-NFA A with n states such that $L(A) = L_f$ is NP-complete.*

Proof. Regarding NP-containment, indeed, we can define a polynomial time verifier for a certificate containing a k-BW-NFA with n states. The maximum size of A lies in $O(n \cdot k \cdot |\Sigma|)$ for the total number of states n, the permissible bandwidth k that implies the number of states a transition can be added to, and the size of the alphabet $|\Sigma|$ that determines the maximum number of transitions between any two given states. To verify this solution, we can construct the minimal DFA representation of L_f in polynomial time, construct its complement and then its intersection with A in polynomial time. If the resulting automaton has a non-empty language, the verifier returns false, otherwise true.

NP-hardness is shown by a reduction from the `Biclique-Cover-Number` problem. We recall its definition. Let $G = (V, E)$ be a bipartite graph over vertices $V = V_1 \cup V_2$ with w.l.o.g. $V_1 = \{a_1, a_2, ..., a_\ell\}$ and $V_2 = \{b_1, b_2, ..., b_r\}$, for some numbers $\ell, r \in \mathbb{N}$, such that for all edges $\{u, v\} \in E$ we have $u \in V_1$ and $v \in V_2$. A biclique-edge-cover $C = \{C_1, C_2, ...\}$ of G is a set of bicliques (i.e., a set of complete bipartite subgraphs of G) that cover all edges in G. For each $C_i \in C$, we assume that it is a subset of V, since these elements uniquely define all edges in a complete subgraph. Deciding whether there exists a biclique-edge-cover with at most c bicliques, for some number $c > 1$, is known to be generally NP-hard [7].

Now, let G be a bipartite graph given as above and let $c \in \mathbb{N}$ with $c > 1$ be some positive number. We proceed by constructing an instance of the above mentioned minimization problem. We set $\Sigma = V$, $L_f = \{\ a_i b_j \mid \{a_i, b_j\} \in E\ \}$, and $n = 4(c-1)$.

($\Rightarrow$): Assume that there exists a biclique-edge-cover $C = \{C_1, \ldots, C_c\}$ of G with $c > 1$. We construct a 2-BW-NFA A with $n = 4(c-1)$ states. Here, we only sketch the idea (see also Fig. 6 for the case $|C| = 3$). For each biclique C_i, we use a gadget with four states $p_0 < p_1 < p_2 < p_3$. For every $\mathtt{a} \in C_i \cap V_1$, we add a transition from p_0 to p_1 reading $\mathtt{a}$, and for every $\mathtt{b} \in C_i \cap V_2$, we add a transition from p_1 to p_3 reading $\mathtt{b}$, where p_3 is a final state. Hence, this gadget accepts exactly the words $\mathtt{ab}$ corresponding to edges covered by C_i. Moreover, we add an ε-transition from p_0 to p_2 and use a ε-transitions from p_2 to chain gadgets, so that the automaton may either continue to a later gadget or accept a word in the current one. To save four states overall, the last two bicliques C_{c-1} and C_c are represented by one combined gadget. It uses a common initial state p_0 and a common final state p_3 but two distinct middle branches for C_{c-1} and C_c. In this way, the whole construction uses exactly $4(c-1)$ states and accepts precisely the language L_f.

($\Leftarrow$): Now assume there exists a 2-BW-NFA A with at most $n = 4(c-1)$ states such that $L(A) = L_f$. It can be shown that the small bandwidth 2 and maximum state space n results in a set of so called maximum middle states M that each can be used to represent one biclique. Each state $q \in M$ represents a state that sits in the middle between a set of vertices from V_1 and a set of vertices from V_2 that can form a biclique (in the example in Fig. 6, we would have $M = \{q_1, q_5, q_6\}$). It can be shown that each state $q \in M$ can be used to construct a biclique in G by bringing together all letters that lead to q with all the letters

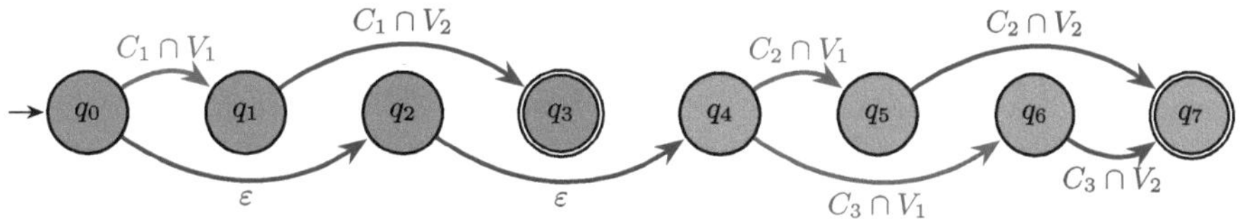

Fig. 6. Construction of a 2-BW-NFA A for a biclique-edge-covering C containing three bicliques. The light blue and orange colored states refer to two distinct gadgets. (Color figure online)

that go out from q to a final state. By showing that A with $n = 4(c-1)$ states can only have at most $|M| \leq c$ many maximum middle states, we obtain that such a 2-BW-NFA A emits a biclique-edge-covering C of G with at most $|C| = c$ bicliques. This direction of the proof, despite its simple intuition, requires a very comprehensive and detailed case distinction to actually confirm that $|M| > c$ is not possible. For verification purposes, the full detailed proof with the specific definition of M can be found in the arXiv version of this paper. □

5 Final Discussion

We introduced the notion of k-bandwidth NFA and analyzed several of its properties. We proved a strict infinite hierarchy of language classes defined by bandwidth (see Fig. 2).

We showed that finite languages can be represented with maximum bandwidth $k = 2$. For $k = 1$, we obtained polynomial time decidability whether a 1-BW-NFA A with $L(A) = L_f$ exists for a given finite language L_f presented as a list of words. The decision procedure implied a minimal size of such an automaton, allowing for a clear answer regarding the minimization problem. However, restricting k to the next smallest constant $k = 2$, we observe that minimization becomes NP-hard, establishing a clear boundary regarding feasibility.

This sets a solid theoretical foundation for the relevant questions of optimizing NFA that are to be encoded on circular DNA for RNA transcription. Some questions remain. First, the hardness result for the minimization problem for $k = 2$ is based on the assumption that the alphabet is unbounded. It would be interesting to know whether this result can be extended to finite alphabets.

Next, most results were established for finite languages, as those are most relevant for our practical motivation. The hardness results clearly translate to unbounded regular languages. Nonetheless, it would be interesting to establish a characterization of all classes in the hierarchy of k-BW-NFA as well as to establish whether the decidability results persist in the case of non-finite regular languages.

Finally, this paper was concerned with exact matching of languages. In practice, covering the target language and allowing for certain unwanted words (given some restrictions) might still be a valid approach. Hence, considering k-BW-NFA in an approximate setting is also a worthwhile direction to consider in the future.

Acknowledgements. This work was supported in part by Institute of Information & Communications Technology Planning & Evaluation (IITP) under the Artificial Intelligence Convergence Innovation Human Resources Development (IITP-2025-RS-2023-00255968) grant funded by the Korea government (MSIT) to D.-J. C., JSPS KAKENHI Grant Number JP23K10976 to S.Z.F., Daiichi-Sankyo "Habataku" Support Program for the Next Generation of Researchers to D. I., "Ambition Internationale" de la Region d'Auvergne-Rhône-Alpes "Rokam: RNA Origami Kinetic Assembly Model" to S. S. In addition, M. W. gratefully acknowledges Dirk Nowotka (Kiel University) for his support and encouragement in participating in this collaboration.

References

1. Chinn, P.Z., Chvátalová, J., Dewdney, A.K., Gibbs, N.E.: The bandwidth problem for graphs and matrices—a survey. J. Graph Theory **6**(3), 223–254 (1982)
2. Cho, D.J., Fazekas, S.Z., Seki, S., Wiedenhöft, M.: A formalization of co-transcriptional splicing as an operation on formal languages. Nat. Comput. **24**(4), 991–1009 (2025)
3. Cho, D.J., Fazekas, S.Z., Seki, S., Wiedenhöft, M.: Programmable co-transcriptional splicing: realizing regular languages via hairpin deletion. In: 31st International Conference on DNA Computing and Molecular Programming (DNA 31). Leibniz International Proceedings in Informatics (LIPIcs), vol. 347, pp. 5:1–5:22. Schloss Dagstuhl – Leibniz-Zentrum für Informatik, Dagstuhl, Germany (2025)
4. Geary, C., Meunier, P.É., Schabanel, N., Seki, S.: Proving the turing universality of oritatami co-transcriptional folding. In: Proceedings of ISAAC 2018. LIPIcs, vol. 123, pp. 23:1–23:13 (2018)
5. Geary, C., Rothemund, P.W., Andersen, E.S.: A single-stranded architecture for cotranscriptional folding of RNA nanostructures. Science **345**(6198), 799–804 (2014)
6. Merkhofer, E.C., Hu, P., Johnson, T.L.: Introduction to Cotranscriptional RNA Splicing, pp. 83–96. Humana Press (2014)
7. Orlin, J.: Contentment in graph theory: covering graphs with cliques. Indagationes Mathematicae (Proceedings) **80**(5), 406–424 (1977)

The Stanat-Weiss Pumping Lemma, Revisited (Extended Abstract)

Hermann Gruber[1], Markus Holzer[2(✉)], and Christian Rauch[3]

[1] BlueMetering GmbH, Höglwörther Str. 1, 82369 München, Germany
hermann.gruber@bluemetering.de
[2] Institut Für Informatik, Universität Giessen, Arndtstr. 2, 35392 Giessen, Germany
holzer@informatik.uni-giessen.de
[3] School of Electrical Engineering and Computer Science, University of Kassel, 34121 Kassel, Germany
christian.rauch@uni-kassel.de

Abstract. We study descriptional and computational properties of the minimal pumping constant induced by the pumping lemma of Stanat and Weiss [D. F. STANAT and S. F. WEISS. A pumping theorem for regular languages. *SIGACT News*, 14(1), 36–37 (Winter 1982)] which provides a necessary and sufficient condition for the regularity of languages.

1 Introduction

Pumping lemmas, e.g., [1,3,8,9,11–13,15], are fundamental tools in the study of formal languages and automata theory. Their primary importance lies in their ability to provide necessary conditions for a language to belong to a specific family, such as regular or context-free languages, allowing researchers to prove non-regularity or non-context-freeness through the contrapositive. Beyond proving non-membership, these lemmas offer deep insights into the expressive power and structural limitations of various language-generating or -accepting mechanisms. Furthermore, they have algorithmic utility; for instance, they imply a decision procedure to determine if a regular language is finite or infinite based on the existence of words that exceed the pumping constant.

Recently, research has focused on the descriptional and computational complexity of these lemmas [4–7], specifically investigating the minimal pumping constant—the smallest value p for which a pumping lemma holds for a given language L. Recent results have established the operational complexity of these constants, determining how they behave under regularity-preserving operations such as the Kleene star, union, intersection, and concatenation. Furthermore, the PUMPING-PROBLEM—deciding whether a given value p satisfies a specific lemma for a language represented by an automaton—has been proven to be computationally intractable. While this problem is coNP-complete for deterministic finite automata (DFAs), it becomes PSPACE-complete for nondeterministic

H. Gruber—The research was carried out while the first author was at Planerio GmbH, Theriesienhöhe 11A, 80538 München, Germany.

T. J. Smith (Ed.): CIAA 2026, LNCS 16695, pp. 47–61, 2026.
https://doi.org/10.1007/978-3-032-31176-4_4

finite automata (NFAs), and even undecidable when considering context-free languages under context-free pumping constraints.

A critical distinction exists between pumping lemmas that provide only necessary conditions and those that are both necessary and sufficient. Most classic lemmas, such as those by Kozen [11] or Bar-Hillel [1], are necessary conditions; a language may satisfy the pumping property but still fail to be regular or context-free. In contrast, Jaffe's pumping lemma [9] was the first to provide a necessary and sufficient condition for regularity. The defining characteristic of Jaffe's lemma is that its pumping property respects the Myhill-Nerode equivalence classes of the language, meaning that pumping a word does not change its equivalence class. This allows the lemma to fully characterize the family of regular languages, linking the pumping constant directly to the deterministic state complexity of the language. The pumping lemma proposed by Stanat and Weiss [15] extends these concepts such that pumping respects the classes of the syntactic monoid of the language under consideration. Thus, the pumping constant induced by Stanat and Weiss' pumping lemma is linked to the size of the syntactic monoid of the language.

We investigate Stanat and Weiss' pumping lemma in detail from the perspectives of automata theory and computational complexity. The key findings presented in this paper include—here $\mathtt{mpm}$ refers to the minimal pumping constant induced by the pumping lemma of Stanat and Weiss:

Robustness of the Constant: It is demonstrated that the minimal pumping constant $\mathtt{mpm}(L)$ is robust against variations in the lemma's conditions, such as changing the word length requirement from $|w| \geq p$ to $|w| = p$ or restricting the pumping iterations to a single case ($t = 1$). These findings align with research showing that Jaffe's pumping lemma remains unchanged under similar modifications.

Descriptional Complexity: The study evaluates $\mathtt{mpm}(L)$ by comparing it to other established measures of descriptional complexity. Beside basic facts, specifically, it establishes that $\mathtt{mpm}(L)$ is generally incomparable to both deterministic state complexity ($\mathtt{sc}$) and nondeterministic state complexity ($\mathtt{nsc}$), as there are language families where $\mathtt{mpm}(L)$ is significantly smaller than these measures and others where it is strictly larger. Furthermore, the paper proves that $\mathtt{mpm}(L)$ is always greater than or equal to the constant induced by Jaffe's pumping lemma ($\mathtt{mpe}(L)$).

Structural Bounds: An upper bound for the minimal constant based on the transformation monoid is provided, showing $\mathtt{mpm}(L) \leq \mathtt{llp}(G) + 1$, where $\mathtt{llp}(G)$ refers to the length of the longest path in the directed Cayley graph of the language's transformation monoid. This bound becomes a precise characterization when the transformation monoid is isomorphic to the syntactic monoid. This nicely resembles a similar result on the $\mathtt{mpe}$-relation to the length of the longest path in the DFA obtained in [6], with a subtle difference, because for $\mathtt{mpe}(L)$ the bound on the longest path may not be exact. For a language accepted by an n-state DFA, the $\mathtt{mpm}$-constant is further bounded by $e \cdot n!$, where e is Euler's number.

Computational Intractability: A significant portion of the research focuses on the PUMPING-PROBLEM deciding if a given value p satisfies the lemma for a specific automaton. The paper establishes that this problem is computationally hard: it is PSPACE-complete for NFAs and coNP-complete for DFAs. This complexity is noteworthy because the problem remains intractable even when dealing with deterministic finite automata.

Thus, the obtained results nicely fit to known results for other minimal pumping constants investigated in the literature. In particular, there are strong parallels between the properties of the minimal pumping constant for the Stanat-Weiss pumping lemma and those of other established constants, most notably Jaffe's pumping constant. These similarities manifest across several domains, as mentioned above. Due to space constraints, most proofs are omitted.

2 Preliminaries

A *nondeterministic finite automaton* (NFA) is a quintuple $A = (Q, \Sigma, \cdot, q_0, F)$, where Q is the finite set of *states*, Σ is the finite set of *input symbols*, $q_0 \in Q$ is the *initial state*, $F \subseteq Q$ is the set of *accepting states*, and the *transition function* $\cdot$ maps $Q \times \Sigma$ to 2^Q, where 2^Q refers to the powerset of Q. The *language accepted* by the NFA A is defined as $L(A) = \{ w \in \Sigma^* \mid (q_0 \cdot w) \cap F \neq \emptyset \}$, where the transition function is recursively extended to a mapping $Q \times \Sigma^* \to 2^Q$ in the usual way, namely for $q \in Q$ let $q \cdot \varepsilon = \{q\}$, where ε refers to empty word, and $q \cdot wa = \bigcup_{p \in (q \cdot w)} p \cdot a$, for every $w \in \Sigma^*$ and $a \in \Sigma$. Let $\mathtt{nsc}(L)$ refer to the *nondeterministic state complexity* of L, i.e., the number of states of the minimal nondeterministic finite automaton accepting L.

An NFA A is *deterministic* (DFA) if $q \cdot a$ is a singleton set, for every $q \in Q$ and $a \in \Sigma$. In this case we simply write $q \cdot a = p$ instead of $q \cdot a = \{p\}$. Similarly we define $\mathtt{sc}(L)$ to be the number of states of the minimal deterministic finite automaton accepting L. We have

$$\mathtt{nsc}(L) \leq \mathtt{sc}(L) \leq 2^{\mathtt{nsc}(L)},$$

for every regular language L. A finite automaton is *minimal* if its number of states is minimal with respect to the accepted language. It is well known that each minimal DFA is isomorphic to the DFA induced by the Myhill-Nerode equivalence relation. The *Myhill-Nerode* equivalence relation $\sim_L$ for a language $L \subset \Sigma^*$ is defined as follows: for $u, v \in \Sigma^*$ let $u \sim_L v$ if and only if $uw \in L \iff vw \in L$, for all $w \in \Sigma^*$. The equivalence class of u is referred to as $[u]_L$ or simply $[u]$ if the language is clear from the context and it is the set of all words that are equivalent to u w.r.t. the relation $\sim_L$, i.e., $[u]_L = \{ v \mid u \sim_L v \}$. Thus $\mathtt{sc}(L)$ is equal to $|\Sigma^*/{\sim_L}|$, which is the number of equivalence classes of $\sim_L$. For nondeterministic finite automata the situation is more involved. Nevertheless, there is a remarkably simple lower bound technique for the nondeterministic state complexity of regular languages, which is due to [2].

Theorem 1. *Let $L \subseteq \Sigma^*$ be a regular language and suppose there exists a set of pairs $S = \{ (x_i, y_i) \mid 1 \leq i \leq n \}$ such that*

1. *$x_i y_i \in L$ for $1 \leq i \leq n$,*
2. *and $i \neq j$ implies $x_i y_j \notin L$ or $x_j y_i \notin L$, for $1 \leq i, j \leq n$.*

Then any nondeterministic finite automaton accepting L has at least n states, i.e., $n \leq \mathtt{nsc}(L)$. Here S is called an extended fooling set *of L.*

Regular languages satisfy a variety of different pumping lemmas. A necessary and sufficient pumping lemma for regular languages, is due to Jaffe [9]:

Lemma 2 (Jaffe). *A language L is regular if and only if there is a constant p (depending on L) such that the following holds: If $w \in \Sigma^*$ and $|w| = p$, then there are words $x \in \Sigma^*$, $y \in \Sigma^+$, and $z \in \Sigma^*$ such that $w = xyz$ and*[1]

$$wv = xyzv \in L \iff xy^t zv \in L$$

for all $t \geq 0$ and each $v \in \Sigma^*$.

For a regular language L let $\mathtt{mpe}(L)$ refer to the minimal number p satisfying the conditions of Lemma 2. An alternative version of Lemma 2 appeared in [16, page 86, Lemma 4.3], where the requirement $|w| = p$ is replaced by the condition $|w| \geq p$. In [7] it was shown that this slight change in condition *does not* effect the value of the minimal pumping constant $\mathtt{mpe}(L)$, for a regular language L. Moreover, the following result is also from [7]:

Theorem 3. *Let L be a regular language over the alphabet Σ. Then*

$$\mathtt{mpe}(L) \leq \mathtt{sc}(L) \leq \sum_{i=0}^{\mathtt{mpe}(L)-1} |\Sigma|^i,$$

while the measures $\mathtt{mpe}$ and $\mathtt{nsc}$ are incomparable in general.

Further properties of $\mathtt{mpe}$ are: (i) $\mathtt{mpe}(L) = 1$ if and only if $L = \emptyset$ or $L = \Sigma^*$ and (ii) for finite language we have $\ell_{\max}(L) + 1 \leq \mathtt{mpe}(L) \leq \ell_{\max}(L) + 2$, where $\ell_{\max}(L) := \max\{ |w| \mid w \in L \}$, and this range of descriptional complexity can be made more precise. We state the following result without proof, since it is a straightforward rewrite of Jaffe's pumping condition—the details are left to the reader.

Lemma 4. *Let $L \subseteq \Sigma^*$ be a finite language. Then $\mathtt{mpe}(L) = \ell_{\max}(L) + 1$ if and only if for every word w with $|w| = \ell_{\max}(L) + 1$ there are words $x \in \Sigma^*$, $y \in \Sigma^+$ and $z \in \Sigma^*$ such that $w = xyz$ and*

$$(xz)^{-1} L = \emptyset,$$

where $u^{-1}L := \{ v \in \Sigma^* \mid uv \in L \}$; otherwise $\mathtt{mpe}(L) = \ell_{\max}(L) + 2$. □

[1] Observe that the words $w = xyz$ and $xy^t z$, for all $t \geq 0$, belong to the same Myhill-Nerode equivalence class of the language L. Thus, one can say that the pumping of the word y in w *respects equivalence classes.*

3 Stanat and Weiss' Pumping Lemma and Variants

As mentioned in the introduction, pumping lemmas generally provide necessary conditions for regularity. However, there are only a few pumping lemmas that are both necessary and sufficient characterizations of regular languages. The most prominent example of such a characterization is Jaffe's pumping lemma [9]. Another characterization, which is less known, is due to Stanat and Weiss [15] and will be discussed below.

Lemma 5 (Stanat & Weiss). *A language $L \subseteq \Sigma^*$ is regular if and only if the following condition holds: there exists a positive integer p such that for every word w of length p or greater, there exist $x, y, z \in \Sigma^*$ with $|y| \geq 1$ such that $w = xyz$ and*

$$uxv \in L \iff uxy^t v \in L,$$

for every $u, v \in \Sigma^*$ and all nonnegative integers $t \geq 0$.

This gives rise to the following definition of a *minimal pumping constant*: for a regular language L, let $\mathtt{mpm}(L)$ denote the minimal number p satisfying the conditions of Lemma 5. As in the case of Jaffe's pumping lemma, where the length condition can be changed from $|w| \geq p$ to $|w| = p$ without changing the minimal pumping constant—see, [7]—such a change by replacing the requirement $|w| \geq p$ by the condition $|w| = p$ results in an alternative version of Lemma 5. It is easy to see that this slight change in condition does not affect the statement. Moreover, it does not affect the minimal pumping constant either. For a regular language L we refer to the minimal number satisfying Lemma 5 with the modified condition $|w| = p$ as $\widetilde{\mathtt{mpm}}(L)$. Obviously, $\widetilde{\mathtt{mpm}}(L) \leq \mathtt{mpm}(L)$. On the other hand, $\mathtt{mpm}(L) \leq \widetilde{\mathtt{mpm}}(L)$, because for every word w that is longer than $\widetilde{\mathtt{mpm}}(L)$ there are words $\widetilde{w} \in \Sigma^*$ and $\widetilde{v} \in \Sigma^*$ with $|\widetilde{w}| = \widetilde{\mathtt{mpm}}(L)$ and $w = \widetilde{w}\widetilde{v}$. But word $\widetilde{w}$ can be pumped, i.e., there are words $\widetilde{x}, \widetilde{y}, \widetilde{z} \in \Sigma^*$ with $|\widetilde{y}| \geq 1$ such that $\widetilde{w} = \widetilde{x}\widetilde{y}\widetilde{z}$ and $u\widetilde{x}v \in L \iff u\widetilde{x}\widetilde{y}^t v \in L$, for every $u, v \in \Sigma^*$ and all nonnegative integers $t \geq 0$. Hence, the word w can be pumped, too. This implies that $\mathtt{mpm}(L) \leq \widetilde{\mathtt{mpm}}(L)$, which proves equality. That is, $\mathtt{mpm}(L) = \widetilde{\mathtt{mpm}}(L)$. Thus, the above given definition of $\mathtt{mpm}(L)$ is robust against this change in the length property.

Moreover, also the pumping property can be simplified to $t = 1$ instead of for every $t \geq 0$, thus leading to a stronger version than the original Stanat and Weiss lemma, without changing the actual minimal pumping constant $\mathtt{mpm}(L)$, for a regular language L.

Lemma 6. *A language $L \subseteq \Sigma^*$ is regular if and only if the following condition holds: there exists a positive integer p such that every word w with $|w| = p$, there exist $x, y, z \in \Sigma^*$ with $|y| \geq 1$ such that $w = xyz$ and*

$$uxv \in L \iff uxyv \in L, \tag{1}$$

for every $u, v \in \Sigma^*$.

Proof. It suffices to show that iterating the equivalence (1) yields

$$uxv \in L \iff uxy^t v \in L$$

for all $u, v \in \Sigma^*$ and all integers $t \geq 0$, which is straight forward, since from $uxv \in L \iff uxyv \in L$ replacing v by yv leads to $uxyv \in L \iff uxy^2v \in L$. Combining the two equivalences gives $uxv \in L \iff uxy^2v \in L$. Repeating this substitution inductively yields the desired original property of Stanat and Weiss. Note that the minimal pumping constant remains the same as in the original pumping lemma, where the condition is required to hold for all $t \geq 0$. □

Note that restricting the number of pumping-iterations to a constant is also possible for Jaffe's pumping lemma like, e.g.,

$$wv = xyzv \in L \iff xzv \in L \tag{2}$$

for each $v \in \Sigma^*$. Here the words w, x, y, and z are required to fulfill the property of Jaffe's pumping lemma, in particular $|w| = p$ is required. Obviously, regularity is equivalent to Jaffe's original pumping lemma which in turn implies the restricted version. Since the restricted version says that there exists p such that for every word w with $|w| = p$, there is a decomposition $w = xyz$ with $y \neq \varepsilon$ such that: $wv \in L \iff xzv \in L$ foreach $v \in \Sigma^*$ This means that w and xz are Myhill-Nerode equivalent (same right quotient). Crucially, $|xz| < |w| = p$. So the restricted version of Jaffe's pumping lemma means that every string of length p is Myhill-Nerode equivalent to a strictly shorter string. This means that the Myhill-Nerode classes are all represented by strings of length less than p, implying there are finitely many classes—hence language L is regular. Thus, the restricted version of Jaffe's pumping lemma also implies regularity, and since every regular language satisfies Jaffe's pumping lemma, we have shown all these three conditions, namely, regularity, Jaffe's pumping lemma, and its restricted form are equivalent. Therefore, we get:

Lemma 7. *A language L is regular if and only if there is a constant p (depending on L) such that the following holds: If $w \in \Sigma^*$ and $|w| = p$, then there are words $x \in \Sigma^*$, $y \in \Sigma^+$, and $z \in \Sigma^*$ such that $w = xyz$ and*

$$wv = xyzv \in L \iff xzv \in L$$

for each $v \in \Sigma^*$. □

Since every decomposition witnessing Jaffe's pumping property also witnesses its restricted version, the minimal pumping constant for a regular language L induced by the restricted version is bounded above by $\texttt{mpe}(L)$. Whether this bound can be strict remains open. Although we have no proof yet, we conjecture that this cannot be the case and thus, both minimal pumping constants coincide.

4 Relation to Other Pumping Constants and State Complexity

First we prove some basic facts for the minimal pumping constant induced by Stanat and Weiss' pumping lemma [15]. From the proof given in [15] we know that

$$\texttt{mpm}(L) \leq \texttt{sc}(L)^{\texttt{sc}(L)}.$$

Later in this section we show that this bound can be significantly improved. We start with the following result that nicely resembles a similar result on the minimal pumping constant induced by Jaffe's pumping lemma [5].

Theorem 8. *Let L be a regular language over the alphabet Σ. Then we have $\texttt{mpm}(L) = 1$ if and only if L is universal or empty, i.e., $L = \Sigma^*$ or $L = \emptyset$.*

As mentioned in the preliminaries, $\ell_{\max}(L)+1 \leq \texttt{mpe}(L) \leq \ell_{\max}(L)+2$ for a finite language L and Lemma 4 provides a more precise characterization of these bounds. In contrast, the case of mpm is slightly simpler, as we see next.

Lemma 9. *Let $L \subseteq \Sigma^*$ be a finite language. Then $\texttt{mpm}(L) = \ell_{\max}(L) + 2$.*

Next we consider the relation to mpe and measures of descriptional complexity of automata such as sc and nsc. We start with the measure mpe, where it is known [5] that $\texttt{mpe}(L) \leq \texttt{sc}(L)$, for every regular language L, and mpe and nsc are incomparable, i.e., (i) there is a family of unary regular languages $(L_i)_{i\geq 2}$ such that the inequality $\texttt{nsc}(L_i) < \texttt{mpe}(L_i)$ holds, and (ii) there is a family of regular languages $(L_i)_{i\geq 4}$, over an alphabet of growing size, such that $3 = \texttt{mpe}(L_i) < i \leq \texttt{nsc}(L_i)$. For the relation between mpe and mpm we find the following situation in general.

Theorem 10. *Let L be any regular language. Then $\texttt{mpe}(L) \leq \texttt{mpm}(L)$.*

Proof. Let p be the minimal pumping constant for the language L w.r.t. Lemma 5. Consider any word w and its decomposition $x, y, z \in \Sigma^*$ with $|y| \geq 1$ such that $w = xyz$ and

$$uxv \in L \iff uxy^tv \in L,$$

for every $u, v \in \Sigma^*$ and all nonnegative integers $t \geq 0$. But then this also holds for $u = \varepsilon$, the empty word, and $v = zv'$, for each $v' \in \Sigma^*$. Thus,

$$uxv = \varepsilon xzv' = xzv' \in L \iff uxy^tv = \varepsilon xy^tzv' = xy^tzv' \in L,$$

all $t \geq 0$ and each $v' \in \Sigma^*$. This in turn is equivalent to

$$wv' = xyzv' \in L \iff xy^tzv' \in L,$$

for all $t \geq 0$ and each $v' \in \Sigma^*$, which is exactly the property of Jaffe's pumping lemma. Therefore, $\texttt{mpe}(L) \leq \texttt{mpm}(L)$ as desired. □

Now what can be said about mpm and the state measures sc and nsc? Both of these measures are incomparable to mpm in general as we shall see next.

Theorem 11. *The following statements hold:*

- *There is a family of regular languages $(L_i)_{i\geq 2}$ such that mpm$(L_i) <$ sc(L_i) holds. The same holds true for the nondeterministic state complexity nsc.*
- *There is a family of regular languages $(L_i)_{i\geq 2}$ such that sc$(L_i) <$ mpm(L_i) holds. The same holds true for the nondeterministic state complexity nsc.*

In [6] a nice relation between the measure mpe and the length of the longest path of the automaton, that is, a simple directed path of maximum length from the initial state of the automaton, was shown, namely

$$\texttt{mpe}(L) \leq \ell_A + 1,$$

where A is a DFA, $L := L(A)$, and ℓ_A refers to the length, i.e., number of transitions, of the longest simple directed path of maximum length in the automaton A starting in the initial state q_0 and ending in any state. While the gap in this inequality can be arbitrarily large, this turns out to be very useful for a restricted class of automata, to be more precise bideterministic finite automata, where the pumping constant can be sandwiched between ℓ_A and $\ell_A + 1$; see [6] for more details.

We can do similar for the measure mpm, by considering the transformation/syntactic monoid of the language accepted by A and the longest path within the associated directed Cayley graph. Let us start with the definition of the transformation/syntactic monoid. The *transformation monoid* $T(A)$ of the DFA $A = (Q, \Sigma, \cdot, q_0, F)$ is the semigroup of all transformations $Q \to Q$ induced by words in Σ^*, i.e.,

$$T(A) = \{\, u^A : Q \to Q \mid u \in \Sigma^* \,\},$$

where u^A is the mapping form $Q \to Q$ which is given by $u^A(q) = q \cdot u$. It should be clear that the mappings $a^A \in \Sigma^A$ for $\Sigma^A = \{a^A \mid a \in \Sigma\}$ are the *generators* of $T(A)$. These mappings can be described in array form, i.e., a transformation $u^A : Q \to Q$ is written as

$$u^A = \begin{pmatrix} q_0 & q_1 & \dots & q_n \\ u^A(q_0) & u^A(q_1) & \dots & u^A(q_n) \end{pmatrix}$$

if $Q = \{q_0, q_1, \dots, q_n\}$. If the DFA A is minimal, then $T(A)$ is isomorphic to the syntactic monoid $M(L)$ of the language L. The *syntactic monoid* of L is the quotient $M(L) = \Sigma^*/\approx_L$, where the *syntactic congruence* $\approx_L$ is defined over Σ^* by $x \approx_L y$ if and only if $uxv \in L \iff uyv \in L$ for every $u, v \in \Sigma^*$; the *congruence class* of x is referred to as $[x]_{\approx_L}$ or simply $[x]_{\approx}$ if the language is clear from the context. As above, define $\Sigma_{\approx_L} = \{\, [a]_{\approx_L} \mid a \in \Sigma \,\}$ as *generator set* of $M(L)$. Finally, we define the *directed Cayley graph* Cay(M, G) of a monoid

$(M, \circ)$ with generator set G as the directed graph whose vertex set is $V := M$ and whose edge set is

$$E := \{ (m, m \circ g) \mid m \in M \text{ and } g \in G \}$$

build by *right multiplication.* If there is no danger of confusion and the set of generators G is clear from the context we simply write $\text{Cay}(M)$ instead of $\text{Cay}(M, G)$. The *length of the longest path* in the Cayley graph $\text{Cay}(M, G)$ is the maximum distance between any two vertices in the graph and it is referred to as $\texttt{llp}(\text{Cay}(M, G))$. We give a small example.

Example 12. Consider the DFA $A = (Q, \Sigma, \cdot, q_0, F)$ with state set $Q = \{1, 2, 3, 4\}$, alphabet $\Sigma = \{a, b\}$, initial state $q_0 = 1$, and set of final states $F = \{3\}$. The transition function can be read from the automaton drawing shown on the left in Fig. 1.

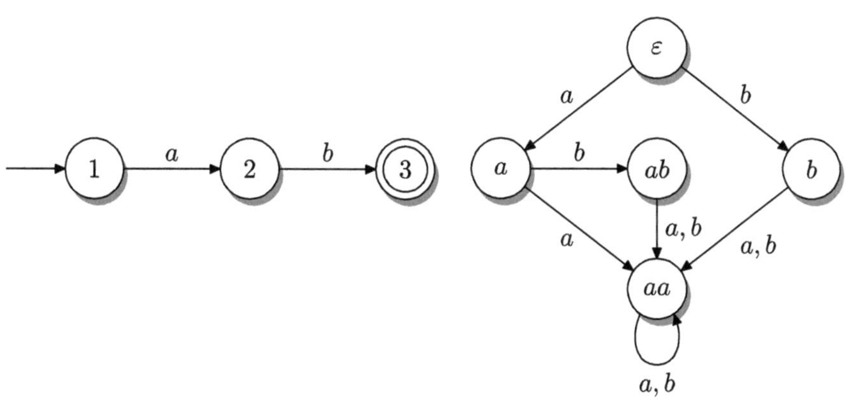

Fig. 1. Left: Minimal DFA A accepting the language $\{ab\}$. The non-accepting sink state is not shown. Right: Cayley graph induced by the transformation monoid of the DFA A, where the superscript A in the vertices and on the edges (labeled by generators) are omitted. In fact $T(A)$ is isomorphic to the syntactic monoid $M(L)$, since A is minimal.

The mappings $\varepsilon^A = id$, a^A, b^A, $(aa)^A$, and $(ab)^A$ form the transformation monoid $T(A)$, which is isomorphic to the syntactic monoid $M(L)$, since A is minimal. In array form these mappings read as follows:

$$a^A = \begin{pmatrix} 1\,2\,3\,4 \\ 2\,4\,4\,4 \end{pmatrix}, \qquad b^A = \begin{pmatrix} 1\,2\,3\,4 \\ 4\,3\,4\,4 \end{pmatrix},$$

and

$$(aa)^A = \begin{pmatrix} 1\,2\,3\,4 \\ 4\,4\,4\,4 \end{pmatrix}, \qquad (ab)^A = \begin{pmatrix} 1\,2\,3\,4 \\ 3\,4\,4\,4 \end{pmatrix}.$$

The Cayley graph $\text{Cay}(T(A), \Sigma^A)$ or simply $\text{Cay}(T(A))$ induced by the DFA A is depicted on the right in Fig. 1. The maximum distance between any of the vertices in Cay($T(A)$) is three. There is only one path of maximal length, namely it starts at vertex ε^A, moving in sequence to vertices a^A and $(ab)^A$, and ending in vertex $(aa)^A$. Hence, $\texttt{llp}(\text{Cay}(T(A))) = 3$. □

Now we are ready for the next theorem, where the proof follows along similar lines as the proof of the pumping lemma by Stanat and Weiss [15]:

Theorem 13. *Let $L \subseteq \Sigma$ be accepted by the DFA $A = (Q, \Sigma, \cdot, q_0, F)$, i.e., $L = L(A)$. Then*

$$\mathit{mpm}(L) \leq \mathtt{llp}(G) + 1.$$

where G is the Cayley graph $Cay(T(A), \Sigma^A)$ with $\Sigma^A = \{ a^A : Q \to Q \mid a \in \Sigma \}$.

This looks very similar to the result for the measure $\texttt{mpe}$ mentioned above. Now the question arises, whether the upper bound shown in Theorem 13 is also a lower bound. For arbitrary DFAs, this is generally not the case; but for the minimal finite automaton and the syntactic monoid of the language, the upper bound also serves as a lower bound. Before we can prove this, we recast Stanat and Weiss' pumping condition in terms of the transition function of a DFA accepting L.

Lemma 14. *Let $A = (Q, \Sigma, \cdot, q_0, F)$ be a DFA and let Q_{reach} refer to the set of all reachable states of A, i.e., $Q_{reach} = \{ q \in Q \mid q = q_0 \cdot w \text{ for } w \in \Sigma^* \}$. Moreover, let $x, y \in \Sigma^*$. Then*

$$uxv \in L(A) \iff uxyv \in L(A) \text{ for every } u, v \in \Sigma^*$$

if and only if

$$(q \cdot x) \cdot y \sim_{L(A)} q \cdot x \text{ for every } q \in Q_{reach}.$$

In case A is the minimal DFA, then the Myhill-Nerode equivalence $\sim_{L(A)}$ is replaced by (ordinary) equality.

Now we are ready for the next theorem:

Theorem 15. *Let $L \subseteq \Sigma^*$ be a regular language with syntactic monoid $M(L)$. Then*

$$\mathit{mpm}(L) = \mathtt{llp}(G) + 1,$$

where G is the Cayley graph $Cay(M(L), \Sigma_{\approx_L})$ with $\Sigma_{\approx_L} = \{ [a]_{\approx_L} \mid a \in \Sigma \}$.

An easy application of Theorem 15 is given next.

Example 16. We continue Example 12. Recall that $L = \{ab\}$. Then the longest path in the directed Cayley graph induced by the transformation monoid of the minimal DFA accepting L is 3. Thus, by Theorem 15 we have $\texttt{mpm}(L) = 3{+}1 = 4$. □

Using Theorem 15 we show how the upper bound on $\mathtt{mpm}(L)$ can be improved. Recall that for a language L accepted by an n-state DFA, it is known from [15] that $\mathtt{mpm}(L) \leq n^n$. To establish the improved bound, we first introduce some additional definitions and notations. Let S_n be the *symmetric group* on n elements, i.e., the group on all permutations on n elements, under composition, and further let T_n be the *full transformation monoid* on n elements, which is the monoid of all mappings on an n-element set under composition. As usual the set of n elements is $\{1, 2, \ldots, n\}$. Obviously $S_n \subseteq T_n$. Notice that S_n is of size $n!$ whereas T_n is of size n^n. For a mapping $t \in T_n$, let the *image* of t be defined as $\mathtt{img}(t) := \{\, t(i) \mid 1 \leq i \leq n \,\}$ and the *rank* of t is the cardinality of its image, i.e., $\mathtt{rank}(t) := |\mathtt{img}(t)|$.

Theorem 17. *For $n \geq 3$ the Cayley graph* $\mathrm{Cay}(T_n, S)$ *has* no *Hamiltonian path for any generating set $S \subseteq T_n$; it follows that the length of the longest path is at most $e \cdot n!$.*

A direct consequence of the previous theorem is the following result on regular languages, which is somehow surprising that although the upper bound on $\mathtt{mpm}$ grows very fast, but n^n grows even faster, i.e., the fraction of $e \cdot n!$ by n^n tends to 0 as n grows to infinity.

Corollary 18. *For $n \geq 3$ it holds $\mathit{mpm}(L) \leq e \cdot n!$ for any language L accepted by an n-state DFA.* □

Now the question arises whether there is a language that is close to the above mentioned bound of $e \cdot n!$. It is known that there is a set of generators of quadratic size containing all transpositions that induces a Hamiltonian path in S_n by applying the Steinhaus-Johnson-Trotter algorithm (see [10, Sec. 7.2.1.2] and [14]) to generate all permutations on n elements in a minimal-change order. In fact, we can improve this further, as shown in the next theorem, by reducing the number of generators to 2—we use the usual cycle notation for permutations.

Theorem 19. *The Cayley graph $\mathrm{Cay}(S_n, S)$ with generating set*

$$S = \{\, (1\,2\, \ldots\, n), (n-1\, n)\}$$

of the permutation inducing one cycle on the n elements and the transposition swapping $n-1$ and n while fixing all other elements has a Hamiltonian path for every $n \geq 2$.

As a consequence of the previous theorem we can show the following.

Theorem 20. *For $n \geq 3$ there is a binary language L accepted by an n-state DFA such that $\mathit{mpe}(L) = n < n! = \mathit{mpm}(L)$.*

Proof. Let $A = (Q, \Sigma, \cdot, q_0, F)$ with the state set $Q = \{1, 2, \ldots, n\}$, input alphabet $\Sigma = \{a, b\}$, initial state $q_0 = 1$, and set of final states $F = \{n\}$. We define its transition function by setting

$$i \cdot a = \sigma(i) \text{ and } i \cdot b = \tau(i) \text{ for } 1 \leq i \leq n,$$

where $\sigma = (1\,2\,\ldots\,n)$ and $\tau = (n-1\,n)$. Clearly the transformation monoid of A induces the Cayley graph for which we have shown in Theorem 19 that it contains a Hamiltonian path. By Theorem 15 we thus obtain that $\mathtt{mpm}(L(A)) = n!$ since the Cayley graph $\mathrm{Cay}(S_n, S)$ contains exactly $n!$ nodes—one for each permutation in the symmetric group S_n.

Therefore, it remains to show that $\mathtt{mpe}(L(A)) = n$. Observe that the word $w = a^{n-1}$ does not contain a subword that is in $L(A)$ and thus *cannot* be pumped by any of its decompositions w.r.t. Lemma 2 which immediately implies that $\mathtt{mpe}(L(A)) > n-1$. By Theorem 3 we have $\mathtt{mpe}(L(A)) \leq n$ and thus $\mathtt{mpe}(L(A)) = n$. □

5 Computational Complexity of the Pumping-Problem

The Pumping-Problem was introduced in [5] and reads as follows, adapted to Stanat and Weiss' pumping lemma:

Language-Pumping-Problem or for short Pumping-Problem:
Input: a finite automaton A and a natural number p, i.e., an encoding $\langle A, 1^p \rangle$.
Output: Yes, if and only if the statement from Lemma 5 holds for the language $L(A)$ w.r.t. the value p.

In case Jaffe's pumping lemma is considered instead, it was shown in [5] that the problem is coNP-complete for DFAs, while it becomes PSPACE-complete for NFAs. Thus, the Pumping-Problem is a rare example of an automaton problem, which is intractable already for a *single* deterministic finite automaton. Moreover, the minimal pumping constant for Jaffe's pumping lemma cannot be approximated within a factor of $o(n^{1-\delta})$ with $0 \leq \delta \leq 1/2$, for a given n-state NFA, unless the Exponential Time Hypothesis (ETH) fails.

Now let us focus on the Pumping-Problem as defined above with Stanat and Weiss' pumping lemma as basis. We show that from the complexity theoretical perspective the Pumping-Problem has the same complexity as the problem w.r.t. Jaffe's pumping lemma. This is expected for NFAs, but somehow surprising for DFAs, since the number of Myhill-Nerode classes can be significantly smaller then the number of syntactic monoid elements. Nevertheless, we first show that the problem under consideration is contained in PSPACE for NFAs.

Theorem 21. *Given an NFA A and a natural number p, it can be decided in* PSPACE *whether for the language $L(A)$ the statement of Lemma 5 holds for the value p.*

Proof. By our investigation of Stanat and Weiss' pumping lemma and its variants, we have shown that the original pumping lemma is equivalent to its strong version stated in Lemma 6 without altering the pumping constant.

Let $\langle A, 1^p \rangle$ be an input instance of the Pumping-Problem, where the NFA is $A = (Q, \Sigma, \cdot, q_0, F)$. We consider the negation of the pumping condition; since PSPACE is closed under complementation, the result follows. The negated pumping property reads as follows: there is a word $w \in \Sigma^*$ of length $|w| = p$, such that for every $x, y, z \in \Sigma^*$ with $|y| \geq 1$ satisfying $w = xyz$, there exist $u, v \in \Sigma^*$ with ($uxv \notin L(A)$ and $uxyv \in L(A)$) or ($uxv \in L(A)$ and $uxyv \notin L(A)$).

A Turing machine M verifies the negated pumping property as follows: first M guesses a word $w \in \Sigma^p$ and stores it; this uses space $O(n)$, since p is encoded in unary. Next the Turing machine iterates deterministically over all $O(n^2)$ choices of split positions defining a factorization $w = xyz$ with $|y| \geq 1$. For each such pair (x, y) the device M checks that there exist $u, v \in \Sigma^*$ with $uxv \in L(A)$ and $uxyv \notin L(A)$, or *vice versa*. To this end, M first guesses a word u letter-by-letter and simulates A on u, by updating the current reachable subset $S \subseteq Q$ using A's transition function when processing each symbol; thus, $S := \{q_0\} \cdot u$. Here, the current reachable subset is encoded using a bit vector of length n. From S, the Turing machine deterministically computes the subsets

$$S_x := S \cdot x \qquad \text{and} \qquad S_{xy} := S \cdot xy.$$

Now observe that there exists a suffix v with either $uxv \in L(A)$ or $uxyv \in L(A)$ if and only if there exists a word v such that exactly one of the subsets $S_x \cdot v$ and $S_{xy} \cdot v$ intersects F. Whether such a v exists is verified by guessing such a word letter-by-letter and deterministically compute S_{xv} and S_{xyv}. Finally, it remains to verify that exactly one of the computed subsets intersects with F, which again can be done in a deterministic fashion. Since the reachable subsets can be represented in $O(n)$ space and the outer universal loop over the factorizations $w = xyz$ uses only polynomial space, by reusing work space for each factorization, the overall space required is polynomial in n. Thus, it follows that the PUMPING-PROBLEM for NFAs is solvable in PSPACE. □

For NFAs it is easy to see that the PUMPING-PROBLEM is even PSPACE-complete.

Theorem 22. *Given an NFA A and a natural number p in unary, it is* PSPACE-*complete to decide whether for the language $L(A)$ the statement from Lemma 5 holds for the value p.*

In the remainder of this section, we consider the PUMPING-PROBLEM where a DFA together with a number p in unary is given as input. We show coNP-completeness in the next theorem.

Theorem 23. *Given a DFA A and a natural number p, it can be decided in* coNP *whether for the language $L(A)$ the statement of Lemma 5 holds for the value p.*

Thus, it remains to prove coNP-hardness.

Theorem 24. *Given a DFA A and a natural number p in unary, it is* coNP-*hard to decide whether for the language $L(A)$ the statement from Lemma 5 holds for the value p.*

Proof. Given a digraph $G = (V, E)$ on n vertices and a designated vertex $v_0 \in V$, our reduction constructs, in polynomial time, a DFA $A = (Q, \Sigma, \cdot, q_0, \{q_f\})$ such that G has *no* directed Hamiltonian path starting in v_0 if and only if Lemma 5 holds for the value $p = n$.

The construction of the automaton A is as follows: the state set Q consists of the vertex set V, together with a dead state q_d not in V, as well as a single accepting state q_f. The input alphabet is $V \cup \{ a_v \mid v \in V \}$ (the union being disjoint). The transition function on reading $v \in V$ is given by: $u \cdot v = v$ if (u, v) is an arc in E, and $u \cdot v = q_d$ otherwise; the transition function on reading a_v, for vertex $v \in V$, is given by $u \cdot a_v = q_f$ if $u = v$ and $u \cdot a_v = q_d$ otherwise. Moreover, let $q_d \cdot a = q_d$ and $q_f \cdot a = q_f$ for all symbols $a \in \Sigma$, thus effectively obtaining a non-accepting sink state as well as an accepting sink state. All other transitions lead to the dead state q_d. The start state q_0 is the designated vertex v_0. See Fig. 2 for a schematic drawing of the construction.

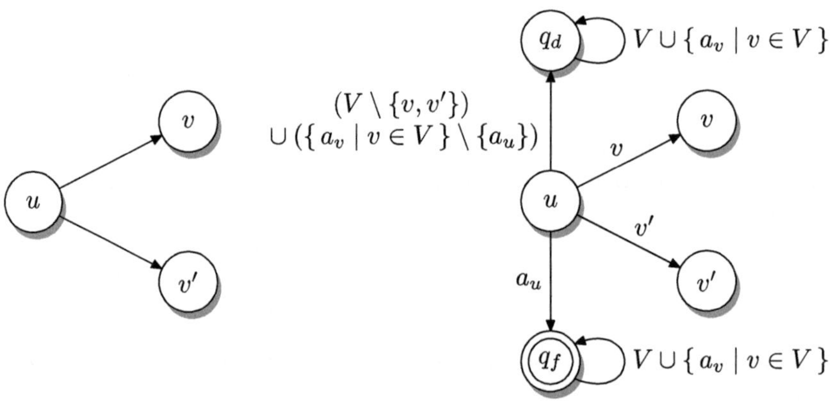

Fig. 2. Left: Schematic drawing of graph $G = (V, E)$ with distinguished vertex v_0 (not shown), depicting vertex u and its edges to v and v'. Right: Schematic drawing of the automaton A with state set $V \cup \{q_d, q_f\}$, input alphabet $V \cup \{ a_v \mid v \in V \}$, initial state v_0 (not shown), and set of final states $\{q_f\}$ constructed from graph G when considering vertex u with its neighboring vertices.

It remains to show that if the graph G has a directed Hamiltonian path starting at v_0, then there is a word w of length n that *cannot* be pumped. Conversely, if no Hamiltonian path exists from v_0, then every word of length n can be pumped, since no such distinguishing structure arises. The proof of these implications are left to the interested reader. □

Thus, we immediately obtain one of the main results of this paper:

Corollary 25. *Given a DFA A and a natural number p in unary, it is* coNP-*complete to decide whether for the language $L(A)$ the statement from Lemma 5 holds for the value p.* □

References

1. Bar-Hillel, Y., Perles, M., Shamir, E.: On formal properties of simple phrase structure grammars. Zeitschrift für Phonetik, Sprachwissenschaft und Kommunikationsforschung **14**, 143–177 (1961). https://doi.org/10.1524/stuf.1961.14.14.143
2. Birget, J.C.: Intersection and union of regular languages and state complexity. Inform. Process. Lett. **43**, 185–190 (1992)
3. Brauer, W.: Automatentheorie: Eine Einführung in die Theorie endlicher Automaten. Leitfäden und Monographien der Informatik, Teubner Stuttgart (1984). https://doi.org/10.1007/978-3-322-92151-2, (in German)
4. Dassow, J., Jecker, I.: Operational complexity and pumping lemmas. Acta Inform. **59**, 337–355 (2022). https://doi.org/10.1007/s00236-022-00431-3
5. Gruber, H., Holzer, M., Rauch, C.: The pumping lemma for regular languages is hard. In: Nagy, B. (ed.) Proceedings of the 27th International Conference on Implementation and Application of Automata, pp. 128–140. No. 14151. LNCS, Springer, Famagusta, Cyprus (2023). https://doi.org/10.1007/978-3-031-40247-0_9
6. Gruber, H., Holzer, M., Rauch, C.: On pumping preserving homomorphisms and the complexity of the pumping problem (extended abstract). In: Fazekas, S. (ed.) Proceedings of the 28th International Conference on Implementation and Application of Automata, pp. 153–165. No. 15015 in LNCS. Springer, Akita, Japan (2024). https://doi.org/10.1007/978-3-031-71112-1_11
7. Holzer, M., Rauch, C.: On Jaffe's pumping lemma, revisited. In: Bordihn, H., Tran, N., Vaszil, G. (eds.) Proceedings of the 25th International Conference on Descriptional Complexity of Formal Systems. pp. 65–78. No. 13918 in LNCS, Springer, Potsdam, Germany (2023). https://doi.org/10.1007/978-3-031-34326-1_5
8. Hopcroft, J.E., Ullman, J.D.: Introduction to Automata Theory, Languages and Computation. Addison-Wesley (1979)
9. Jaffe, J.: A necessary and sufficient pumping lemma for regular languages. SIGACT News **10**(2), 48–49 (1978). https://doi.org/10.1145/990524.990528
10. Knuth, D.E.: Combinatorial Algorithms, Part 1, The Art of Computer Programming, vol. 4A. Addison-Wesley (2011)
11. Kozen, D.C.: Automata and Computability. Undergraduate Texts in Computer Science, Springer (1997). https://doi.org/10.1007/978-1-4612-1844-9
12. Rabin, M.O., Scott, D.: Finite automata and their decision problems. IBM J. Res. Dev. **3**, 114–125 (1959). https://doi.org/10.1147/rd.32.0114
13. Savitch, W.J.: Abstract Machines and Grammars. Little, Brown and Company (1982)
14. Sedgewick, R.: Permutation generation methods. ACM Comput. Surv. **9**(2), 137–164 (1977)
15. Stanat, D.F., Weiss, S.F.: A pumping theorem for regular languages. SIGACT News **14**(1), 36–37 (1982)
16. Yu, S.: Chapter 2: Regular languages. In: Rozenberg, G., Salomaa, A. (eds.) Handbook of Formal Languages, vol. 1, pp. 41–110. Springer (1997)

Decidability of Regularity for Families of Languages

Oscar H. Ibarra[1] and Ian McQuillan[2(✉)]

[1] Department of Computer Science, University of California, Santa Barbara, CA 93106, USA
ibarra@cs.ucsb.edu
[2] Department of Computer Science, University of Saskatchewan, Saskatoon, SK S7N 5A9, Canada
mcquillan@cs.usask.ca

Abstract. We study decidability of the regularity problem, "given a machine M of a given type, is $L(M)$ a regular language? If so, construct a finite automaton accepting $L(M)$." The problem is easily shown to be undecidable for one-way nondeterministic machines with only one counter that does not increase after decreasing. In contrast, we prove that the problem is decidable for one-way deterministic reversal-bounded multi-counter machines. This result also holds for deterministic Parikh automata which are less powerful. To help with this decidability proof, we used a variant of this problem we call the bounded-regularity problem, "given a machine M of a given type, is $L(M)$ a bounded regular language?". We find that bounded-regularity is often easier to decide than regularity and it is decidable for one-way nondeterministic pushdown automata augmented with reversal-bounded multi-counters.

1 Introduction

It is common to study decision problems for classes of machines or grammars. Examples of problems that are often studied for classes of machines $\mathcal{M}$ (where each machine is over Σ) are: the **emptiness problem** ("given $M \in \mathcal{M}$, is $L(M) = \emptyset$?"), the **membership problem** ("given $M \in \mathcal{M}$ and $w \in \Sigma^*$, is $w \in L(M)$?"), the **finiteness problem** ("given $M \in \mathcal{M}$, is $L(M)$ finite?"), the **universe problem** ("given $M \in \mathcal{M}$, is $L(M) = \Sigma^*$?), the **equality problem** ("given $M_1, M_2 \in \mathcal{M}$, is $L(M_1) = L(M_2)$"), the **containment problem** ("given $M_1, M_2 \in \mathcal{M}$, is $L(M_1) \subseteq L(M_2)$"), and the **regular equality problem** ("given $M_1 \in \mathcal{M}$ and a finite automaton M_2, is $L(M_1) = L(M_2)$?"). Decidability of these problems have been studied for the regular languages and the context-free languages [12].

Another decision problem that is often studied is:

Supported, in part, by Natural Sciences and Engineering Research Council of Canada Grant 2022-05092 (Ian McQuillan).

T. J. Smith (Ed.): CIAA 2026, LNCS 16695, pp. 62–77, 2026.
https://doi.org/10.1007/978-3-032-31176-4_5

regularity problem for $\mathcal{M}$: "Given $M \in \mathcal{M}$, is $L(M)$ a regular language? If so, construct a finite automaton accepting $L(M)$."

It is known that this problem is undecidable for the context-free languages [10]. In contrast, it is known that the problem is decidable for deterministic pushdown automata [17] using quite a complicated procedure. The complexity of this procedure was improved in [18], but the problem is still difficult to decide.

In this paper, we are interested in decidability of regularity for other classes of automata that are more powerful than regular languages. One of the principle machine models we use in this paper is one-way nondeterministic k-counter machines. These machines have a finite control, and k data stores that each holds a non-negative integer. At each step, a machine can read from the input, test whether a specific counter i contains zero or not, and in response to that and the state, it can change states, and add either -1, 0, or 1 to counter i. While even deterministic 2-counter machines accept all recursively enumerable languages [12], restrictions can limit their power to make certain properties be decidable. Such a machine is r-reversal-bounded if, in every accepting computation, each counter switches from increasing to decreasing, or vice versa, at most r times. Denote by $\mathsf{NCM}(k, r)$ (respectively $\mathsf{DCM}(k, r)$), the class of one-way nondeterministic (resp. deterministic) r-reversal-bounded k-counter machines; and $\mathsf{NCM} = \bigcup_{k,r\geq 0} \mathsf{NCM}(k, r)$ and $\mathsf{DCM} = \bigcup_{k,r\geq 0} \mathsf{DCM}(k, r)$. Also, we study one-way nondeterministic (resp. deterministic) pushdown automata augmented by some number of reversal-bounded counters, and let NPCM (resp. DPCM) be this class. Many fundamental properties of these machines, including decidability of emptiness, membership, finiteness, universe, equality, containment, and regular equality, were studied in [14]. However, one problem that was not studied is decidability of the regularity problem. Certainly, it is beneficial for a machine model to have a procedure to convert a machine to an equivalent finite automaton if possible.

Using the same technique that shows regularity is undecidable for pushdown automata, we first show that it is undecidable for $\mathsf{NCM}(1, 1)$. While not surprising, the authors could not find the result in the literature. This class $\mathsf{NCM}(1, 1)$ accepts the smallest full trio (a family closed under homomorphism, inverse homomorphism, and intersection with regular languages) containing $\{a^n b^n \mid n \geq 0\}$, and therefore regularity is undecidable for any class of machines accepting full trios that contains this language. Thus, other restrictions are needed for decidability.

To study new classes where regularity is decidable, two key tools are used. The first tool involves bounded languages, which are any language $L \subseteq \Sigma^*$ such that $L \subseteq w_1^* w_2^* \cdots w_n^*$ for some $w_1, \dots, w_n \in \Sigma^*$. We define another problem:

bounded-regularity problem for $\mathcal{M}$: "Given $M \in \mathcal{M}$, is $L(M)$ a bounded regular language?"

Related to this problem is the boundedness problem: "given $M \in \mathcal{M}$, is $L(M)$ a bounded language? If so, determine $w_1, \dots, w_n$ such that $L(M) \subseteq w_1^* \cdots w_n^*$." This boundedness problem is notoriously tricky to determine decidability, but it is decidable for context-free languages [9], and for vector addition systems with states [6], and for NPCM [3]. Here, we show that for every semilinear full trio (with all properties providing effective constructions), if the boundedness problem is decidable, then so is the bounded-regularity problem. Therefore, this problem is decidable for NPCM. Hence, the bounded-regularity problem can be easier to decide than the regularity problem.

The second tool is an interesting known property for the regular languages. For every regular language L, there is a bounded language $R \subseteq L$ such that R is regular language and the Parikh image of L and R are equal [7,16].

Both of these results are key to proving the main result of the paper, that the regularity problem is decidable for DCM. This proof takes a given DCM machine M, and constructs a bounded subset of $L(M)$, which is proven to be non-regular if and only if $L(M)$ is non-regular. From this, the bounded-regularity problem can be used to decidable the regularity problem for DCM.

Techniques used in this paper also provide a set of tools to study the regularity problem for other classes of machines or grammars.

2 Preliminaries

Let $\mathbb{N}$ be the set of positive integers, and $\mathbb{N}_0$ be the set of non-negative integers. For $k \in \mathbb{N}$, we denote $\mathbb{N}(k) = \{1, \ldots, k\}$. For $k \in \mathbb{N}_0$, let $\pi(k)$ be $\hat{0}$ if $k = 0$ and $\hat{1}$ otherwise. For a set X, and $k \in \mathbb{N}$, let $[X]^k$ denote the set of k-tuples over X, and $[X]^+$ be the union of $[X]^k$ over all $k \in \mathbb{N}$. Given $B = \{\vec{v}_1, \ldots, \vec{v}_n\} \subseteq [\mathbb{N}_0]^k$ let $B^{\oplus} = \{x_1\vec{v}_1 + \cdots + x_n\vec{v}_n \mid x_1, \ldots, x_n \in \mathbb{N}_0\}$. Let $S \subseteq [\mathbb{N}_0]^k$. Then:

- S is a *linear set* if $S = \vec{c} + B^{\oplus}$, for some $\vec{c} \in [\mathbb{N}_0]^k$ and finite $B \subseteq [\mathbb{N}_0]^k$. Here, $\vec{c}$ is called the constant, B is called the set of periods, and $(\vec{c}, B)$ is called a representation of S, and S is said to be generated by $(\vec{c}, B)$.
- S is a *semilinear set* if $S = S_1 \cup \cdots \cup S_m$, for some linear sets $S_1, \ldots, S_m$. If $(\vec{c}_i, B_i)$ is a representation of S_i, then $\{(\vec{c}_i, B_i) \mid 1 \leq i \leq m\}$ is a representation of S, and S is generated by this representation.

We refer the reader to [12] for an introduction to the theory of automata and formal languages. We assume knowledge of finite automata, pushdown automata, closure properties, and decision problems. We describe some necessary notation. An alphabet is a finite set of symbols. Given an alphabet Σ, a string over Σ is a finite sequence of symbols of Σ, usually denoted by writing the symbols juxtaposed together. We denote by Σ^* the set of all words over Σ which includes the empty word λ. A language L over Σ is any subset of Σ^*. Given Σ, if $\Gamma \subseteq \Sigma$, π_Γ is the homomorphism from Σ^* to Γ^* that maps $a \in \Gamma$ to a, and a to λ otherwise.

Let $\Sigma = \{a_1, \ldots, a_k\}$ be an alphabet of distinct symbols. Given a word w, $|w| \in \mathbb{N}_0$ is its length, and $|w|_a$ is the number of a's in w, for $a \in \Sigma$. The *Parikh image* of $w \in \Sigma^*$, $\psi(w) = (|w|_{a_1}, \ldots, |w|_{a_k})$, and for $L \subseteq \Sigma^*$, $\psi(L) = \{\psi(w) \mid w \in L\}$. A language $L \subseteq \Sigma^*$ is *semilinear* (resp. *linear*) if $\psi(L)$ is a semilinear (resp. linear) set. A language $L \subseteq \Sigma^*$ is *letter-bounded* if there exists $b_1, \ldots, b_n \in \Sigma$ such that $L \subseteq b_1^* \cdots b_n^*$; L is called *distinct-letter-bounded* if $b_1, \ldots, b_n$ are distinct. A language $L \subseteq \Sigma^*$ is *bounded* if there exists $w_1, \ldots, w_n \in \Sigma^+$ such that $L \subseteq w_1^* \cdots w_n^*$; and if $L = w_1^* \cdots w_n^*$, we call L *elementary bounded*. The commutative closure of L, $\text{comm}(L) = \{w \in \Sigma^* \mid \psi(w) = \psi(v), v \in L\}$.

Definition 1. *A* one-way k-counter machine *is a tuple* $M = (Q, \Sigma, \delta, q_0, F)$, *where* Q, Σ, q_0, F *are the finite set of states, input alphabet, initial state in* Q, *and final state set that is a subset of* Q *respectively. The transition function* δ *is a partial function from* $Q \times (\Sigma \cup \{\lambda, \lhd\}) \times \mathbb{N}(k) \times \{\hat{0}, \hat{1}\}$ *to finite subsets of* $Q \times \{-1, 0, +1\}$, *where each transition* $(q', e) \in \delta(q, a, i, \tau)$ *is often denoted as* $\delta(q, a, i, \tau) \to (q', e)$. *The symbol* $\lhd$ *is the fixed input end marker. The machine* M *is* deterministic *if, for all* $q \in Q, a \in \Sigma \cup \{\lhd\}, i \in \mathbb{N}(k), \tau \in \{\hat{0}, \hat{1}\}$,

$$|\delta(q, a, i, \tau) \cup \delta(q, \lambda, i, \tau)| \leq 1,$$

and for all $q \in Q, a \in \Sigma \cup \{\lhd\}$, *and there is at most one counter* i *such that* $\delta(q, a, i, \tau) \cup \delta(q, \lambda, i, \tau) \neq \emptyset$ *for some* $\tau \in \{\hat{0}, \hat{1}\}$.

Here, a transition $\delta(q, a, i, \tau) \to (q', e)$ can be applied if q is the current state, $a \in \Sigma \cup \{\lhd, \lambda\}$ is the current input letter, which can be empty, or the right end marker, and counter i contains 0 if $\tau = \hat{0}$, and contains a positive value if $\tau = \hat{1}$. Hence, τ is a counter test that can be used to tell if a counter is empty or not.

For a machine to be deterministic, given the state and input letter, only one counter can have its status tested and have its value changed, and given its counter status, only one transition is possible.

Definition 2. *A* configuration *of* M *is a tuple* $(q, w, \vec{u}), u \in [\mathbb{N}_0]^k$ *for describing where* M *is in state* q, *with* $w \in \Sigma^* \lhd \cup \Sigma^*$ *the remaining input, and* $\vec{u}$ *is the contents of the* k *counters. We will associate labels* T_δ *bijectively with each transition of* δ. *The computation relation* $\vdash_M^\rho$ *is defined between configurations (often leaving off* M *if it is clear), with*

$$(q, aw, (e_1, \ldots, e_k)) \vdash_M^\rho (q', w, (e_1, \ldots, e_{i-1}, e_i + e, e_{i+1}, \ldots, e_k)), \qquad (1)$$

if ρ *is the transition label of , is that okay? It was messing up syntax highlighting* $\delta(q, a, i, \pi(e_i)) \to (q', e)$ *where* $a \in \Sigma \cup \{\lambda, \lhd\}$. *A* computation *of* M *on* $w \in \Sigma^* \cup \Sigma^* \lhd$ *is a sequence* $(q_0, w_0, \vec{u}_0) \vdash_M^{\rho_1} \cdots \vdash_M^{\rho_n} (q_n, w_n, \vec{u}_n)$ *where* $w_0 = w, \vec{u}_0 = \vec{0}$. *This computation is an* accepting computation *if* $q_n \in F, w \in \Sigma^* \lhd$, *and* $w_n = \lambda$.

Extended computations are defined by $\vdash_M^\gamma, \gamma \in T_\delta^*$, *where* $\vdash_M^\lambda$ *is from any configuration to itself, and inductively where* ρ *is a transition label by* $\vdash_M^{\rho\gamma}$ *is* $\vdash_M^\rho$ *followed by* $\vdash_M^\gamma$. *We also use the more standard notation of* $\vdash_M$ *to mean* $\vdash_M^\rho$ *for some* $\rho \in T_\delta$, *and* $\vdash_M^*$ *to mean* $\vdash_M^\gamma$ *for some* $\gamma \in T_\delta^*$.

A word $w \in \Sigma^*$ *is accepted by* M *if there is an accepting computation of* M *on* $w\lhd$. *The language accepted by* M *is denoted by* $L(M)$, *and is the set of all words in* Σ^* *accepted by* M.

The counters are r*-reversal-bounded if, during any accepting computation, each counter makes at most* r *changes between performing maximal sequences of non-decreasing and non-increasing the counter (and vice versa).*

To note here, a word $w \in \Sigma^*$ is accepted if there is a sequence of transitions that reads all of w, followed by $\lhd$, and then enters a final state; there can be a sequence of λ-transitions at the beginning of the computation, between any two input letters, between the last input letter and $\lhd$, and after $\lhd$. The end marker $\lhd$ does not increase the languages accepted for nondeterministic machines as machines defined without the end marker can simulate those with an end marker, by guessing that they are at the end of the input and then simulating a transition on $\lhd$ using λ. However, for deterministic machines, it is known that the end marker strictly increases the family of languages accepted [13]; that is, DCM defined with $\lhd$ accepts strictly more languages than DCM defined without it. The decidability results in this paper therefore use the more powerful definition with the end marker. This is in contrast to deterministic pushdown automata where the languages accepted are the same with or without the end marker [13].

A language family is a *full trio* if it is closed under homomorphism, inverse homomorphism, and intersection with regular languages. Many standard families, especially those that are accepted by one-way nondeterministic machine models, form full trios, including the context-free languages, and the languages accepted by $\mathsf{NCM}(k, r)$, for each k, r.

If we say that semilinearity is *effective* (for a class of machines or another finite representation accepting a family of languages), that means that there is an effective procedure to calculate, from a representation of L of this type, a semilinear representation of $\psi(L)$.

3 Bounded-Regularity in Full Trios

First, we note the following which follows easily from Greibach's Theorem [10] which we will state next. Consider a language family $\mathcal{L}$ that is effectively closed under union, and concatenation with regular sets, which has an undecidable universe problem. If P is any property that is defined on $\mathcal{L}$ where a) it is false for at least one language in $\mathcal{L}$, b) it is true for all regular languages, and c) it is preserved by inverse gsm, union with $\{\lambda\}$, and intersection with regular set, then P is undecidable for $\mathcal{L}$. This property was used to prove that regularity is undecidable for context-free languages. This can easily be strengthened to $\mathsf{NCM}(1, 1)$, as follows. First, the languages accepted by $\mathsf{NCM}(1, 1)$ are closed under union, and concatenation by regular sets [14], and it has an undecidable universe problem [2]. Second, we use the property P of being a regular language, and a) there is at least one language accepted by $\mathsf{NCM}(1, 1)$ that is not a regular language (say $\{a^n b^n \mid n > 0\}$), b) every regular language is a regular language, c) the regular languages are closed under inverse gsm (we omit the definition of gsm here and refer to [10,12] because we only require the known fact that the regular languages are closed under them), union with $\{\lambda\}$, and intersection. It is also known that $\mathsf{NCM}(1, 1)$ is the smallest full trio containing $\{a^n b^n \mid n > 0\}$ [2] and therefore:

Proposition 3. *The regularity problem is undecidable for* $\mathsf{NCM}(1,1)$. *Thus, the regularity problem is undecidable for any full trio containing* $\{a^n b^n \mid n > 0\}$.

Hence, we will study decidability of regularity for subsets of the families of interest. In this section, we will consider only families consisting of bounded languages. We need some existing concepts from the literature.

Given any semilinear set $S \subseteq [\mathbb{N}_0]^n$ and $w_1, \ldots, w_n \in \Sigma^*$, let $\phi_{w_1,\ldots,w_n}(S) = \{w_1^{i_1} \cdots w_n^{i_n} \mid (i_1, \ldots, i_n) \in S\}$. In the literature, all languages $L \subseteq w_1^* \cdots w_n^*$ such that there is a semilinear set S with $\phi_{w_1,\ldots,w_n}(S) = L$ are called bounded Ginsburg semilinear languages (or simply bounded semilinear) [5]. Conversely, given $L \subseteq w_1^* \cdots w_n^*$, then $\phi^{-1}_{w_1,\ldots,w_n}(L) = \{(i_1, \ldots, i_n) \mid w_1^{i_1} \cdots w_n^{i_r} \in L\} \subseteq [\mathbb{N}_0]^n$.

A representation of a linear set S is *reg-linear* if each periodic vector $\vec{v}_i = (v_1, \ldots, v_k)$ has at most one non-zero coordinate. A linear set S is *reg-linear* if there exists a reg-linear representation of S. Similarly, a representation $\{(\vec{c}_i, B_i) \mid 1 \leq i \leq m\}$ of a semilinear set is *reg-semilinear* if each $(\vec{c}_i, B_i)$ is reg-linear, for $1 \leq i \leq m$. And, a semilinear set S is *reg-semilinear* if there exists some reg-semilinear representation of S. This concept was studied in [8] but we give it a name here. Using this notation, the following was shown in that paper.

Proposition 4. *[8] Let* $a_1, \ldots, a_n$ *be distinct letters and* $L \subseteq a_1^* a_2^* \cdots a_n^*$. *Then* L *is regular if and only if* $\mathrm{comm}(L)$ *is regular.*

Proposition 5. *[8] Let* $L \subseteq w_1^* \cdots w_n^*, w_i \in \Sigma^*$. *Then* L *is regular if and only if* $\phi^{-1}_{w_1,\ldots,w_n}(L)$ *is reg-semilinear.*

Corollary 6. *[8] It is decidable to determine, for an arbitrary semilinear set* $S \subseteq [\mathbb{N}_0]^k$, *whether* $\psi^{-1}(S)$ *is regular.*

We can put these together as follows:

Proposition 7. *Let* $a_1, \ldots, a_n$ *be distinct letters and* $L \subseteq a_1^* \cdots a_n^*$. *The following are equivalent*

- L *is regular,*
- $\psi(L)$ *is reg-semilinear,*
- $\mathrm{comm}(L)$ *is regular.*

Proof. The first and third are equivalent by Proposition 4. And the first and second are equivalent by Proposition 5 as ψ and ϕ^{-1} coincide when $a_1, \ldots, a_n$ are distinct. □

Corollary 6 directly implies the following, because in any effectively semilinear representation, we can effectively construct a semilinear representation of $\psi(L)$ and $\psi^{-1}(\psi(L)) = \mathrm{comm}(L)$.

Corollary 8. *For any effectively semilinear family* $\mathcal{L}$, *it is decidable, given* $L \in \mathcal{L}$ *whether* $\mathrm{comm}(L)$ *is regular.*

As an aside, a reg-semilinear set is a special case of a stratified semilinear set, which is defined as follows. First, a set $S \subseteq [\mathbb{N}_0]^n$ is *stratified* if the following two conditions are satisfied: 1) each element in S has at most two non-zero coordinates, 2) there are no integers i, j, k, and m, and $x = (x_1, \ldots, x_n)$ and $x' = (x'_1, \ldots, x'_n)$ in S such that $1 \leq i < j < k < m \leq n$, and $x_i x'_j x_k x'_m \neq 0$. And a set $S \subseteq [\mathbb{N}_0]^n$ is *stratified semilinear* if S is expressible as a finite union of linear sets, each with a stratified set of periods. It is a long-standing open problem as to whether it is decidable if a given semilinear set is a stratified semilinear set [9].

Next, we establish the following, which is key for this paper.

Proposition 9. *Let $\mathcal{L}$ be any family that is effectively semilinear and effectively closed under inverse homomorphism and intersection with regular languages. It is decidable, given words $w_1, \ldots, w_n$ and $L \subseteq w_1^* \cdots w_n^*$, $L \in \mathcal{L}$, whether L is regular.*

Proof. Let $a_1, \ldots, a_n$ be new distinct letters, and let h be a homomorphism that maps a_i to w_i for each i. Let $L' = h^{-1}(L) \cap a_1^* \cdots a_n^*$. It is immediate that $L' \in \mathcal{L}$.

We will first show that

$$L' = \{a_1^{i_1} \cdots a_n^{i_n} \mid w_1^{i_1} \cdots w_n^{i_n} \in L\}. \tag{2}$$

Let $w \in h^{-1}(L) \cap a_1^* \cdots a_n^*$. Then there exists $u \in L$ such that $h(w) = u, w = a_1^{i_1} \cdots a_n^{i_n}$ and so $u = w_1^{i_1} \cdots w_n^{i_n}$ and hence w is in $\{a_1^{i_1} \cdots a_n^{i_n} \mid w_1^{i_1} \cdots w_n^{i_n} \in L\}$. Conversely, let $w \in \{a_1^{i_1} \cdots a_n^{i_n} \mid w_1^{i_1} \cdots w_n^{i_n} \in L\}$. So $w = a_1^{i_1} \cdots a_n^{i_n}$ where $w_1^{i_1} \cdots w_n^{i_n} \in L$. Thus, $w \in h^{-1}(w_1^{i_1} \cdots w_n^{i_n}) \cap a_1^* \cdots a_n^*$.

Next we see that $h(L') = L$. Let $w \in h(L')$. Then there exists $u \in L', h(u) = w, u = a_1^{i_1} \cdots a_n^{i_n}, w_1^{i_1} \cdots w_n^{i_n} \in L$ by (2), and this latter word equals w, and thus $w \in L$. Conversely, let $w \in L$. So $w = w_1^{i_1} \cdots w_n^{i_n}$ for some $i_1, \ldots, i_n$. Thus, $u = a_1^{i_1} \cdots a_n^{i_n} \in h^{-1}(L) \cap a_1^* \cdots a_n^*$, and in L', and $h(u) = w$. Hence, $w \in h(L')$.

Next, we see that L' is regular if and only if L is regular. Indeed, assume L is regular, then $h^{-1}(L) \cap a_1^* \cdots a_n^* = L'$ is regular since regular languages are closed under inverse homomorphism and intersection with regular languages [12]. Conversely, assume L' is regular. Then $h(L') = L$ is as well since regular languages are closed under homomorphism.

Hence it is sufficient to decide if L' is regular. But L' is regular if and only if $\text{comm}(L')$ is regular by Proposition 7. But this is decidable by Corollary 8. Hence, we can decide if L is regular. □

Therefore, the regularity problem is decidable for bounded languages in any semilinear family closed under inverse homomorphism and intersection with regular languages (assuming effective closure properties and that the words $w_1, \ldots, w_n$ of the bounded language are given).

We define the following problem variant.

Bounded-Regularity Problem for $\mathcal{M}$: "Given $M \in \mathcal{M}$, is $L(M)$ a bounded regular language? If so, construct a finite automaton accepting $L(M)$?"

We see the following general result:

Proposition 10. *Let $\mathcal{L}$ be any family that is effectively semilinear and effectively closed under inverse homomorphism. If the boundedness problem is decidable, then the bounded-regularity problem is decidable.*

Proof. Since the boundedness problem is decidable for $\mathcal{L}$, we can determine if $L \in \mathcal{L}$ is bounded, and if so, to determine words $w_1, \ldots, w_n$ such that $L \subseteq w_1^* \cdots w_n^*$ (the boundedness problem includes determining the words $w_1, \ldots, w_n$ as defined in Sect. 1). If the answer to this is 'no', then clearly the bounded-regularity problem is false. Assume L is bounded. Then by Proposition 9, it is decidable whether L is regular and hence bounded-regularity is decidable. □

Corollary 11. *The bounded-regularity problem is decidable for* NPCM.

This follows from the known effective semilinearity and closure under inverse homomorphism and intersection with regular languages [14] as well as decidability of the boundedness problem [3]. This also clearly implies decidability for NCM (and other equivalent models such as Parikh automata [4] and BLIND multicounter automata [11]).

It is also known that the boundedness problem is decidable for other semilinear full trios, such as one-way Turing machines with a single two-way worktape with a bound on the number of switches between switching directions on the worktape, and for one-way pushdown automata with the ability to flip their pushdown a bounded number of times [15]. Therefore, bounded-regularity will be decidable for these classes as well.

This can be made more general. In [19], a property called *downward closures* is defined and it was shown that in each semilinear full trio (with all properties effective), downward closures are computable. And in [1], another property was studied called concatenative amalgamation (we refer there for the definition). It was shown that for every full trio with all properties effective (equivalent to effective closure under rational transductions) that supports concatenative amalgamation, downward closures are computable if and only if boundedness is decidable. Combining these two together, we obtain that for any semilinear full trio (with all properties effective) that supports concatenative amalgamation, bounded-regularity is decidable. Some examples of classes that support concatenative amalgamation are given in [1].

4 Regularity for DCM

In this section, we consider decidability of regularity for one-way deterministic reversal-bounded multicounter machines (DCM).

We require the following fundamental known result regarding regular languages. First, given an alphabet Σ and a sequence of words $\alpha = (w_1, \ldots, w_n)$ where $w_i \in \Sigma^*, 1 \leq i \leq n$, define $\alpha^* = w_1^* \cdots w_n^*$. This must be a bounded elementary and regular language.

Proposition 12. *[7,16] For every regular language, $R \subseteq \Sigma^*$, there is a sequence $\alpha \in [\Sigma^*]^+$ that can be effectively determined, such that $\psi(R) = \psi(R \cap \alpha^*)$ and $R \cap \alpha^*$ is regular.*

In this proposition, it is clear that $R \cap \alpha^*$ is regular; the more difficult fact is that R and its bounded subset $R \cap \alpha^*$ have the same Parikh image. While the same thing was proven for context-free languages, we only require this result for regular languages here.

We need the well-known *Myhill-Nerode relation* [12]. Given a language $L \subseteq \Sigma^*$, $x \approx_L y$ if, $xz \in L \iff yz \in L$ for all $z \in \Sigma^*$. Hence, $x \not\approx_L y$ if and only if there exists a distinguishing string $z \in \Sigma^*$ such that exactly one of xz and yz is in L. The Myhill-Nerode Theorem states that a language L is regular if and only if $\approx_L$ has a finite number of equivalence classes. Equivalently, a language is non-regular if and only if there are infinitely many words $x_1, x_2, \ldots$ such that, for each $\iota \neq \rho$, $x_\iota \not\approx_L x_\rho$.

We need the following normal form. A k-counter DCM $M = (Q, \Sigma, \delta, q_0, F)$ is in *normal form* if (referring to counter i by C_i),

1. all counters are 1-reversal-bounded,
2. in every computation, for each C_i, it possibly increases, then possibly decreases, and then possibly returns to zero, and it must do all three before accepting,
3. in every computation, it can only enter a final state after reading $\lhd$, and there are no transitions from any final state,
4. directly after any transition that decreases C_i, it tests if C_i is zero or not,
5. every input w can be fully read, and M will reach a configuration where no transitions can be applied,
6. Q is partitioned into subsets Q_Y where $Y \in \{1^\uparrow, 1^\downarrow, 1^0\} \times \cdots \times \{k^\uparrow, k^\downarrow, k^0\}$, and if M is in a state $q \in Q_Y$ with $Y = (1^{z_1}, \ldots, k^{z_k})$, then in any computation that reaches q,

$$z_i = \begin{cases} \uparrow & \text{if } C_i \text{ has not changed, or the last change to } C_i \text{ was an increase,} \\ \downarrow & \text{if last change to } C_i \text{ was decrease and has not yet detected zero,} \\ 0 & \text{if } C_i \text{ has changed, and later detected zero.} \end{cases}$$

Lemma 13. *Every* DCM M *can be effectively converted into another* DCM M' *in normal form such that $L(M) = L(M')$.*

Proof. To note, this conversion can increase the number of counters. First, it is known that we can convert DCM machines into another where all counters are 1-reversal-bounded that accepts the same language [14]. Next, we can use the finite control to keep track of each counter, and whether each has not yet increased, increased but not yet decreased, decreased but not yet detected zero, or decreased then detected zero, verifying conditions 4 and 6. Clearly also, we can add additional states to make sure that it reads $\lhd$ before entering a final state. Therefore if it reaches a final state, the computation must already be accepting, and so further transitions can be removed. Finally, to make sure it reads the entire input word, we add new "dead" states (possibly more than one to preserve property 6) so that a transition on an input letter is always available if there is one. To detect and remove infinite sequences of λ transitions, it detects cycles of states on λ where within the cycle no counter has decreased across the entire cycle, and if this occurs, it switches to a dead state. If a counter decreases across each cycle, then that counter will eventually reach 0, and after it does, it can repeat this procedure (with fewer counters if any still above zero, and therefore the procedure can terminate) to detect and stop infinite loops. □

Most of the proof that regularity is decidable is in Lemma 14. Intuitively, the proof examines computations on words x of the DCM machine M. But it treats computations as finite automata without using counters, and instead by reading letters associated with counters without actually counting. This language is then regular which allows for the application of Proposition 12. We can then instead find alternate computations over a bounded language that changes each counter by the same amount because the letters associated with counter changes appear in dedicated components of the Parikh image. This results in a new computation not on x but on a different word $\overline{x}$ that reaches the same configuration in M as does x (and makes the same counter changes) but $\overline{x}$ is within a bounded language. Furthermore, if we start with a non-regular language $L = L(M)$, and therefore we have an infinite number of words $x_1, x_2, \ldots$ such that $x_\iota \not\approx_L x_\rho$ for $\iota \neq \rho$, we also show that $\overline{x_\iota} \not\approx_{L'} \overline{x_\rho}$ where L' is a bounded subset of L. Thus, L' is non-regular exactly when L is non-regular.

Lemma 14. *Given a* DCM$(k,1)$ $M = (Q, \Sigma, \delta, q_0, F)$ *in normal form, there is a sequence* $\alpha \in [\Sigma^*]^+$ *such that* $L = L(M)$ *is non-regular if and only if* $L \cap \alpha^*$ *is non-regular. Moreover, we can effectively construct* α *from* M.

Proof. Assume $L = L(M)$ is non-regular. Let $M = (Q, \Sigma, \delta, q_0, F) \in$ DCM$(k,1)$ in normal form accepting a non-regular language, and therefore Q is partitioned into subsets Q_Y as per the normal form. We refer to counter i by C_i. Without loss of generality, we treat $\lhd$ as a normal character of Σ which M verifies is at the end before accepting.

We define the relation $\vdash^{\rightarrowtail}$ so that $(q, w, \vec{v}) \vdash^{\rightarrowtail} (p, \lambda, \vec{u})$ when $(q, w, \vec{v}) \vdash^* (p, \lambda, \vec{u})$ but no λ transitions are possible from $(p, \lambda, \vec{u})$. Given $w \in \Sigma^*$, let δ^* be a function where $\delta^*(w) = (q, \vec{u})$, such that $(q_0, w, \vec{0}) \vdash^{\rightarrowtail} (q, \lambda, \vec{u})$. It is clear that $\delta^*(w)$ must exist, because M can fully read every word and cannot use infinite sequences of λ transitions by the normal form. Due to determinism, there cannot be two such configurations.

Consider the computation of $w \in \Sigma^*$,

$$(q_0, w_0, \vec{u}_0) \vdash \cdots \vdash (q_n, w_n, \vec{u}_n), w_0 = w, w_n = \lambda, \vec{u}_0 = \vec{u}_n = \vec{0}, \tag{3}$$

where no λ transitions are possible from $(q_n, w_n, \vec{u}_n)$. From this computation, we say that this computation *uses*

$$\beta = (p_1, p_1', a_1, p_2, p_2', a_2, \ldots, p_l, p_l') \tag{4}$$

if there exists distinct $Y_1, \ldots, Y_l$, such that for each i, $1 \le i \le l$, $p_i, p'_i \in Q_{Y_i}$, $p_1 = q_0, p'_l = q_n$, and in the computation, for each i, $1 \le i \le l$, from p_i, it reaches p'_i, and if $i < l$, in the next transition it switches to p_{i+1} while reading $a_i \in \Sigma \cup \{\lambda\}$. Therefore, we can rewrite (3) as

$$\begin{aligned}
&(p_1, t_1 a_1 t_2 \cdots a_{l-1} t_l, \vec{v}_1) \vdash^{\gamma_1} (p'_1, a_1 t_2 \cdots a_{l-1} t_l, \vec{v}'_1) \vdash^{\epsilon_1} \\
&(p_2, t_2 a_2 \cdots a_{l-1} t_l, \vec{v}_2) \vdash^{\gamma_2} (p'_2, a_2 t_3 \cdots a_{l-1} t_l, \vec{v}'_2) \vdash^{\epsilon_2} \\
&\vdots \\
&(p_l, t_l, \vec{v}_l) \vdash^{\gamma_l} (p'_l, \lambda, \vec{v}'_l),
\end{aligned} \tag{5}$$

such that each γ_i is a sequence of transitions within Q_{Y_i} that reads $t_i \in \Sigma^*$, and when $i < l$, ϵ_i is a transition from a state in Q_{Y_i} to $Q_{Y_{i+1}}$ that reads $a_i \in \Sigma \cup \{\lambda\}$. There are only finitely many such vectors β. Let $Z_\beta(M)$ be all computations (not necessarily accepting) of M that use β.

Let $\Delta = \{c_1, \ldots, c_k, d_1, \ldots, d_k\}$ and $\Gamma = \Sigma \cup \Delta$. For each β of the form (4) we create an NFA M_β accepting the regular language L_β over Γ^* as follows: M_β operates just like M, switching states just like M, but only using the sequence β. In M_β, p'_l is the accepting state, so that there is a word in L_β for every computation on w in $Z_\beta(M)$, whether w is accepting in M or not. Furthermore, instead of actually counting, it reads symbols associated with the counter instructions in Δ by reading c_i for each increase of C_i and d_i for each decrease of C_i. Formally, L_β contains all strings w' where $w' = w_1 y_1 \cdots w_m y_m, w_i \in \Sigma \cup \{\lambda\}, y_i \in \Delta \cup \{\lambda\}, m \ge 0$, and for $1 \le i \le m$ there exists transitions of M, $\tau_1, \ldots, \tau_m$ that defines a computation in $Z_\beta(M)$ starting at $p_1 = q_0$ and ending at p'_l, where τ_i reads w_i and

$$y_i = \begin{cases} c_i & \text{if } \tau_i \text{ increases } C_i, \\ d_i & \text{if } \tau_i \text{ decreases } C_i, \\ \lambda & \text{if } \tau_i \text{ does not change any counter.} \end{cases} \tag{6}$$

Further, for each i, $1 \le i \le k$, M_β guesses when it switches from simulating a transition with C_i positive but decreasing to transitions where C_i is zero, which is why nondeterminism is needed in M_β (and otherwise the behaviour of M_β is deterministic). Certainly, M_β can accept words where counting is not "properly followed" as in M; e.g. where more d_i's are read than c_i's, or where it guesses C_i is zero even though more c_i's have been read than d_i's.

Next, we will argue by using L_β that each $w \in \Sigma^*$ read by a computation in $Z_\beta(M)$ has a corresponding word $\overline{w} \in \Sigma^*$ with $\delta^*(w) = \delta^*(\overline{w})$ where $\overline{w}$ is over a bounded language.

Because L_β is regular, there is $\alpha'_\beta \in [\Gamma^*]^+$ such that $\psi(L_\beta) = \psi(L_\beta \cap (\alpha'_\beta)^*)$ by Proposition 12. Notice that this subset of L_β, $L_\beta \cap (\alpha'_\beta)^*$, has each word including symbols of Δ, and for each i, all c_i's occur before any d_i because all increases of C_i occur before any decreases via the states in M, and hence, this must be true for all words in L_β. Then given $w \in \Sigma^*$, where there is a computation in $Z_\beta(M)$ reading w on transitions $\tau_1, \ldots, \tau_m$ say, and so we can create a string $\overset{\sqcap}{w} = w_1 y_1 \cdots w_m y_m \in L_\beta$ where τ_i reads w_i and y_i changes as in (6). Define a new equivalence relation $\equiv \subseteq \Sigma^* \times \Sigma^*$ where $[z]$ is the equivalence class of z, such that $z_1 \equiv z_2$, $z_1, z_2 \in \Sigma^*$ if there is a β with a computation on $z_1, z_2 \in Z_\beta(M)$, and $\psi(\overset{\sqcap}{z_1}) = \psi(\overset{\sqcap}{z_2})$; hence, $\overset{\sqcap}{z_1}$ and $\overset{\sqcap}{z_2}$ have the same number of copies of each symbol of Δ. For each w, each equivalence class $[w]$ has a $\overline{w} \in [w]$, $\overset{\sqcap}{\overline{w}} \in L_\beta \cap (\alpha'_\beta)^*$ with the same Parikh image, and therefore have the same number of each symbol of Δ. Let $\alpha_\beta = \pi_\Sigma(\alpha'_\beta)$. Then given w, $\delta^*(w) = \delta^*(\overline{w})$, because the computation on both end in the same state (the last state of β which is the only possible state that can be accepting), and has the same number of each symbol of Δ, hence counter increases and decreases, but $\overline{w} \in \alpha_\beta^*$. Hence $w \approx_L \overline{w}$ (the Myhill-Nerode relation), because both words reach the same configuration of M, and so for any $z \in \Sigma^*$, either $wz \in L$ and $\overline{w}z \in L$, or neither are in L. Let α_1 be the concatenation of α_β over all β. Clearly α_1 is a finite sequence. Also, any word in α_β^* for any β will be in α_1^* as well.

Since L is not regular, there exists an infinite sequence of words $x_1, x_2, \ldots$ such that $x_\iota \not\approx_L x_\rho$, for all $\iota \neq \rho$. We now know that for this sequence, there are words $\overline{x_1}, \overline{x_2}, \ldots$ over a bounded language α_1^* such that $x_i \equiv \overline{x_i}$ for all i, and hence $\delta^*(x_i) = \delta^*(\overline{x_i})$. Because $x_\iota \not\approx_L x_\rho$ and $\overline{x_\iota} \not\approx_L \overline{x_\rho}$, there must exist z such that exactly one of $\overline{x_\iota}z, \overline{x_\rho}z \in L$. Even though $\overline{x_\iota}, \overline{x_\rho}$ come from within a bounded language, all the distinguishing strings z that must exist from L being non-regular do not need to also come from within a bounded language. But our goal is to create a bounded subset that is provably non-regular. Thus, we wish to next show that there is another sequence $\alpha_2 \in [\Sigma^*]^+$ such that, for all $x_\iota \not\approx_L x_\rho$, there is a string $z' \in \alpha_2^*$ such that exactly one of $\overline{x_\iota}z', \overline{x_\rho}z'$ is in L. This involves a similar technique as we used above.

Let ι and ρ be such that $x_\iota \not\approx_L x_\rho$, and let z be such that exactly one of $x_\iota z, x_\rho z$ is in L, which must exist. Assume without loss of generality that $x_\iota z \in L$ and $x_\rho z \notin L$. Even though it is straightforward to take this z and create another word z' in a bounded language such that $\overline{x_\iota}z' \in L$, we need to make sure that this same z' does not accept after reading it from the configuration $\delta^*(x_\rho)$, which will start in a different configuration than $\delta^*(x_\iota)$ (it does not need to change counters from $\delta^*(x_\rho)$ as it would after reading x_ι). Therefore, additional steps are required.

Consider the accepting computation on $x_\iota z$, $(q_0, x_\iota z, \vec{0}) \vdash^{\rightarrowtail} (p, z, \vec{v}) \vdash^{\rightarrowtail} (p_f, \lambda, \vec{0})$. The latter part can be rewritten as

$$\begin{aligned}
&(p_1, t_1 a_1 t_2 \cdots a_{l-1} t_l, \vec{v}_1) \vdash^{\gamma_1} (p'_1, a_1 t_2 \cdots a_{l-1} t_l, \vec{v}'_1) \vdash^{\epsilon_1} \\
&(p_2, t_2 a_2 \cdots a_{l-1} t_l, \vec{v}_2) \vdash^{\gamma_2} (p'_2, a_2 t_3 \cdots a_{l-1} t_l, \vec{v}'_2) \vdash^{\epsilon_2} \\
&\vdots \\
&(p_l, t_l, \vec{v}_l) \vdash^{\gamma_l} (p'_l, \lambda, \vec{v}'_l),
\end{aligned} \tag{7}$$

such that there exists distinct $Y_1, \ldots Y_l, p_i, p'_i \in Y_i, p_1 = p, \vec{v}_1 = \vec{v}, p'_l = p_f \in F, \vec{v}'_l = \vec{0}, t_i \in \Sigma^*$, and if $i < l$ each ϵ_i is a transition from a state in Q_{Y_i} to $Q_{Y_{i+1}}$ that reads $a_i \in \Sigma \cup \{\lambda\}$. Let $\beta = (p_1, p'_1, a_1, p_2, p'_2, a_2, \ldots, p_l, p'_l)$. By contrast, consider the maximal computation of M on $x_\rho z$ (that fully reads every letter and applies as many transitions as possible for which no more are possible): $(q_0, x_\rho z, \vec{0}) \vdash^{\rightarrowtail} (q, z, \vec{u}) \vdash^{\rightarrowtail} (q', \lambda, \vec{u}')$, $q' \notin F$, and it does not pass over any final states since $x_\rho z \notin L$. We can again rewrite the latter half:

$$\begin{aligned}
&(q_1, s_1 b_1 s_2 b_2 \cdots b_{m-1} s_m, \vec{u}_1) \vdash^{\mu_1} (q_1', b_1 s_2 b_2 s_3 \cdots b_{m-1} s_m, \vec{u}_1') \vdash^{\theta_1} \\
&(q_2, s_2 b_2 \cdots b_{m-1} s_m, \vec{u}_2) \vdash^{\mu_2} (q_2', b_2 s_3 \cdots b_{m-1} s_m, \vec{u}_2') \vdash^{\theta_2} \\
&\vdots \\
&(q_m, s_m, \vec{u}_m) \vdash^{\mu_m} (q_m', \lambda, \vec{u}_m'),
\end{aligned} \tag{8}$$

such that there exists distinct $Z_1, \ldots, Z_m, q_i, q_i' \in Z_i, q_1 = q, \vec{u}_1 = \vec{u}, q' = q_m', \vec{u}' = \vec{u}_m', s_i \in \Sigma^*$ and if $i < m$, each θ_i is a transition from a state in Q_{Z_i} to $Q_{Z_{i+1}}$ that reads $b_i \in \Sigma \cup \{\lambda\}$. We can construct a vector $\gamma = (q_1, q_1', b_1, \ldots, b_{m-1}, q_m, q_m')$.

Let $\Delta_1 = \{c_1^1, \ldots, c_k^1, d_1^1, \ldots, d_k^1\}$ and $\Delta_2 = \{c_1^2, \ldots, c_k^2, d_1^2, \ldots, d_k^2\}$ be two new alphabets, and $\Gamma' = \Sigma \cup \Delta_1 \cup \Delta_2$. Corresponding to both β and γ, we create an NFA $M_{\beta,\gamma}$ accepting a language $L_{\beta,\gamma} = L(M_{\beta,\gamma})$ over $(\Gamma')^*$ that is similar to M_β above, but now it intuitively runs two such NFAs in parallel. That is, states are ordered pairs (p', q') where $p', q' \in Q$, and the first component switches states according to β while the second switches according to γ; and instead of actually counting, it reads symbols associated with the counter instructions. We make (p_l', q_m') the accepting state in $M_{\beta,\gamma}$. Therefore, $L_{\beta,\gamma}$ consists of all strings w' where

$$\pi_{\Sigma \cup \Delta_1}(w') = w_1 y_1 \cdots w_s y_s, \pi_{\Sigma \cup \Delta_2}(w') = g_1 z_1 \cdots g_r z_r, \tag{9}$$

$w_i \in \Sigma \cup \{\lambda\}, y_i \in \Delta_1 \cup \{\lambda\}, 1 \le i \le s, g_i \in \Sigma \cup \{\lambda\}, z_i \in \Delta_2 \cup \{\lambda\}, 1 \le i \le r$, there exists a sequence of transitions $\tau_1, \ldots, \tau_s$ of M that switch states according to β using the first coordinate of the states reading $w_1, \ldots, w_s$ respectively, and then reading y_i instead of changing counters as in (6) using Δ_1; and there exists a sequence of transitions $\kappa_1, \ldots, \kappa_r$ of M that switch states according to γ using the second coordinate of the states, reading $g_1, \ldots, g_r$ respectively, and then reading z_i instead of changing counters in a similar fashion using Δ_2. Because $L_{\beta,\gamma}$ is a regular language, by Proposition 12, there is a $\alpha_{\beta,\gamma}' \in [(\Gamma')^*]^+$ such that $\psi(L_{\beta,\gamma}) = \psi(L_{\beta,\gamma} \cap (\alpha_{\beta,\gamma}')^*)$.

Specifically consider the string z (that distinguishes x_ι and x_ρ). There is a string $\underset{\sqcup}{z} \in (\Gamma')^*$ obtained as above in (9), where the computation (7) reads w_i while y_i follows the counters of that computation, and the computation (8) reads g_i while z_i follows the counters of that computation. To note here, $\underset{\sqcup}{z}$ depends on z, w_ι, and w_ρ, and the sequence β and γ determines when counters become empty, which is reflected in the transitions that can be applied. That means $\underset{\sqcup}{z} \in L_{\beta,\gamma}$. Furthermore, there exists $z' \in L_{\beta,\gamma} \cap (\alpha_{\beta,\gamma}')^*$ with the same Parikh image as $\underset{\sqcup}{z}$. This means the number of each character of $\Delta_1 \cup \Delta_2$ is the same. Let $\widetilde{z} = \pi_\Sigma(z')$, and $\alpha_{\beta,\gamma} = \pi_\Sigma(\alpha_{\beta,\gamma}')$.

Thus, given the computation (7) and (8), at the end of reading x_ι and x_ρ, if we replace z with reading $\widetilde{z}$, it will increase and decrease counters exactly the same from both configurations $\delta^*(x_\iota)$ and $\delta^*(x_\rho)$, and still obey β in the first and γ in the second, and reach the same configuration as reading z. Let us concatenate all sequences $\alpha_{\beta,\gamma}$ to obtain α_2. Finally, let α be obtained by concatenating α_1 with α_2.

Claim. Let $L' = L \cap \alpha^*$. Then:

1. $\delta^*(x_\iota) = \delta^*(\overline{x_\iota})$ for all ι,
2. $x_\iota \not\approx_L x_\rho$ implies $\overline{x_\iota} \not\approx_{L'} \overline{x_\rho}$,
3. L' is non-regular.

Proof. We have already determined the first item. For 2, consider x_ι and x_ρ, $\iota \neq \rho$ from the list, and so $x_\iota \not\approx_L x_\rho$ and therefore $\delta^*(x_\iota) \neq \delta^*(x_\rho)$ and then $\overline{x_\iota} \neq \overline{x_\rho}$. We know $\delta^*(\overline{x_\iota}) = \delta^*(x_\iota)$ and $\delta^*(\overline{x_\rho}) = \delta^*(x_\rho)$. So there exists z such that, without loss of generality, $x_\iota z \in L$ but $x_\rho z \notin L$. But consider $\widetilde{z}$ obtained as above. It follows that $\overline{x_\iota}\widetilde{z} \in L$ and $\overline{x_\rho}\widetilde{z} \notin L$, but $\overline{x_\iota}, \overline{x_\rho} \in \alpha_1^*, \widetilde{z} \in \alpha_2^*$, and so $\overline{x_\iota}\widetilde{z} \in L'$ but $\overline{x_\rho}\widetilde{z} \notin L'$. Hence, for all pairwise distinct $x_\iota, x_\rho, \overline{x_\iota} \not\approx_{L'} \overline{x_\rho}$, and thus L' is non-regular, and so 3 is true. □

Conversely, if $L(M) \cap \alpha^*$ with α as constructed above is non-regular, then $L(M)$ must be non-regular as well since α^* is a regular language, and regular languages are closed under intersection. □

Theorem 15. *The regularity problem is decidable for* DCM.

Proof. Let M be a DCM machine, which we assume is in normal form. By Lemma 14, we can effectively determine $\alpha \in [\Sigma^*]^+$ such that $L(M)$ is non-regular if and only if $L \cap \alpha^*$ is non-regular. The latter language can be accepted by a DCM since the family of DCM languages is closed under intersection with regular languages; and it can be effectively constructed. By Corollary 11, and because the languages accepted by DCM are a subset of NPCM, we can decide given α whether $L(M) \cap \alpha^*$ is regular, thus whether $L(M)$ is regular. □

5 Conclusions and Future Directions

We showed that the regularity problem is decidable for DCM. Because we also showed it to be undecidable for NCM(1, 1), the obvious open problems would be to consider regularity for more powerful deterministic acceptors such as DPCM, or deterministic stack automata. We also defined and studied the bounded-regularity problem, which is to determine if a given automaton accepts a bounded regular language. We showed this problem to be decidable for NPCM, and for any semilinear full trio with a decidable boundedness problem; hence, the bounded-regularity problem is often easier to decide than the regularity problem.

Acknowledgements. We thank the anonymous reviewers for a careful reading of this paper and for correcting an error in a previous version of Proposition 9.

References

1. Anand, A., Schmitz, S., Schütze, L., Zetzsche, G.: Verifying unboundedness via amalgamation. In: Proceedings of the 39th Annual ACM/IEEE Symposium on Logic in Computer Science. LICS '24. Association for Computing Machinery, New York (2024)
2. Baker, B.S., Book, R.V.: Reversal-bounded multipushdown machines. J. Comput. Syst. Sci. **8**(3), 315–332 (1974)
3. Baumann, P., D'Alessandro, F., Ganardi, M., Ibarra, O.H., McQuillan, I., Schütze, L., Zetzsche, G.: Unboundedness problems for machines with reversal-bounded counters. In: Lecture Notes in Computer Science. Proceedings of the 26th International Conference on Foundations of Software Science and Computation Structures, FoSSaCS 2023, vol. 13992, pp. 240–264 (2023)
4. Cadilhac, M., Finkel, A., McKenzie, P.: Bounded Parikh automata. Int. J. Found. Comput. Sci. **23**(08), 1691–1709 (2012)
5. Carpi, A., D'Alessandro, F., Ibarra, O.H., McQuillan, I.: Relationships between bounded languages, counter machines, finite-index grammars, ambiguity, and commutative regularity. Theoret. Comput. Sci. **862**, 97–118 (2021)
6. Czerwinski, W., Hofman, P., Zetzsche, G.: Unboundedness problems for languages of vector addition systems. In: Chatzigiannakis, I., Kaklamanis, C., Marx, D., Sannella, D. (eds.) 45th International Colloquium on Automata, Languages, and Programming (ICALP 2018). Leibniz International Proceedings in Informatics (LIPIcs), vol. 107, p. 119. Schloss Dagstuhl–Leibniz-Zentrum fuer Informatik, Dagstuhl, Germany (2018)
7. Ganty, P., Majumdar, R., Monmege, B.: Bounded underapproximations. Formal Methods Syst. Des. **40**, 206–231 (2012)
8. Ginsburg, S., Spanier, E.H.: Bounded regular sets. Proceedings of the American Mathematical Society **17**, 1043–1049 (1966)
9. Ginsburg, S.: The Mathematical Theory of Context-Free Languages. McGraw-Hill Inc., New York (1966)
10. Greibach, S.: A note on undecidable properties of formal languages. Math. Syst. Theory **2**(1), 1–6 (1968)
11. Greibach, S.: Remarks on blind and partially blind one-way multicounter machines. Theoret. Comput. Sci. **7**, 311–324 (1978)
12. Hopcroft, J.E., Ullman, J.D.: Introduction to Automata Theory, Languages, and Computation. Addison-Wesley, Reading, MA (1979)
13. Ibarra, O., McQuillan, I.: The effect of end-markers on counter machines and commutativity. Theoret. Comput. Sci. **627**, 71–81 (2016)
14. Ibarra, O.H.: Reversal-bounded multicounter machines and their decision problems. J. ACM **25**(1), 116–133 (1978)
15. Ibarra, O.H., McQuillan, I.: Techniques for showing the decidability of the boundedness problem of language acceptors. In: Lecture Notes in Computer Science. Proceedings of the 28th International Conference on Developments in Language Theory, DLT 2024, vol. 14791, pp. 156–172 (2024)
16. Latteux, M., Leguy, J.: Une propriété de la famille GRE. In: Proceedings of the Conference on Algebraic, Arithmetic, and Categorial Methods in Computation Theory. FCT '79, Akademie-Verlag, Berlin (1979)
17. Stearns, R.: A regularity test for pushdown machines. Inf. Control **11**, 323–340 (1967)

18. Valiant, L.G.: Regularity and related problems for deterministic pushdown automata. J. ACM **22**(1), 1–10 (1975)
19. Zetzsche, G.: An approach to computing downward closures. In: Halldórsson, M.M., Iwama, K., Kobayashi, N., Speckmann, B. (eds.) ICALP 2015. LNCS, vol. 9135, pp. 440–451. Springer, Heidelberg (2015). https://doi.org/10.1007/978-3-662-47666-6_35

Impartial Derivation Games on Context-Free Grammars

Yusuke Inoue[1(✉)], Yuji Komatsu[1], Hiyu Inoue[2], and Shun-ichi Kimura[2]

[1] Nagoya University, Nagoya, Japan
y-inoue@sqlab.jp
[2] Hiroshima University, Hiroshima, Japan

Abstract. Combinatorial games are two-player perfect-information games without chance moves. In a specific class of such games, each position is assigned a nonnegative integer, called its Grundy number, which determines whether it is winning or losing. In particular, the boundedness and periodicity of Grundy numbers are central topics in the study of heap-based games such as subtraction games and Kayles, and their analysis often shows intriguing parallels with formal language theory. In this paper, we introduce impartial normal-play games in which players alternately rewrite words according to a context-free grammar. In this setting, we show that the resulting Grundy numbers are bounded when the grammar is k-bounded. In addition, we prove that given a CFG G and an integer c, it is undecidable whether (i) there exists a word whose Grundy number is c, and (ii) every target word is winning for the first player. Our framework provides a uniform perspective on a broad family of heap-based games while introducing undecidability phenomena into their study. We also observe that the Grundy number of a given word can be computed in polynomial time by a dynamic-programming method analogous to weight computation for weighted context-free grammars.

Keywords: Context-free grammars · Grundy numbers · Impartial games · Undecidability

1 Introduction

Combinatorial game theory studies two-player games with perfect information and no chance moves [21]. A particularly well-studied class consists of *impartial*, *loop-free* games under the *normal-play* convention, where impartial means that both players have the same legal moves, loop-free means that every play is finite, and normal play means that the player who makes the last move wins. Many classical examples are heap-based games played on heaps of stones, such as *subtraction games* and *Kayles*, which are instances of *octal games*, and more generally of *code games*.

An important aspect in the study of these games is the notion of *Grundy numbers* [9,22]. A Grundy number indicates whether a position is winning or losing:

T. J. Smith (Ed.): CIAA 2026, LNCS 16695, pp. 78–91, 2026.
https://doi.org/10.1007/978-3-032-31176-4_6

the previous player has a winning strategy for the positions with Grundy number zero, while the next player has a winning strategy for those with a nonzero Grundy number. In code games, one is typically interested in the sequence of Grundy numbers $\mathcal{G}(n)$ indexed by the initial heap size n. For many code games, this sequence is known to exhibit simple or arithmetic periodicity. For example, in the case of Kayles, the Grundy sequence becomes periodic with period 12 from $n = 71$ [1]. However, whether every finite octal game yields a periodic Grundy sequence remains an open problem [18], and a large body of work is devoted to case studies.

In the analysis of Grundy sequences, a natural question is whether the values are bounded, that is, whether the sequence takes only finitely many distinct values. For example, subtraction games always exhibit simple periodicity [21]. A standard proof first establishes that the Grundy numbers are bounded and then applies the pigeonhole principle to deduce periodicity. This type of reasoning is very familiar in formal language theory. In fact, in the study of more general octal games—where a move may split a heap into several smaller heaps—the arguments used to analyze periodicity closely resemble those appearing in the theory of context-free grammars. This resemblance stems from the fact that the *disjunctive sum* of games is based on essentially the same principle as context-freeness.

In this paper, we propose a new point of contact between combinatorial game theory and formal language theory by introducing *successful derivation games* on CFGs. Given a CFG G and a word w, the successful derivation game (G, w) is an impartial normal-play game in which two players alternately apply rewriting steps. The player who completes a derivation to the target word w wins the game. In this setting, we establish the following results:

- If a CFG G is k-bounded with some $k \in \mathbb{N}$, then the associated Grundy numbers are bounded (Theorem 3),
- Given a linear CFG G, the supremum of the associated Grundy numbers is not computable (Theorem 4), and
- Given a 2-bounded CFG G, the problem of deciding whether every target word is winning for the first player (resp. the second player) is undecidable (Theorem 5).

These results contribute to both combinatorial game theory and formal language theory. From the perspective of combinatorial game theory, our framework provides a unifying setting for a wide range of heap-based games, while also clarifying their connection to CFGs. The framework further gives rise to many new classes of games that have not previously been studied, extending the scope of combinatorial game theory beyond the traditional setting of code games. More importantly, our work brings undecidability into the study of heap-based games. Although the complexity of computing Grundy numbers has been studied extensively [10,20], general questions of decidability have received much less attention, except for games that explicitly encode arithmetic [13,16]. In this respect, our decidability results open up a new direction in the study of combinatorial games.

From the viewpoint of formal language theory, our results motivate an extension of the usual notion of weight structures [5,7]. As we will discuss later, the Grundy numbers of derivation games can be computed in polynomial time. This is because the two operations involved in Grundy-number computation—the *mex* operation and the *nim-sum*—form an algebraic structure similar to a semiring, which allows us to apply a CYK-style algorithm analogous to the one used for weighted context-free grammars [7]. This perspective is reminiscent of earlier work that extends semiring-based weighted models to more general algebraic structures, such as multioperator weighted tree automata [6,15,23] and tree valuation weighted tree automata [4,24].

Finally, we briefly comment on games that have been studied in formal language theory. A standard setting in this area is the so-called *regular games*, which include Büchi, parity, and Muller games [3,8]. In these games, two players interactively construct an infinite word on a finite automaton or graph, and the first player aims to ensure that the resulting word belongs to a given ω-language. Closer to the setting of this paper are games played on CFGs, such as *Romeo-Juliet games* [17] and *context-free games* [11]. In Romeo-Juliet games, one player chooses a position in the current sentential form to rewrite, while the other selects and applies a production rule. In context-free games, each player has a separate set of rules and they alternate applying them during a derivation.

Despite being played on CFGs, these games differ substantially from the impartial, loop-free combinatorial games considered in this paper. The classical grammar-based games are inherently asymmetric, may allow infinite plays, and are primarily motivated by applications such as verification and reactive synthesis. For these reasons, no direct connection between these grammar-based games and the derivation games studied here has yet been established. Nevertheless, our framework could provide a natural bridge between these approaches.

2 Derivation Games

Let $\mathbb{N}$ denote the set of all non-negative integers. For a set X, let $\mathcal{P}_{\text{fin}}(X)$ denote the set of all finite subsets of X. For a word $w \in \Sigma^+$, let $|w|$ denote the length of w.

Definition 1 (context-free grammar). *A* context-free grammar *(or* CFG*) is a 4-tuple $G = (\Gamma, \Sigma, R, I)$ where*

- *Γ is a finite set of nonterminal symbols,*
- *Σ is a finite set of terminal symbols,*
- *$I \in \Gamma$ is the initial symbol, and*
- *$R \subseteq \Gamma \times (\Gamma \cup \Sigma)^+$ is a finite set of rules such that if $A \to B \in R$, then there is no derivation of the form $B \Rightarrow^* A$.* ∎

Here, a derivation $\alpha \Rightarrow^* \beta$ with $\alpha, \beta \in (\Gamma \cup \Sigma)^+$ is defined in the usual way. Note that grammars in Chomsky normal form satisfy the condition in the last item above, so the restrictions imposed in our definition do not reduce the generative power. Moreover, because there are no ε-rules and no loops consisting of unit rules, every derivation has finite length. These assumptions are standard in the study of weighted CFGs, as infinite derivations may lead to ill-defined weights. As we will discuss later, our games can be viewed as a form of weighted CFGs, and similar assumptions are therefore necessary.

A derivation $\alpha \Rightarrow^* \beta$ is said to be *successful* if $\beta \in \Sigma^+$, and said to be *incomplete* otherwise. For each nonterminal $A \in \Gamma$, let $L(G, A)$ be the language defined as

$$L(G, A) = \{w \in \Sigma^+ : \text{ there is a derivation of the form } A \Rightarrow^* w.\}$$

The language generated by $G = (\Gamma, \Sigma, R, I)$ is defined as $L(G) = L(G, I)$.

Definition 2 (successful derivation game). *For a CFG $G = (\Gamma, \Sigma, R, I)$ and a word $w \in L(G)$, the* successful derivation game (G, w) *is the two-player combinatorial game defined as follows.*

- *A* position *is a finite multiset of the form*

$$\{(A_1, w_1), (A_2, w_2), \ldots, (A_\ell, w_\ell)\}$$

where $(A_i, w_i) \in \Gamma \times \Sigma^+$ and $w_i \in L(G, A_i)$ for each $1 \le i \le \ell$.
- *A legal move from a position Q is defined as follows. First, the current player chooses an element (A, w) of Q. Next, the player chooses a sequence $(A_1, w_1), \ldots, (A_\ell, w_\ell) \in \Gamma \times \Sigma^+$ with $\ell \ge 0$ such that*
  - *$w_i \in L(G, A_i)$ for each i, and*
  - *$A \to u_0 A_1 u_1 \cdots u_{\ell-1} A_\ell u_\ell \in R$ where $w = u_0 w_1 u_1 \cdots u_{\ell-1} w_\ell u_\ell$.*

  Finally, the position Q is updated to

$$Q' = (Q \setminus \{(A, w)\}) \cup \{(A_1, w_1), \ldots, (A_\ell, w_\ell)\}.$$

If there exists a legal move from Q to Q', we write $Q \rightsquigarrow Q'$.
- *The game starts from the initial position $Q_0 = \{(I, w)\}$, and the players alternate making legal moves $Q_i \rightsquigarrow Q_{i+1}$ for each $i \ge 0$. The player who is unable to move loses.* ■

This game has finitely many options and always terminates in finitely many moves. Therefore, by a standard result in combinatorial game theory, exactly one of the two players has a winning strategy (see e.g., [21]). Positions where the first player has a winning strategy are called $\mathcal{N}$*-positions*, and those where the second player has a winning strategy are called $\mathcal{P}$*-positions*.

Example 1 (subtraction game). Let us consider the CFG **Sub** $= (\{A\}, \{a\}, \{A \to aaA, A \to aA, A \to aa, A \to a\}, A)$. In this case, the derivation game $(\mathbf{Sub}, a^n)$ is the game in which players alternately remove one or two stones from a heap of n stones. For example, the derivation

$$A \Rightarrow aaA \Rightarrow aaaaA \Rightarrow aaaaa$$

corresponds to the legal play

$$\{(A, a^5)\} \rightsquigarrow \{(A, a^3)\} \rightsquigarrow \{(A, a^1)\} \rightsquigarrow \emptyset.$$

In this play, the first player wins. ■

Example 2 (Kayles). Let **Kayles** $= (\{A\}, \{a\}, R, A)$ be the CFG where

$$R = \{A \to AaaA, A \to AaA, A \to aaA, A \to aA, A \to aa, A \to a\}.$$

In this case, the derivation game $(\mathbf{Kayles}, a^n)$ is the *Kayles* (or **0.77**) with a heap of n stones. For example, the derivation

$$A \Rightarrow A\,a\,A \Rightarrow aaA\,a\,A \Rightarrow aaA\,a\,aaA \Rightarrow aaa\,a\,aaA \Rightarrow aaa\,a\,aaa$$

corresponds to the legal play

$$\{(A, a^7)\} \rightsquigarrow \{(A, a^3), (A, a^3)\} \rightsquigarrow \{(A, a^1), (A, a^3)\}$$
$$\rightsquigarrow \{(A, a^1), (A, a^1)\} \rightsquigarrow \{(A, a^1)\} \rightsquigarrow \emptyset.$$

In this play, the first player wins. ■

Example 3 (**4.3**). The derivation game $(\mathbf{4.3}, a^n)$ with $\mathbf{4.3} = (\{A\}, \{a\}, \{A \to AA, A \to aA, A \to a\}$ is the code game called **4.3**. ■

As these examples illustrate, successful derivation games encompass many classical combinatorial games. In addition, one can define simple derivation games that nevertheless exhibit nontrivial behavior.

Example 4 (subtraction game with colored stones). The derivation game $(\mathbf{CSub}, w)$ with $\mathbf{CSub} = (\{A\}, \{a, b\}, \{A \to aaA, A \to aA, A \to bA, A \to aa, A \to a, A \to b\}, A)$ and $w \in \Sigma^+$ is the game in which a player can remove one a, remove two a, or remove one b from left to right. ■

2.1 Grundy Numbers

Let us discuss *Grundy numbers.* In combinatorial game theory, the Grundy number of a game is a standard tool for deciding whether a position is winning or losing.

Let $\mathsf{mex} : \mathcal{P}_{\mathrm{fin}}(\mathbb{N}) \to \mathbb{N}$ be the *mex* operation defined as $\mathsf{mex}(X) = \mathsf{min}(\mathbb{N} \setminus X)$ for each $X \in \mathcal{P}_{\mathrm{fin}}(\mathbb{N})$. In addition, let $\oplus : \mathbb{N} \times \mathbb{N} \to \mathbb{N}$ denote the *nim-sum* operation defined as the bitwise exclusive OR on binary representations. For example, $5 \oplus 3 = 101_{(2)} \oplus 011_{(2)} = 110_{(2)} = 6$.

Definition 3 (Grundy number). *For a position Q of a successful derivation game on a CFG $G = (\Gamma, \Sigma, R, I)$, define the* Grundy number $\mathcal{G}_G(Q) \in \mathbb{N}$ *as*

$$\mathcal{G}_G(\{(A_1, w_1), \ldots, (A_n, w_n)\}) = \mathcal{G}_G(\{(A_1, w_n)\}) \oplus \cdots \oplus \mathcal{G}_G(\{(A_n, w_n)\}),$$

and

$$\mathcal{G}_G(\{(A,w)\}) = \mathsf{mex}(\{ \bigoplus_{0\le i\le k} \mathcal{G}_G(\{(A_i,w_i)\}) : A \to u_0A_1u_1\cdots u_{\ell-1}A_\ell u_\ell \in R,$$
$$w = u_0w_1u_1\cdots u_{\ell-1}w_\ell u_\ell,\ w_i \in L(G,A_i).\})$$

In particular, $\mathcal{G}_G(\emptyset) = 0$. Because the derivation length of a word is finite, $\mathcal{G}_G(Q)$ is well-defined. ■

For each nonterminal $A \in \Gamma$, define the function $\mathcal{G}_{G,A} : \Sigma^+ \to \mathbb{N} \cup \{\bot\}$ as

$$\mathcal{G}_{G,A}(w) = \begin{cases} \mathcal{G}_G(\{(A,w)\}) & \text{if } w \in L(A), \\ \bot & \text{otherwise} \end{cases}$$

for each $w \in \Sigma^+$. Note that $\mathcal{G}_{G,A}(w) \neq \bot$ if and only if $w \in L$.

Because $\mathcal{G}_G(\{(A,w)\})$ is defined inductively along derivations, the value $\mathcal{G}_{G,A}(w)$ can be computed by a CYK-style dynamic programming algorithm, similarly to the weight computation problem for weighted CFGs [7].

Theorem 1. *Let $G = (\Gamma, \Sigma, R, I)$ be a CFG, and let $A \in \Gamma$ be a nonterminal symbol. Given a word $w \in \Sigma^+$, $\mathcal{G}_{G,A}(w)$ can be computed in polynomial time with respect to $|w|$.*

Proof Sketch. One can compute $\mathcal{G}_{G,A}(w)$ in the following way. First, $w = a_1\cdots a_n$ is recursively decomposed into smaller subwords $a_i\cdots a_j$, and for each nonterminal B and each subword $a_i\cdots a_j$, the value $\mathcal{G}_{G,B}(a_i\cdots a_j)$ is computed. The computed values are stored in a memory in the form of tuples $(B,i,j,x) \in \Gamma \times \mathbb{N} \times \mathbb{N} \times \mathbb{N}$ where $x = \mathcal{G}_{G,B}(a_i\cdots a_j)$. These values are then reused when computing values for larger subwords. This algorithm is basically the same as the CYK-algorithm [14,25], and runs in polynomial time. □

The following is a consequence of the Sprague-Grundy theorem [9,22]:

Theorem 2. *Let Q be a position of a successful derivation game (G,w). Then, a position Q is a $\mathcal{N}$-position if and only if $\mathcal{G}_{G,I}(Q) \neq 0$.* ■

By Theorem 1, whether Q is a $\mathcal{N}$-position can be decided in polynomial time.

Example 5 (subtraction game). It holds that $\mathcal{G}_{\mathbf{Sub},A}(a^n) = n \bmod 3$ for the subtraction game defined in Example 1. For example,

$$\begin{aligned} \mathcal{G}_{\mathbf{Sub},A}(aaa) &= \mathcal{G}_{\mathbf{Sub}}(\{(A,aaa)\}) \\ &= \mathsf{mex}(\{\mathcal{G}_{\mathbf{Sub}}(\{(A,aa)\}), \mathcal{G}_{\mathbf{Sub}}(\{(A,a)\})\}) \\ &= \mathsf{mex}(\{\mathsf{mex}(\{\mathcal{G}_{\mathbf{Sub}}(\{(A,a)\})\}), 0\}), \mathsf{mex}(\{0\})\}) \\ &= \mathsf{mex}(\{\mathsf{mex}(\{1,0\}), 1\}) = \mathsf{mex}(\{2,1\}) = 0 \end{aligned}$$

holds. Therefore, the first player can win for $(\mathbf{Sub}, a^n)$ if and only if n is not a multiple of 3. ■

Example 6 (Kayles). The sequence $\{\mathcal{G}_{\mathbf{Kayles},A}(a^n)\}_{n\geq 1}$ is

$$1, 2, 3, 1, 4, 3, 2, 1, 4, 2, 6, 4, 1, 2, 7, 1, 4, 3, 2, 1, 4, 6, 7, \ldots$$

with the Kayles defined in Example 2. It is known that $\mathcal{G}_{\mathbf{Kayles},A}(a^n) \neq 0$ for each $n \geq 1$, that is, the first player always wins for Kayles (see Chap. 4 of [1] for details). ■

Example 7 (subtraction game with colored stones). Let us consider the game $(\mathbf{CSub}, w)$ discussed in Example 4. Let $w = a^{d_1}b^{e_1}a^{d_2}b^{e_2}\cdots a^{d_k}b^{e_k}$ with $d_2, \ldots, d_k, e_1, \ldots, e_{k-1} > 0$ and $d_1, e_k \geq 0$, and let $\{0, 1, 2\} \ni \overline{d_i} \equiv d_i \mod 3$. Then, $\mathcal{G}_{\mathbf{CSub},A}(w) = 0$ if and only if

(1) $\overline{d_i} \neq 2$ for each $1 \leq i \leq k$ and $\sum_{i=1}^{k} \overline{d_i} + \sum_{i=1}^{k} e_i$ is even, or
(2) there exists $1 \leq m \leq k$ such that $\overline{d_1}, \ldots, \overline{d_{m-1}} \neq 2, \overline{d_m} = 2$, and $\sum_{i=1}^{m-1} \overline{d_i} + \sum_{i=1}^{m-1} e_i$ is odd

holds. The proof of this fact was completed with assistance from Hiroki Inazu, Takahiro Yamashita, and Kosaku Watanabe (Hiroshima University). ■

3 Boundedness of Grundy Numbers

As mentioned in Sect. 1, boundedness of Grundy numbers plays a crucial role in their analysis. For example, it is known that in subtraction games—equivalently, derivation games over unary linear context-free grammars—the Grundy numbers are bounded. In this section, we prove that the Grundy numbers of successful derivation games are bounded whenever the underlying grammar is k-bounded.

3.1 Examples of Grundy Sequences

First, let us consider several grammars that will be used in the subsequent sections.

Lemma 1. *For any context-free language $L \subseteq \Sigma^+$, there is a CFG $G = (\Gamma, \Sigma, R, I)$ such that $\mathcal{G}_{G,I}(w) \geq 1$ if $w \in L$ and $\mathcal{G}_{G,I}(w) = \perp$ otherwise.*

Proof. Let $G' = (\Gamma', \Sigma, R', I')$ be a CFG such that $L(G') = L$. Define $G = (\Gamma, \Sigma, R, I)$ as $\Gamma = \Gamma' \cup \{I, J\}$, $R = R' \cup \{I \to I', I \to J, J \to I'\}$. Then, for each $w \in L$,

$$\begin{aligned}
\mathcal{G}_{G,I}(w) &= \mathcal{G}_G(\{(I, w)\}) \\
&= \mathsf{mex}\{\mathcal{G}_{G'}(\{(I', w)\}), \mathcal{G}_G(\{(J, w)\})\} \\
&= \mathsf{mex}\{\mathcal{G}_{G'}(\{(I', w)\}), \mathsf{mex}\{\mathcal{G}_{G'}(\{(I', w)\})\}\}
\end{aligned}$$

holds. If $\mathcal{G}_{G'}(\{(I', w)\}) = 0$, then $\mathcal{G}_{G,I}(w) \neq 0$ by the definition of mex. In addition, if $\mathcal{G}_{G'}(\{(I', w)\}) \neq 0$, then $\mathsf{mex}(\{\mathcal{G}_{G'}(\{(I', w)\})\}) = 0$. Thus, $\mathcal{G}_{G,I}(w) \neq 0$. Consequently, we have $\mathcal{G}_{G,I}(w) \neq 0$ for each $w \in L$. □

Lemma 2. *For any context-free language $L \subseteq \Sigma^+$ and any positive integer $c \in \mathbb{N}$, there is a CFG $G = (\Gamma, \Sigma, R, I)$ such that $\mathcal{G}_{G,I}(w) = c$ if $w \in L$ and $\mathcal{G}_{G,I}(w) = \bot$ otherwise.*

Proof. We first show for the case where $c = 0$. Let $G' = (\Gamma', \Sigma, R', I')$ be a CFG such that $L(G') = L$. Define $G = (\Gamma, \Sigma, R, I_0)$ as $\Gamma = \Gamma' \cup \{I_0, J, K\}$, $R = R' \cup \{I_0 \to J, J \to I', J \to K, K \to I'\}$. Then, for each $w \in L$, $\mathcal{G}_{G,J}(w) \neq 0$ by the same reason as Lemma 1. Thus, $\mathcal{G}_{G,I_0}(w) = \mathsf{mex}(\{\mathcal{G}_{G,J}(w)\}) = 0$ holds for each $w \in L$.

Next, by letting $G_c = (\Gamma' \cup \{J, K\} \cup \{I_0, I_1, I_2, \ldots, I_c\}, \Sigma, R' \cup \{I_i \to I_j : 0 \leq j < i \leq c\}, I_0)$, it holds that $\mathcal{G}_{G_c,I_c}(w) = c$ if $w \in L$ and $\mathcal{G}_{G,I}(w) = \bot$ otherwise. This is because, $\mathcal{G}_{G_c,I_c}(w) = \mathsf{mex}(\{\mathcal{G}_{G_j,I_j}(w) : 0 \leq j \leq c\}) = \mathsf{mex}(\{0, 1, \ldots, c-1\}) = c$ holds inductively. □

Using these constructions, one can discuss grammars such as the following.

Proposition 1. *There is a CFG $G = (\Gamma, \Sigma, R, I)$ such that $\mathcal{G}_{G,I}(w) = |w|$.*

Proof. Let $G' = (\Gamma', \Sigma, R', I')$ be a grammar such that $\mathcal{G}_{G,I'}(w) = 0$ for each $w \in \Sigma^+$. By Lemma 2, such grammar G' indeed exists because Σ^+ is a context-free language. Define $G = (\Gamma, \Sigma, R, I)$ as $\Gamma = \Gamma' \cup \{I\}$ and

$$R = R' \cup \{I \to I', I \to II'\}.$$

Note that $L(G) = \Sigma^+$, and thus, $\mathcal{G}_{G,I}(w) \neq \bot$ for each $w \in \Sigma^+$. We show that $\mathcal{G}_{G,I}(w) = |w|$ holds for each $w \in \Sigma^+$ by induction on the length of w. First, $\mathcal{G}_{G,I}(a) = \mathsf{mex}(\{\mathcal{G}_{G,I'}(a)\}) = \mathsf{mex}(\{0\}) = 1$ for each $a \in \Sigma$. Next, as the induction hypothesis, assume that $\mathcal{G}_{G,I}(w_1) = |w_1|$ holds for any word w_1 such that $|w_1| < k$. Then, for a word $w \in \Sigma^+$ such that $|w| = k$, we have

$$\begin{aligned}
\mathcal{G}_{G,I}(w) &= \mathsf{mex}(\{\mathcal{G}_{G,I'}(w)\} \cup \{\mathcal{G}_{G,I}(w_1) \oplus \mathcal{G}_{G,I'}(w_2) : w = w_1w_2 \text{ with } w_1, w_2 \in \Sigma^+\}) \\
&= \mathsf{mex}(\{0\} \cup \{|w_1| \oplus 0 : w = w_1w_2 \text{ with } w_1, w_2 \in \Sigma^+\}) \\
&= \mathsf{mex}(\{0\} \cup \{\ell \oplus 0 : 1 \leq \ell \leq k-1\}) \\
&= \mathsf{mex}(\{0, 1, 2, \ldots, k-1\}) = k
\end{aligned}$$

holds. This concludes the proof. □

This construction is essentially equivalent to the single-heap Nim[1]. The player splits a heap of $|w|$ stones into two heaps of sizes $|w_1|$ and $|w_2|$, and assigns them to heaps I and I', respectively. However, because $\mathcal{G}_{G,I'}(w) = 0$ for each $w \in \Sigma^+$, the I'-heap is always equivalent to the empty heap. Consequently, the player is merely choosing how many stones to "discard" from the original heap of size $|w|$.

Although single-heap Nim is a very simple one-heap game, the number of stones that can be removed in a single move is unbounded, and consequently its Grundy numbers can grow arbitrarily large. In the next subsection, we revisit this phenomenon from the perspective of grammatical boundedness, and show how this viewpoint unifies the boundedness of heap branching and the boundedness of the number of stones that can be removed in a single move.

[1] Single-heap Nim refers to the game in which a player may remove any positive number of stones from a single heap, and its Grundy number satisfies $\mathcal{G}(n) = n$ for all n [2].

3.2 k-Bounded CFGs

For a positive integer k, a context-free grammar $G = (\Gamma, \Sigma, R, I)$ is *k-bounded* if for each $A \in \Gamma$, there is no derivation of the form

$$A \Rightarrow^* \alpha_0 A_1 \alpha_1 \cdots \alpha_k A_{k+1} \alpha_{k+1}$$

with $\alpha_0, \alpha_1 \ldots, \alpha_{k+1} \in (\Gamma \cup \Sigma)^*$ and $A_1, \ldots, A_{k+1} \in \Gamma$. In particular, a 1-bounded CFG is also called a *linear context-free grammar.* For a k-bounded CFG $G = (\Gamma, \Sigma, R, I)$ and a nonterminal symbol $A \in \Gamma$, the *rank* of A is defined as

$$\mathsf{rank}(A) = \max\{d : \text{there is a derivation of the form } A \Rightarrow^* \alpha_0 A_1 \cdots A_d \alpha_{d+1}\}.$$

That is, the rank of $A \in \Gamma$ is the largest number of nonterminals that can appear in a sentential form derived from A. Note that $1 \leq \mathsf{rank}(A) \leq k$ for each A. In addition, for each rule $A \to u_0 A_1 u_1 \cdots u_{\ell-1} A_\ell u_\ell$, it holds that $\mathsf{rank}(A_1) + \cdots + \mathsf{rank}(A_\ell) \leq \mathsf{rank}(A)$.

Theorem 3. *Let $G = (\Gamma, \Sigma, R, I)$ be a k-bounded CFG with some $k \in \mathbb{N}$. Then, there is an upper-bound $N \in \mathbb{N}$ such that $\mathcal{G}_{G,I}(w) \leq N$ for each $w \in L(G)$.*

Proof. It suffices to show that for each $A \in \Gamma$, there is an upper-bound $N_A \in \mathbb{N}$ such that $\mathcal{G}_{G,A}(w) \leq N_A$ for each $w \in L(G, A)$. Let us show this by induction on $\mathsf{rank}(A) \geq 1$.

Assume that $\mathsf{rank}(A) = 1$. Then, for each $w \in L(G, A)$, we have

$$\mathcal{G}_{G,A}(w) = \mathsf{mex}(\{\mathcal{G}_{G,A_1}(w) : A \to u_0 A_1 u_1 \in R,\ w = u_0 w_1 u_1,\ w_1 \in L(G, A_1)\} \cup \{0\})$$

holds if $(A \to w) \in R$, and

$$\mathcal{G}_{G,A}(w) = \mathsf{mex}(\{\mathcal{G}_{G,A_1}(w) : A \to u_0 A_1 u_1 \in R,\ w = u_0 w_1 u_1,\ w_1 \in L(G, A_1)\})$$

if $(A \to w) \notin R$. Here, the size of the set $\{\mathcal{G}_{G,A_1}(w) : A \to u_0 A_1 u_1 \in R,\ w = u_0 w_1 u_1,\ w_1 \in L(G, A_1)\}$ is bounded by the number of rules of the form $A \to u_0 A_1 u_1 \in R$. Because $\mathsf{mex}(X) \leq |X|$ holds for any finite subset $X \subseteq \mathbb{N}$, we can conclude that $\mathcal{G}_{G,A}(w) \leq N_A$ with $N_A = |R| + 1$.

Next, assume that there is an upper-bound $N_{A_i} \in \mathbb{N}$ for each $A_i \in \Gamma$ with $\mathsf{rank}(A_i) \leq k - 1$. Let A be a nonterminal such that $\mathsf{rank}(A) = k$. Then, for each $w \in L(G, A)$, it holds that

$$\begin{aligned}\mathcal{G}_{G,A}(w) = \mathsf{mex}(& \bigcup_{\substack{A \to u_0 A_1 \cdots A_\ell u_\ell \in R \\ 2 \leq \ell \leq k}} \{\bigoplus_{1 \leq i \leq \ell} \mathcal{G}_{G,A_i}(w_i) : w = u_0 w_1 \cdots w_\ell u_\ell,\ w_i \in A_i\} \\ & \cup \{\mathcal{G}_{G,A_1}(w_1) : A \to u_0 A_1 u_1 \in R, w = u_0 w_1 u_1,\ w_1 \in L(G, A_1)\} \cup \{0\})\end{aligned}$$

or

$$\begin{aligned}\mathcal{G}_{G,A}(w) = \mathsf{mex}(& \bigcup_{\substack{A \to u_0 A_1 \cdots A_\ell u_\ell \in R \\ 2 \leq \ell \leq k}} \{\bigoplus_{1 \leq i \leq \ell} \mathcal{G}_{G,A_i}(w_i) : w = u_0 w_1 \cdots w_\ell u_\ell,\ w_i \in A_i\} \\ & \cup \{\mathcal{G}_{G,A_1}(w_1) : A \to u_0 A_1 u_1 \in R, w = u_0 w_1 u_1,\ w_1 \in L(G, A_1)\})\end{aligned}$$

by the definition of $\mathcal{G}_{G,A}(w)$. However, for each $A \to u_0A_1 \cdots A_\ell u_\ell \in R$, the value of $\bigoplus_{1\le i\le \ell} \mathcal{G}_{G,A_i}(w_i)$ can take at most $\prod_i N_{A_i}$ distinct values regardless of the partition $w = u_0w_1 \cdots w_\ell u_\ell$. In addition, the size of $\{\mathcal{G}_{G,A_1}(w_1) : A \to u_0A_1u_1 \in R, w = u_0w_1u_1, w_1 \in L(G, A_1)\} \cup \{0\}$ is at most $|R|$ in the same reason as the base case. Thus, the size of the input set of mex is at most N_A with

$$N_A = |R| + \sum_{A \to A \to u_0A_1 \cdots A_\ell u_\ell \in R} \Big(\prod_{1\le i\le \ell} N_{A_i}\Big),$$

and this depends only on G. □

For example, there is no k-bounded CFG $G = (\Gamma, \Sigma, R, I)$ such that $\mathcal{G}_{G,I}(w) = |w|$ (see also Proposition 1). Note that the converse of the theorem does not hold. For instance, the grammar encoding Kayles in Example 2 is not bounded, while its sequence of Grundy numbers remains bounded.

4 Undecidability of Ranges

In the previous section, we showed that k-bounded CFGs can take only finitely many Grundy numbers. Moreover, the proof is constructive and allows us to obtain a coarse upper bound on the Grundy numbers. However, determining the exact upper bound is an undecidable problem, even in the linear case.

Lemma 3. *Given a linear CFG $G = (\Gamma, \Sigma, R, I)$, it is undecidable whether there exists a word $w \in \Sigma^+$ such that $\mathcal{G}_{G,I}(w) = 2$.*

Proof. We prove the claim by a reduction from the emptiness problem for the intersection of two linear CFGs, which is known to be undecidable (see e.g., Theorem 6 in [12]). Let G_0 and G_1 be linear CFGs over an alphabet Σ. Then, by the construction explained in Lemma 2, one can define the following linear CFGs:

- $G'_0 = (\Gamma_0, \Sigma, R_0, I_0)$ such that $\mathcal{G}_{G'_0,I_0}(w) = 0$ if $w \in L(G_0)$ and $\mathcal{G}_{G'_0,I_0}(w) = \bot$ otherwise, and
- $G'_1 = (\Gamma_1, \Sigma, R_1, I_1)$ such that $\mathcal{G}_{G'_1,I_1}(w) = 1$ if $w \in L(G_1)$ and $\mathcal{G}_{G'_1,I_1}(w) = \bot$ otherwise.

Finally, let us consider the grammar $G = (\Gamma_0 \cup \Gamma_1 \cup I, \Sigma, R_0 \cup R_1 \cup \{I \to I_0, I \to I_1\}, I)$. Then, for each $w \in \Sigma^+$,

$$\mathcal{G}_{G,I}(w) = \begin{cases} \mathsf{mex}(\{0,1\}) = 2 & \text{if } w \in L(G_0) \wedge w \in L(G_1) \\ \mathsf{mex}(\{0\}) = 1 & \text{if } w \in L(G_0) \wedge w \notin L(G_1) \\ \mathsf{mex}(\{1\}) = 0 & \text{if } w \notin L(G_0) \wedge w \in L(G_1) \\ \bot & \text{otherwise} \end{cases}$$

holds because $\mathcal{G}_{G,I}(w) = \mathsf{mex}(\{\mathcal{G}_{G',I'}(w) : (G', I') \in \{(G'_0, I_0), (G'_1, I_1)\},\ w \in L(G', I')\})$. Thus, there is a word $w \in \Sigma^+$ such that $\mathcal{G}_{G,I}(w) = 2$ if and only if $L(G_0) \cap L(G_1) \neq \emptyset$. Therefore, if one could decide the existence of w satisfying $\mathcal{G}_{G,I}(w) = 2$, then one could also decide the emptiness of the intersection of two CFGs, which is a contradiction. □

Theorem 4. *Given a context-free grammar $G = (\Gamma, \Sigma, R, I)$, there is no algorithm that computes the strict upper bound* $\max(\mathcal{G}_{G,I}(L(G)))$ *of Grundy numbers. This remains true even when the grammar is restricted to be linear.*

Proof. Consider the same grammar G defined in Lemma 3. Then, $\max(\mathcal{G}_{G,I}(L(G))) = 2$ if and only if $L(G_0) \cap L(G_1) \neq \emptyset$ in the same reason as Lemma 3. □

This theorem implies that there is no algorithm that computes the range $\mathcal{G}_{G,I}(L(G)) \subseteq \mathbb{N}$ of the function $\mathcal{G}_{G,I}$. For similar reasons, it is also undecidable whether the game is always winning or always losing.

Theorem 5. *Given a context-free grammar $G = (\Gamma, \Sigma, R, I)$, it is undecidable whether*

(1) $\mathcal{G}_{G,I}(w) = 0$ for all $w \in L(G)$, that is, $\{(I, w)\}$ is always $\mathcal{P}$-position,
(2) $\mathcal{G}_{G,I}(w) \neq 0$ for all $w \in L(G)$, that is, $\{(I, w)\}$ is always $\mathcal{N}$-position, or
(3) given an integer $c \in \mathbb{N}$, there exists a word $w \in \Sigma^+$ such that $\mathcal{G}_{G,I}(w) = c$.

This remains true even when the grammar is restricted to be 2-bounded.

Proof. We first show that (1) is undecidable. Given a linear CFG $G' = (\Gamma', \Sigma, R', I')$, construct the 2-bounded CFG $G = (\Gamma' \cup \{I, J_1, J_2\}, \Sigma \cup \{\#\}, R, I)$ where

$$R = R' \cup \{I \to J_2\#I', J_2 \to J_1, J_2 \to a, J_1 \to a\}$$

with a fixed letter $a \in \Sigma^*$. Note that $L(G) = a\#L(G')$. In addition, $\mathcal{G}_{G,J_2}(a) = \mathsf{mex}(\{0, \mathcal{G}_{G,J_1}(a)\}) = \mathsf{mex}\{0, 1\} = 2$. Therefore, for each $w \in L(G')$

$$\begin{aligned}\mathcal{G}_{G,I}(a\#w) &= \mathsf{mex}(\{\mathcal{G}_{G,J_2}(a) \oplus \mathcal{G}_{G,I'}(w)\}) \\ &= \mathsf{mex}(\{2 \oplus \mathcal{G}_{G',I'}(w)\})\end{aligned}$$

holds. Here, we have that for each $w \in \Sigma^+$, $\mathcal{G}_{G,I}(a\#w) \neq 0 \iff 2 \oplus \mathcal{G}_{G',I'}(w) = 0 \iff \mathcal{G}_{G',I'}(w) = 2$ holds. Therefore, if one could decide the existence of w satisfying $\mathcal{G}_{G,I}(a\#w) \neq 0$, then one could also decide the existence of w satisfying $\mathcal{G}_{G',I'}(w) = 2$, which contradicts to Lemma 3.

Next, (2) follows by an easy reduction from (1). Given a CFG $G' = (\Gamma', \Sigma, R', I')$, construct the CFG $G = (\Gamma' \cup \{I\}, \Sigma, R' \cup \{I \to I'\}, I)$. Then, for each $w \in L(G')$, $\mathcal{G}_{G,I}(w) = \mathsf{mex}(\mathcal{G}_{G',I'}(w))$ holds. Thus, $\mathcal{G}_{G,I}(w) \neq 0$ iff $\mathcal{G}_{G',I'}(w) = 0$. Therefore, (2) is also undecidable because the existence of w satisfying $\mathcal{G}_{G',I'}(w) = 0$ is undecidable by (1).

Finally, let us show that (3) is undecidable by induction on $c \geq 0$. The base case where $c = 0$ follows from (2). Assume that given a CFG $G' = (\Gamma', \Sigma, R', I')$, the existence of a word $w \in \Sigma^+$ satisfying $\mathcal{G}_{G,I}(w) = c - 1$ is undecidable. For a non-negative integer $i \in \mathbb{N}$, let $G_i = (\Gamma_i, \Sigma, R_i, I_i)$ be the CFG such that $\mathcal{G}_{G',I'}(w) = i$ for each $w \in L(G')$ (see also Lemma 2). Then, by constructing $G = (\Gamma_0 \cup \dots \cup \Gamma_{c-2} \cup \Gamma' \cup \{I\}, \Sigma, R, I)$ where

$$R = R \cup \dots \cup R_{c-2} \cup R' \cup \{I \to I_i : 0 \leq i \leq c - 2\} \cup \{I \to I'\},$$

we have $\mathcal{G}_{G,I}(w) = \mathsf{mex}(\{0, 1, \ldots, c-2, \mathcal{G}_{G',I'}(w)\})$ holds for each $w \in L(G')$. Thus, there is a word $w \in L(G')$ such that $\mathcal{G}_{G,I}(w) = c$ if and only if there is w such that $\mathcal{G}_{G',I'}(w) = c - 1$. This concludes the proof by the induction hypothesis. □

The proof relies on the undecidability of the emptiness problem for the intersection of two linear CFGs [12], which in turn follows from the Post Correspondence Problem [19]. Consequently, the argument essentially depends on the combinatorial nature of words over alphabets with at least two letters. For this reason, it is not immediate that a similar phenomenon would arise in the usual heap-based games.

5 Future Work

5.1 Incomplete Derivation Games

In the successful derivation games considered in this paper, all pairs (A, w) appearing in a position satisfy $w \in L(A)$. In contrast, one can consider the following variant of the game:

Definition 4 (incomplete derivation game). *For a CFG $G = (\Gamma, \Sigma, R, I)$ and a word $w \in \Sigma^+$, the* incomplete derivation game (G, w) *is defined as follows.*

- *A* position *is a finite multiset over $\Gamma \times \Sigma^+$.*
- *A legal move from a position Q is defined as follows. First, the current player chooses an element (A, w) of Q. Next, the player chooses a sequence $(A_1, w_1), \ldots, (A_\ell, w_\ell) \in \Gamma \times \Sigma^+$ such that $A \to u_0 A_1 u_1 \cdots u_{\ell-1} A_\ell u_\ell \in R$ with $w = u_0 w_1 u_1 \cdots u_{\ell-1} w_\ell u_\ell$. Finally, the position Q is updated to $Q' = (Q \setminus \{(A, w)\}) \cup \{(A_1, w_1), \ldots, (A_\ell, w_\ell)\}$.*
- *The game starts from the initial position $Q_0 = \{(I, w)\}$, and the players alternate making legal moves $Q_i \rightsquigarrow Q_{i+1}$ for each $i \geq 0$. The player who is unable to move loses.* ■

That is, in an incomplete derivation game, players are not required to guarantee that the derivation eventually reaches the target word. As a result, a derivation may get stuck. This setting has the following advantages.

- In general subtraction games, it often happens that some stones remain in the position, but no legal move is available and the game terminates. This situation corresponds exactly to a derivation that gets stuck before completion. Thus, the framework of incomplete derivation games naturally generalizes a wide range of games, including in particular finite code games.
- When the computation of Grundy numbers is viewed as a weight computation for weighted CFGs, incomplete derivation games fit the framework of weighted CFGs more naturally.

On the other hand, since incomplete derivation games allow derivations to stop before completion, Grundy numbers are defined even for words outside the language generated by the grammar. Consequently, the analysis becomes more difficult, and several of the techniques developed in this paper no longer apply. In particular, undecidability questions for incomplete derivation games, including those considered in Theorems 4 and 5, constitute a major direction for future research.

5.2 Other Directions

In addition to incomplete derivation games, several further problems arise. A natural starting point is the following fundamental question:

Problem 1. For a general CFG $G = (\Gamma, \Sigma, R, I)$, what kind of periodic behavior does the function $\mathcal{G}_{G,I}$ exhibit? ■

At present, even for finite octal games it is unknown whether Grundy sequences are always ultimately periodic [18]. From this viewpoint, Problem 1 appears to be at least as challenging. A more approachable problem is the following.

Problem 2. Given a CFG $G = (\Gamma, \Sigma, R, I)$, is the boundedness of the function $\mathcal{G}_{G,I}$ decidable? ■

In the k-bounded case, $\mathcal{G}_{G,I}$ is always bounded (Theorem 3). In contrast, as shown in Proposition 1, boundedness does not hold in general. From the perspective of Grundy-number analysis, understanding the decidability of boundedness is a particularly intriguing problem.

From a language-theoretic viewpoint, it is natural to ask for a CFG $G = (\Gamma, \Sigma, R, I)$, how the language $L(G)$ relates to the languages $L_{\mathcal{N}} = \{w \in \Sigma^+ : \mathcal{G}_{G,I}(w) \neq 0\}$ and $L_{\mathcal{P}} = \{w \in \Sigma^+ : \mathcal{G}_{G,I}(w) = 0\}$. The authors have already identified several examples of grammars G for which $L_{\mathcal{P}}$ is not context-free. As discussed in Sect. 2.1, this direction is closely connected to weighted CFGs.

Finally, Theorem 4 demonstrates that even linear grammars already give rise to substantial difficulties for the analysis of Grundy numbers. This naturally motivates a closer study of games on right-linear (regular) grammars. Even in the right-linear—and even unary—case, the presence of nonterminal symbols provides a form of finite memory, making the resulting games more expressive than standard heap-based games. For this reason, such restricted settings still appear to merit further investigation. We note that in the right-linear case, it has recently been shown that the range of Grundy numbers is computable, and that the languages $L_{\mathcal{N}}$ and $L_{\mathcal{P}}$ are regular; these results will be reported elsewhere.

References

1. Elwyn, R., Berlekamp, J.H., Conway: Winning ways for your mathematical plays, vol. 1, CRC Press (2001). 2nd edition
2. Bouton, C.L.: Nim, a game with a complete mathematical theory. Annals Math. **3**(1/4), 35–39 (1901)
3. Bruyère, V.: Computer aided synthesis: a game-theoretic approach. In: International Conference on Developments in Language Theory, pp. 3–35. Springer (2017)
4. Droste, M., Götze, D., Märcker, S., Meinecke, I.: Weighted tree automata over valuation monoids and their characterization by weighted logics. In: Algebraic Foundations in Computer Science: Essays Dedicated to Symeon Bozapalidis on the Occasion of His Retirement, pp. 30–55. Springer (2011)
5. Droste, M., Meinecke, I.: Weighted automata and weighted mso logics for average and long-time behaviors. Inf. Comput. **220**, 44–59 (2012)
6. Fülöp, Z., Stüber, T., Vogler, H.: A büchi-like theorem for weighted tree automata over multioperator monoids. Theory Comput. Syst. **50**(2), 241–278 (2012)
7. Goodman, J.: Semiring parsing. Comput. Linguist. **25**(4), 573–606 (1999)
8. Grädel, E., Thomas, W., Wilke, T.: Automata, logics, and infinite games: a guide to current research, Springer (2003)
9. Patrick, M.: Grundy: Mathematics and games. Eureka **2**, 6–8 (1939)
10. Guy, R.K., Smith, C.A.B.: The g-values of various games. Math. Proc. Cambridge Philos. Soc. **52**, 514–526 (1956) Cambridge University Press
11. Holík, L., Meyer, R., Muskalla, S.: Summaries for context-free games, (2016). arXiv:1603.07256 arXiv preprint
12. Hoogeboom, H.J.: Undecidable problems for context-free grammars. Preprinthttps://liacs.leidenuniv.nl/hoogeboomhj/second/codingcomputations.pdf (2015)
13. James, P.: Jones: some undecidable determined games. Internat. J. Game Theory **11**(2), 63–70 (1982)
14. Kasami, T.: An efficient recognition and syntax-analysis algorithm for context-free languages. Coordinated Science Laboratory Report no. R, 257 (1966)
15. Kuich, W.: Linear systems of equations and automata on distributive multioperator monoids. Contributions General Algebra **12**, 247–256 (1999)
16. Larsson, U., Wästlund, J.: From heaps of matches to the limits of computability, (2012). arXiv:1202.0664 arXiv preprint
17. Muscholl, A., Schwentick, T., Segoufin, L.: Active context-free games. Theory Comput. Syst. **39**(1), 237–276 (2006)
18. Nowakowski, R.J.: Unsolved problems in combinatorial games. Games no Chance **5**(70), 125–168 (2019)
19. Emil, L.: Post: a variant of a recursively unsolvable problem, (1946)
20. William, L., Sibert, J.H., Conway: Mathematical kayles, (1992)
21. Aaron, N.: Siegel: Combinatorial game theory. In: Graduate studies in mathematics, vol. 146, American Mathematical Society (2013)
22. Sprague, R.: Über mathematische kampfspiele. Tohoku Math. J. **41**, 438–444 (1935)
23. Stüber, T., Vogler, H., Fülöp, Z.: Decomposition of weighted multioperator tree automata. Int. J. Found. Comput. Sci. **20**(02), 221–245 (2009)
24. Teichmann, M., Osterholzer, J.: A link between multioperator and tree valuation automata and logics. Theoret. Comput. Sci. **594**, 106–119 (2015)
25. Younger, D.H.: Recognition and parsing of context-free languages in time n^3. Inf. Control **10**(2), 189–208 (1967)

Decomposing Regular Languages Under Shuffle Along Trajectories

Sungmin Kim[1], Hyunki Hong[1], Taeryung Lim[1], Yo-Sub Han[1(✉)], and Kai Salomaa[2]

[1] School of Computing, Yonsei University, 50 Yonsei-Ro, Seodaemun-Gu, Seoul 03722, Republic of Korea
{rena_rio,hhong,taeryunglim,emmous}@yonsei.ac.kr
[2] School of Computing, Queen's University, Kingston, ON K7L 3N6, Canada
salomaa@queensu.ca

Abstract. We study the decomposition problem for regular languages under the shuffle along trajectories. Given a language, the problem is to decide whether or not there exists a pair of languages whose shuffle along the trajectory is equal to the given language. We prove that for regular languages, the decomposition problem is decidable under the perfect shuffle and the 2-comet shuffle. Furthermore, for the case when there exists a decomposition, we provide algorithms that construct a pair of automata where the shuffle of the languages they recognize along each trajectory equals the given language.

Keywords: Shuffle along trajectories · Language decomposition · Regular languages · Büchi automata

1 Introduction

For two strings x and y, the *shuffle* $x \sqcup\!\sqcup y$ of x and y is the set of strings that can be obtained by interleaving x and y [14]. Extending the shuffle operation to languages, the shuffle of languages X and Y is defined as the union of the shuffle of two strings, each in X and Y. In other words, $X \sqcup\!\sqcup Y = \bigcup_{x \in X, y \in Y} x \sqcup\!\sqcup y$. The shuffle of two languages is widely studied in language theory [4,6,15,17] with applications in formal verification [2,12], XML database systems [13] and natural language processing [18]. Especially in settings where we need a finer control over how the two strings are interleaved, we equip the operation with a set of *trajectories*, which are strings over $\{0, 1\}$. Consider two strings x and y over Σ and a trajectory t with $|t| = |x| + |y|$, where the length $|w|$ of w denotes the number of characters in a string w. The *shuffle along trajectory* t of x and y is defined as $x \sqcup\!\sqcup_t y = a_1 a_2 \cdots a_{|t|}$, where $a_i = x[j_0]$ if $t[i] = 0$ and $a_i = y[j_1]$ otherwise [19], where $w[\ell]$ denotes the ℓth character of w, for $\ell = 1, 2, \ldots, |w|$, and j_0 and j_1 are the number of 0's and 1's in the prefix of t with length i. Finally, for a (not necessarily finite) set T of trajectories, the shuffle of two languages X and Y along the trajectory set T is defined as the union of all $x \sqcup\!\sqcup_t y$, where

T. J. Smith (Ed.): CIAA 2026, LNCS 16695, pp. 92–107, 2026.
https://doi.org/10.1007/978-3-032-31176-4_7

$x \in X$, $y \in Y$, $t \in T$ satisfy $|t| = |x| + |y|$. Then, it is easy to see that the operation $\sqcup\!\sqcup$ is equivalent to the operation $\sqcup\!\sqcup_{\{0,1\}^*}$.

On the other hand, for a given language R, the *decomposition problem* with respect to a binary operator $\circ$ on languages is to compute two languages X and Y satisfying $X \circ Y = R$. For instance, for the language concatenation operation and regular R, the problem of testing whether or not there is a solution to the decomposition problem with $X \neq R$ and $Y \neq R$ is PSPACE-complete [9]. It is also known that such a solution always exists when R is unary [16]. Our interest lies in the decomposition problem with respect to the shuffle along trajectories. We are given a language R and a trajectory set T, and the objective is to compute a pair (X, Y) of languages that satisfy $X \sqcup\!\sqcup_T Y = R$. In this case, we say that the pair (X, Y) is a *decomposition* with respect to T, and if such a decomposition exists, then R can be *decomposed* with respect to T. Furthermore, if a pair (X, Y) where X or Y is $\{\lambda\}$ is a decomposition of R with respect to T, then we call the pair a *trivial decomposition*. For example, for the trajectory set $\{0,1\}^*$, the pair $(R, \{\lambda\})$ and the pair $(\{\lambda\}, R)$ are always trivial decompositions of R. Note that, for certain types of trajectories, there may not be a trivial decomposition. For the trajectory set $\{01\}^*$, for instance, the language R cannot be decomposed into $(R, \{\lambda\})$ or $(\{\lambda\}, R)$, unless $R = \{\lambda\}$.

Problem 1 (Trajectory Shuffle Decomposition). Given a regular language R and a trajectory language T, decide whether or not there exists a non-trivial decomposition (X, Y) of R with respect to T.

Domaratzki and Salomaa proved that Problem 1 is decidable for 1-thin regular trajectory sets [8] and letter-bounded regular trajectory sets [7]. Here, a language R is *1-thin* if there is at most 1 string in R for any given length, and R is *letter-bounded* if there exists a finite sequence of characters $\sigma_1, \sigma_2, \ldots, \sigma_k$ that are not necessarily distinct and satisfies $R \subseteq \sigma_1^* \sigma_2^* \cdots \sigma_k^*$. For example, the trajectory set $(01)^*$ is 1-thin, and thus Problem 1 is decidable for this trajectory set. Our goal is to solve Problem 1 for the following trajectories. The first case is the perfect shuffle trajectory set $T_P = (01)^* + (10)^*$ and the second case is the k-comet trajectory set $T_{\leq k} = (01)^* 0^{\leq k}$ when $k = 2$. Note that these trajectory sets are not 1-thin nor letter-bounded. We use the name "k-comet" for $T_{\leq k}$ as a reference to the left-sided comet languages, which is the family of languages that can be expressed in the form of AB^* for some regular languages A and B [3].

Problem 2 (Perfect Shuffle Decomposition). Given a regular language R, decide whether or not there exists a decomposition (X, Y) of R with respect to T_P, where $T_P = L((01)^* + (10)^*)$.

Problem 3 (k-Comet Shuffle Decomposition). Given a regular language R and an integer $k \geq 2$, decide whether or not there exists a decomposition (X, Y) of R with respect to $T_{\leq k} = L((01)^* 0^{\leq k})$.

Our Contributions. We present algorithms that solve the perfect shuffle decomposition problem and the k-comet shuffle decomposition problem for $k = 2$. The algorithm for Problem 2 applies set operations on the given language R to obtain a candidate decomposition (X, Y), after which we verify by testing $X \sqcup\!\sqcup_{T_P} Y = R$. On the other hand, the algorithm for Problem 3 under $k = 2$ first computes an alternating Büchi automaton whose language is non-empty if and only if R has a non-trivial decomposition with respect to $T_{\leq 2}$. Then, if the language of the automaton is non-empty, we further compute a pair of finite-state automata whose shuffle between the languages along $T_{\leq 2}$ equal R. Since our approaches explicitly construct the automata for X and Y verifying $X \sqcup\!\sqcup_T Y = R$, we obtain as corollaries that if R is non-trivially decomposable with respect to either trajectory, then it also has a non-trivial regular decomposition.

2 Preliminaries

For a string w and two integers i, j with $1 \leq i \leq j \leq |w|$, let $w[i:j]$ denote the substring $w[i]w[i+1] \cdots w[j]$. For an alphabet Σ and an integer n, let Σ^n denote the set of all strings over Σ with length n. The set $\Sigma^* \setminus \Sigma^n$ is denoted by $\Sigma^{\neq n}$. The *deletion along trajectory operation* $\rightsquigarrow_T$ [5] is defined as

$$L_1 \rightsquigarrow_T L_2 = \{x \in \Sigma^* \mid \exists w_1 \in L_1, w_2 \in L_2 : x \sqcup\!\sqcup_T w_2 = w_1\},$$

where the resulting language consists of the strings obtained by deleting strings in L_2 from strings in L_1, where the deletion is controlled by some trajectory $t \in T$. Also, the *toggle* function `tog` is a morphism that maps 0 to 1 and 1 to 0. The *density function* of a language L is a function $\phi_L : \mathbb{N} \to \mathbb{N}$, where $\phi_L(n) = |L \cap \Sigma^n|$. A language L is *k-thin* if there exists an $n_0 \in \mathbb{N}$ such that $\phi_L(n) \leq k$ for all $n \geq n_0$. The language is *strictly k-thin* if $n_0 = 1$. A subset L_1 of L is *length-complete* if $\phi_{L_1}(n) \neq 0$ whenever $\phi_L(n) \neq 0$. Furthermore, if no subset of L_1 is length-complete of L, then L_1 is *minimal length-complete* (MLC) [21]. For a language L, the language $L_{\min}$ is the minimal length-complete subset of L consisting of the lexicographically least strings in L for every length. Note that $L_{\min}$ is strictly 1-thin. A language L is *prefix-closed* if every prefix u of $w \in L$ is also in L. Finally, for a language L, let $\texttt{even}(L)$ and $\texttt{odd}(L)$ be the set of strings constructed by deleting for each string in L the characters in odd positions and even positions, respectively.

Proposition 1 (Andrasiu et al. [1]). *For a regular language L, the language $L_{\min}$ is regular and effectively constructable from L.*

We recall the definitions for infinite strings and infinite trees as follows. For a finite set Q, an infinite Q-tree Ψ is an infinite subset of Q^* that is prefix-closed. A node in a Q-tree Ψ is an element of Ψ, and the root node of Ψ is denoted as ε to disambiguate the context with the empty string λ. Intuitively, each node n is a sequence of elements of Q which we need to observe while moving from the root to n. For a node $n \in \Psi$, the *depth* of the node is $|n|$, with $|\varepsilon| = 0$. The node $n[1 : |n| - 1]$ is the *parent* of n, and all nodes $n_1, \ldots, n_k \in \Psi$ with $n_i = nq_i$ for some $q_i \in Q$ are the *children* of n. Let $\Delta_\Psi(n) = \{q \in Q \mid nq \in \Psi\}$ be the set of *extension elements* of n, which is the elements q in Q for which nq is a child of n. A *path* of Ψ is an infinite string π over Q satisfying $\pi[1 : i] \in \Psi$ for all integers $i \in \mathbb{N}_0$. For a path π of Ψ, let $\mathtt{Inf}(\pi) = \{q \in Q \mid |\{i \in \mathbb{N}_0 \mid \pi[i] = q\}| \geq \omega\}$ be the subset of Q that appears infinitely often in π.

We recall the definition of the Büchi automaton on infinite strings with the additional power of alternation. For a given ground set X, let $\mathcal{B}^+(X)$ be the set of positive Boolean formulae over $X \cup \{\mathtt{True}, \mathtt{False}\}$, which are Boolean formulae constructed using only conjunction ($\wedge$) and disjunction ($\vee$). A subset $S \subseteq X$ *satisfies* a given Boolean formula ϕ if the formula evaluates to `True` when elements in S are assigned `True` and elements not in S are assigned `False`.

Definition 1. *An* alternating Büchi automaton *is specified by a 5-tuple $A = (Q, \Gamma, \delta, s, F)$, where Q is the set of states, Γ is the alphabet, $\delta : Q \times \Gamma \to \mathcal{B}^+(Q)$ is the transition function, $s \in Q$ is the start state, and $F \subseteq Q$ is the set of accepting states.*

The automaton is *nondeterministic* if the Boolean formulae in the image of every transition function does not use conjunction. A *run* of an infinite string w over Γ on an alternating Büchi automaton $A = (Q, \Gamma, \delta, s, F)$ is an infinite Q-tree Ψ, where, for every node $n \in \Psi$ and for the last state q of $s \cdot n$, the set of extension elements $\Delta_\Psi(n)$ satisfies the positive Boolean formula $\delta(q, w[|n|])$. A run Ψ is *accepting* if all paths π of Ψ satisfy $\mathtt{Inf}(\pi) \cap F \neq \emptyset$. The automaton *accepts* an infinite string w if there exists a run of w that is accepting. Otherwise, the automaton *rejects* w. The *language of* A is the set of all infinite strings that are accepted by A, which we denote as $L(A)$. It is worth noting that there are several variants of automata that describe sets of infinite strings, and the main difference is the acceptance condition of runs. Finally, the following results imply that the emptiness of a given Büchi automaton is decidable.

Proposition 2 (Miyano and Hayashi [20]). *For an alternating Büchi automaton with n states, there exists a nondeterministic Büchi automaton with $2^{O(n)}$ states which describes the same language.*

Proposition 3 (Emerson and Lei [10,11]). *The language described by a given nondeterministic Büchi automaton is non-empty if and only if it contains a reachable cycle including an accepting state, and thus emptiness of the language can be decided in linear time.*

3 The Perfect Shuffle Decomposition

Since we examine Problem 2, we write $T = L((01)^* + (10)^*)$ instead of T_P for simplicity. Given a regular language R, assume that there exists a non-trivial decomposition for R with respect to T, and let (X, Y) be a decomposition of R. Because $X \sqcup\!\sqcup_T Y = Y \sqcup\!\sqcup_T X$, we have the following lemma.

Lemma 1. *If R has a decomposition (X, Y) with respect to T, then* $\mathtt{odd}(R) = \mathtt{even}(R) = X \cup Y$.

Furthermore, for any non-negative integer i, each string in R with length $2i$ only depends on strings in X and Y with length i. Thus, we can freely swap the entire subset of X and Y of a particular length and still obtain the same perfect shuffle as illustrated in Fig. 1. This lead us to establish the following statement.

Proposition 4. *For languages X, Y and a non-negative integer i, let $X_i = (X \cap \Sigma^{\neq i}) \cup (Y \cap \Sigma^i)$ and $Y_i = (X \cap \Sigma^i) \cup (Y \cap \Sigma^{\neq i})$. Then, we have $X \sqcup\!\sqcup_T Y = X_i \sqcup\!\sqcup_T Y_i$.*

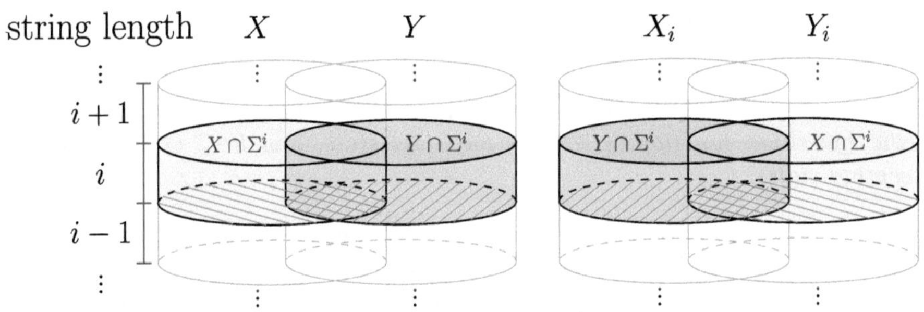

Fig. 1. An illustration of Proposition 4. Each cylinder corresponds to the set of strings of a particular length in X or Y. The sets X_i and Y_i are obtained by swapping the strings of length i from X and Y.

For example, consider the language

$$R_P = L((aabb)^* + (aaab)^* + (aaba)^* + (abac)^* + (abbc)^* + (baca)^* + (bacb)^*)$$

over $\Sigma = \{a, b, c\}$; R_P is our running example in Sect. 3. Then, for

$$X = L((aa)^* + (ab)^*) \text{ and } Y = L((ab)^* + (bc)^*),$$

the pair (X, Y) forms a decomposition of R_P. Next, let us consider the languages

$$X_2 = L(bc + aaaa(aa)^* + (ab)^*) \text{ and } Y_2 = L(aa + bcbc(bc)^* + (ab)^*).$$

Note that X_2 and Y_2 are the languages obtained by exchanging strings of length 2 between X and Y. Then, it is easy to see that $X_2 \sqcup\!\sqcup_T Y_2 = R_P = X \sqcup\!\sqcup_T Y$ as in Proposition 4. Furthermore, it is immediate that if R can be decomposed into the trivial decomposition, then $R = \{\lambda\}$ since the shuffle operation must shuffle strings of the same length. We are now ready to design a decomposition algorithm for Problem 2, which we present in Algorithm 1.

We remark that any pair of languages returned by Algorithm 1 is a decomposition of R with respect to T. This is because Algorithm 1 checks whether or not the shuffle of the returned language pair equals R on Line 8, before returning a pair of languages. Therefore, the return value of Algorithm 1 is always a decomposition of R. We formalize this result in Lemma 2.

Algorithm 1. Perfect shuffle decomposition

Require: R is a language over Σ

if $\texttt{even}(R) \neq \texttt{odd}(R)$, **return** $\texttt{False}$

$\texttt{Intersection} \leftarrow \texttt{even}(R \cap \{\sigma\sigma \mid \sigma \in \Sigma\}^*)$

$\texttt{DiffShuff} \leftarrow R \setminus (\texttt{Intersection} \sqcup\!\sqcup_T \Sigma^*)$

$\texttt{ThinSymDiff} \leftarrow (\texttt{even}(R) \setminus \texttt{Intersection})_{\text{min}}$

$P = (01)^*$

$\texttt{DiffY} \leftarrow \texttt{DiffShuff} \rightsquigarrow_P \texttt{ThinSymDiff}$

$\texttt{DiffX} \leftarrow \texttt{even}(R) \setminus (\texttt{DiffY} \cup \texttt{Intersection})$

if $(\texttt{DiffX} \cup \texttt{Intersection}) \sqcup\!\sqcup_T (\texttt{DiffY} \cup \texttt{Intersection}) \neq R$, **return** $\texttt{False}$

return $(\texttt{DiffX} \cup \texttt{Intersection}, \texttt{DiffY} \cup \texttt{Intersection})$

Lemma 2. *If Algorithm 1 does not reject the input, then the returned value is always a decomposition of R with respect to T.*

It remains to prove that the algorithm returns a decomposition whenever R is decomposable. The idea of computing a decomposition (X, Y) of R is to first compute the intersection and symmetric difference of X and Y by a series of set operations on R. Lemmas 3 and 4 guarantee that the values for $\texttt{Intersection}$, $\texttt{even}(R) \setminus \texttt{Intersection}$ and $\texttt{DiffShuff}$ are exactly the respective values for the intersection, symmetric difference, and the shuffle between $X \setminus Y$ and $Y \setminus X$. Note that the result stands for any decomposition (X, Y) of R. Thus, the intersection, symmetric difference and the shuffle of difference are invariant for every decomposition of R.

Lemma 3. *The variable* $\texttt{Intersection}$ *in Algorithm 1 is exactly $X \cap Y$ for all decompositions (X, Y) if R has a decomposition with respect to T.*

Lemma 4. *If R has a decomposition with respect to T, then in Algorithm 1, we have* $\texttt{even}(R) \setminus \texttt{Intersection} = (X \cup Y) \setminus (X \cap Y)$ *and* $\texttt{DiffShuff} = (X \setminus Y) \sqcup\!\sqcup_T (Y \setminus X)$ *for all decompositions (X, Y).*

Returning to our running example R_P, we observe that

$$R_P \cap \{\sigma\sigma \mid \sigma \in \Sigma\}^* = L((aabb)^*) \text{ and } \texttt{Intersection} = L((ab)^*).$$

Now because $\texttt{even}(R_P) = \texttt{odd}(R_P) = L((aa)^* + (ab)^* + (bc)^*)$, it follows that

$$\texttt{even}(R_P) \setminus \texttt{Intersection} = L((aa)^* + (bc)^*) = (X \cup Y) \setminus (X \cap Y).$$

Furthermore,

$$\texttt{DiffShuff} = R_P \setminus (\texttt{Intersection} \sqcup\!\sqcup_T \Sigma^*)$$

is computed as $L((abac)^* + (baca)^*)$, which is equivalent to the shuffle of the disjoint components

$$(X \setminus Y) \sqcup\!\sqcup_T (Y \setminus X) = L((aa)^*) \sqcup\!\sqcup_T L((bc)^*).$$

For the symmetric difference of X and Y obtained in Lemma 4, we choose a single string for each length by computing $[(X \setminus Y) \cup (Y \setminus X)]_{\min}$, which is exactly `ThinSymDiff` in Algorithm 1. Then, we apply Proposition 4 for each length where the chosen string is in $Y \setminus X$. Recall that this swapping does not affect the shuffle of the two languages. If we swap the strings in X and Y for every such length, then we obtain a pair (X', Y') of languages where every string in `ThinSymDiff` is in X' but not in Y'. We can then extract all the strings that interacted with the chosen string for each length by deleting the string along the trajectory T from `DiffShuff`. Note that $\texttt{DiffShuff} = (X' \setminus Y') \sqcup\!\sqcup_T (Y' \setminus X')$ since `DiffShuff` is invariant for all decompositions of R. (See Fig. 2 for an example.)

Lemma 5. *Let (X, Y) be a decomposition of R with respect to T. There exists a pair of languages (X', Y') such that $X' \sqcup\!\sqcup_T Y' = X \sqcup\!\sqcup_T Y = R$ and* $\texttt{ThinSymDiff} \subseteq X' \setminus Y'$ *in Algorithm 1.*

Lemma 6. *Let (X, Y) be a decomposition of R with respect to T, where the set* `ThinSymDiff` *in Algorithm 1 is contained in $X \setminus Y$. Then,* $\texttt{DiffY} = Y \setminus X$ *and* $\texttt{DiffX} = X \setminus Y$.

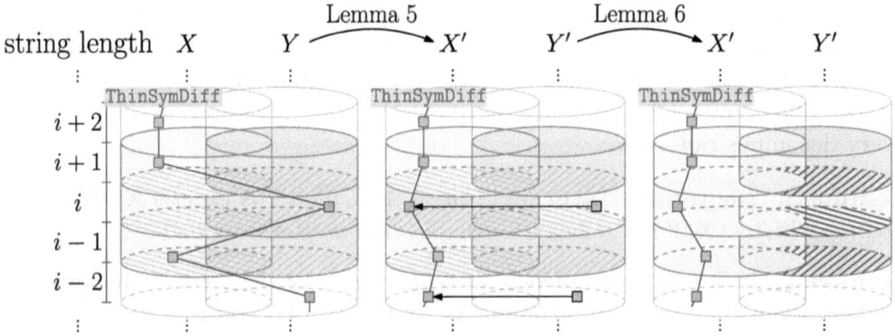

Fig. 2. An illustration of Lemmas 5 and 6. We first swap the strings between X and Y using Proposition 4 to obtain X' and Y' so that elements of `ThinSymDiff` are all included in $X' \setminus Y'$. Then we obtain $Y' \setminus X'$ through deletion of `DiffShuff` by `ThinSymDiff` along T.

For our example language R_P, since $(\texttt{even}(R_P) \setminus \texttt{Intersection}) = L((aa)^* + (bc)^*)$, we have $\texttt{ThinSymDiff} = L((aa)^*)$. Then, we have

$$\texttt{DiffShuff} \rightsquigarrow_P \texttt{ThinSymDiff} = L((bc)^*),$$

which is equivalent to $\texttt{DiffY} = Y \setminus X$. Symmetrically, we find $\texttt{DiffX} = L((aa)^*)$, which leads to the final decomposition (X, Y), where

$$X = L((aa)^* + (ab)^*) \text{ and } Y = L((ab)^* + (bc)^*).$$

These lead us to establish the following results.

Lemma 7. *If R has a decomposition with respect to T, then*

$$(\texttt{DiffX} \cup \texttt{Intersection}) \sqcup\!\sqcup_T (\texttt{DiffY} \cup \texttt{Intersection}) = R$$

in Algorithm 1.

Theorem 1. *Given a regular language R, we can decide whether or not R can be decomposed with respect to the perfect shuffle and, if yes, effectively compute a decomposition (X, Y) of R, where X and Y are regular.*

In other words, if a regular language R can be decomposed into two languages, then we can compute a decomposition whose components are regular. Therefore, we have the following corollary.

Corollary 1. *If a regular language can be decomposed with respect to T, then it can be decomposed into two regular languages.*

4 The 2-Comet Shuffle Decomposition

We now consider the shuffle decomposition problem of a regular language R under the 2-comet trajectory set $T_{\leq 2}$. Our approach is to first construct a Büchi automaton that is non-empty if and only if R has a decomposition with respect to $T_{\leq 2}$. Each ω-string accepted by the Büchi automaton corresponds to a decomposition of R. Then, we extract a single ω-string from the Büchi automaton to construct a pair of DFAs for the corresponding decomposition of R.

For an integer n, let $T_n = L((01)^*0^n)$. Lemma 8 characterizes the strings of a particular length in each decomposition of R with respect to $T_{\leq 2}$ (Fig. 3).

Lemma 8. *Let (X, Y) be a decomposition of R with respect to $T_{\leq 2}$. For an integer k, the pair $(X \cap \Sigma^{k+1}, Y \cap \Sigma^k)$ is either $(\texttt{odd}(R \cap \Sigma^{2k+1}), \texttt{even}(R \cap \Sigma^{2k+1}))$, $((R \rightsquigarrow_{T_0} \Sigma^*) \cap \Sigma^{k+1}, \emptyset)$, or $(\emptyset, ((R \rightsquigarrow_{tog(T_2)})\Sigma^*) \cap \Sigma^k)$.*

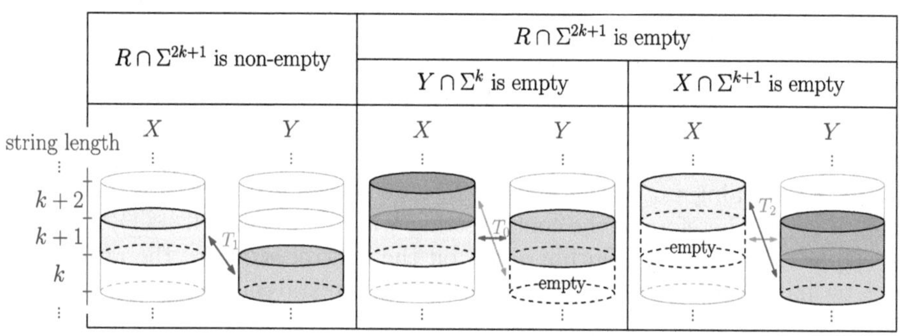

Fig. 3. An illustration of Lemma 8. There are three choices for $X \cap \Sigma^{k+1}$ and $Y \cap \Sigma^k$.

We now use a different running example

$$R_C = \{(abba)^k u \mid k \geq 0, u \in \{\lambda, a, ab\}\}$$

in Sect. 4. Note that $(\{(ab)^i a \mid i \leq k-1\}, \{(ba)^i(\lambda + b) \mid i \leq k-1\})$ is a decomposition of R_C. For Lemma 8, we have $R_C \cap \Sigma^{2k+1} = \{abbaa\}$ when $k = 2$. This gives us $X \cap \Sigma^{k+1} = \texttt{odd}(R_C \cap \Sigma^{2k+1}) = \{aba\}$, and $Y \cap \Sigma^k = \texttt{even}(R_C \cap \Sigma^{2k+1}) = \{ba\}$. Meanwhile, for $k = 3$, we have $R_C \cap \Sigma^{2k+1} = \emptyset$. Therefore, we have two cases, depending on which of the following two sets is $\emptyset$ between $X \cap \Sigma^{k+1}$ or $Y \cap \Sigma^k$. Indeed, we have $X \cap \Sigma^4 = \emptyset$, and therefore $Y \cap \Sigma^3 = (R_C \rightsquigarrow_{\texttt{tog}(T_2)} \Sigma^*) \cap \Sigma^3 = \{bab\}$.

In light of Lemma 8, we can now represent any decomposition (X, Y) of R with respect to $T_{\leq 2}$ as an infinite string γ over the alphabet $\Gamma = \{①, ②, ③\}$, where $\gamma[1] = ①$ if and only if $\lambda \in X$, and for all $k \geq 0$, $\gamma[k+2]$ represents which case $(X \cap \Sigma^{k+1}, Y \cap \Sigma^k)$ corresponds to in Lemma 8. We call such an infinite string γ a *control string*. Likewise, we call Γ the *control alphabet* and any character in Γ a *control character*. Finally, for a control string $\gamma \in \Gamma^\omega$, let us define two languages X_γ and Y_γ, where for each integer $k \geq 0$, the strings of lengths $k+1$ and k in X_γ and Y_γ respectively are defined as follows.

Cases	$X_\gamma \cap \Sigma^{k+1}$	$Y_\gamma \cap \Sigma^k$
$\gamma[k+2] = ①$	$\texttt{odd}(R \cap \Sigma^{2k+1})$	$\texttt{even}(R \cap \Sigma^{2k+1})$
$\gamma[k+2] = ②$	$(R \rightsquigarrow_{T_0} \Sigma^*) \cap \Sigma^{k+1}$	$\emptyset$
$\gamma[k+2] = ③$	$\emptyset$	$(R \rightsquigarrow_{\texttt{tog}(T_2)} \Sigma^*) \cap \Sigma^k$

For the remaining case $X_\gamma \cap \Sigma^0$, we define that $\lambda \in X_\gamma$ if and only if $\gamma[1] = ①$. We call the pair (X_γ, Y_γ) of languages the *decomposition candidate* of R with respect to γ. Moreover, for a finite string g over Γ, let us define $X_g = X_\gamma \cap \Sigma^{\leq |g|+1}$ and $Y_g = Y_\gamma \cap \Sigma^{\leq |g|}$, where γ is an infinite string over Γ that satisfies $\gamma[1 : |g|] = g$. We define the decomposition candidate of R with respect to g as (X_g, Y_g). The control string γ is *valid* with respect to R if $X_\gamma ⧢_{T_{\leq 2}} Y_\gamma = R$. In other words, there exists a valid control string with respect to R if and only if R is decomposable.

Our strategy is to construct a Büchi automaton that recognizes exactly the valid control strings of a given language, and to check emptiness on that machine. The idea is to simultaneously simulate five finite automata, each for R, $\texttt{odd}(R \cap \Sigma(\Sigma\Sigma)^*)$, $\texttt{even}(R \cap \Sigma(\Sigma\Sigma)^*)$, $R \rightsquigarrow_{T_0} \Sigma^*$, and $R \rightsquigarrow_{\texttt{tog}(T_2)} \Sigma^*$. Returning to our running example language R_C, we observe that $R_C \cap \Sigma^5 \neq \emptyset$ and $R_C \cap \Sigma^7 = \emptyset$. Therefore, if the control string γ is *valid*, then $\gamma[4] = ①, \gamma[5] = ②$ or $\gamma[4] = ①, \gamma[5] = ③$. During a run on the control string γ, the automaton tries to find a counterexample for either of the two statements:

1. **Soundness:** for all $x \in X_\gamma$ and $y \in Y_\gamma$, we have $x ⧢_{T_{\leq 2}} y \subseteq R$,
2. **Completeness:** for all $w \in R$, there exists an $x_w \in X_\gamma$ and $y_w \in Y_\gamma$ such that $w \in x_w ⧢_{T_{\leq 2}} y_w$.

If the automaton indeed finds a counterexample, then it enters a rejecting state with no chance of visiting another accepting state. This does not allow the run to be accepting since the path corresponding to that counterexample only makes a finite number of visit to accepting states. Note that, if there are no counterexamples for both statements, then we have $X_\gamma \sqcup_{T_{\leq 2}} Y_\gamma = R$, and therefore we can decide the decomposability of R by deciding the emptiness of this automaton. We also note that a control string γ is valid if and only if it satisfies both soundness and completeness. We say that the decomposition candidate (X_γ, Y_γ) for a control string γ is *sound* if $X_\gamma \sqcup_{T_{\leq 2}} Y_\gamma \subseteq R$, and *complete* if $X_\gamma \sqcup_{T_{\leq 2}} Y_\gamma \supseteq R$.

Before giving an explicit construction of the alternating automaton, we define some necessary notations for the construction. Let $A_m = (Q_m, \Sigma, \delta_m, s_m, F_m)$ be the minimal DFA for R, $\mathtt{odd}(R \cap \Sigma(\Sigma\Sigma)^*)$, $\mathtt{even}(R \cap \Sigma(\Sigma\Sigma)^*)$, $R \rightsquigarrow_{T_0} \Sigma^*$, and $R \rightsquigarrow_{\mathtt{tog}(T_2)} \Sigma^*$ for $m = R, \mathtt{odd}, \mathtt{even}, T_0$ and T_2, respectively. Let us call the 5-tuple of states in $Q_R \times Q_{\mathtt{odd}} \times Q_{\mathtt{even}} \times Q_{T_0} \times Q_{T_2}$ a *snapshot*, where, for a snapshot $\sharp q$, we denote the state in Q_m as $\sharp q[m]$ for $m \in \{R, \mathtt{odd}, \mathtt{even}, T_0, T_2\}$. Let us denote the set of all snapshots as $\sharp Q$, and let us define the *snapshot transition* $\sharp\delta : \sharp Q \times \Sigma^2 \to \sharp Q$ as $\sharp\delta(\sharp q, \sigma_1\sigma_2) =$

$$(\delta_R(\sharp q[R], \sigma_1\sigma_2), \delta_{\mathtt{odd}}(\sharp q[\mathtt{odd}], \sigma_1), \delta_{\mathtt{even}}(\sharp q[\mathtt{even}], \sigma_2), \delta_{T_0}(\sharp q[T_0], \sigma_1), \delta_{T_2}(\sharp q[T_2], \sigma_2)).$$

In other words, the snapshot transition simulates each of the five components on a string $\sigma_1\sigma_2$ of length 2, where A_R is simulated on $\sigma_1\sigma_2$, $A_{\mathtt{odd}}$ and A_{T_0} are simulated on σ_1 and $A_{\mathtt{even}}$ and A_{T_2} are simulated on σ_2. Finally, we define four Boolean formulae $\varphi_{s1}, \varphi_{c1}, \varphi_{s2}, \varphi_{c2} : \sharp Q \times \{\mathtt{odd}, \mathtt{even}, T_0, T_2\}^2 \times \Sigma^2 \to \{\mathtt{True}, \mathtt{False}\}$ and one additional Boolean formula $\varphi_o : \sharp Q \times \Gamma \times \Sigma \to \{\mathtt{True}, \mathtt{False}\}$ as follows.

1. $\varphi_{s1}(\sharp q, m_1, m_2, \sigma_1\sigma_2) = \mathtt{True}$ when the following holds:
$$[(\delta_{m_1}(\sharp q[m_1], \sigma_1\sigma_2) \in F_{m_1}) \wedge (\sharp q[m_2] \in F_{m_2})] \Rightarrow [(\delta_R(\sharp q[R], \sigma_1\sigma_2)) \in F_R],$$

2. $\varphi_{c1}(\sharp q, m_1, m_2, \sigma_1\sigma_2) = \mathtt{True}$ when the following holds:
$$[(\delta_R(\sharp q[R], \sigma_1\sigma_2)) \in F_R] \Rightarrow [(\delta_{m_1}(\sharp q[m_1], \sigma_1\sigma_2) \in F_{m_1}) \wedge (\sharp q[m_2] \in F_{m_2})],$$

3. $\varphi_{s2}(\sharp q, m_1, m_2, \sigma_1\sigma_2) = \mathtt{True}$ when the following holds:
$$[(\delta_{m_1}(\sharp q[m_1], \sigma_1) \in F_{m_1}) \wedge (\delta_{m_2}(\sharp q[m_2], \sigma_2) \in F_{m_2})] \Rightarrow [(\delta_R(\sharp q[R], \sigma_1\sigma_2)) \in F_R],$$

4. $\varphi_{c2}(\sharp q, m_1, m_2, \sigma_1\sigma_2) = \mathtt{True}$ when the following holds:
$$[(\delta_R(\sharp q[R], \sigma_1\sigma_2)) \in F_R] \Rightarrow [(\delta_{m_1}(\sharp q[m_1], \sigma_1) \in F_{m_1}) \wedge (\delta_{m_2}(\sharp q[m_2], \sigma_2) \in F_{m_2})],$$

5. $\varphi_o(\sharp q, g, \sigma) = \texttt{True}$ when the following holds:

$$[g = ①] \wedge [(\delta_{\texttt{odd}}(\sharp q[\texttt{odd}], \sigma) \in F_{\texttt{odd}}) \wedge (\sharp q[\texttt{even}] \in F_{\texttt{even}})] \Rightarrow [(\delta_R(\sharp q[R], \sigma)) \in F_R] .$$

The idea is to use the Boolean formulae to check the soundness and completeness of the decomposition candidate of R with respect to the input control string. Consider a control string γ and two strings x and y with the same length. Let $\sharp q$ be the snapshot obtained by feeding $x \sqcup_T y$ to the snapshot transition, starting from the tuple $(s_R, s_{\texttt{odd}}, s_{\texttt{even}}, s_{T_0}, s_{T_2})$. For any character $\sigma \in \Sigma$, the string $(x\sigma) \sqcup_T y$ can only be obtained by shuffling $x\sigma$ and y along T_1. Therefore, if (X_γ, Y_γ) is sound and both $x\sigma \in X_\gamma$ and $y \in Y_\gamma$, then the string $(x\sigma) \sqcup_T y$ should appear in R; we check this using φ_o with the parameters $\sharp q$, $\gamma[|x| + 2]$ and σ. The completeness of odd-length strings in R are immediate from the choice of $g = \gamma[|x| + 2] = ①$, since for any odd-length string $w \in R$, there must exist two strings $x \in L(A_{\texttt{odd}})$ and $y \in L(A_{\texttt{even}})$ that satisfy $x \sqcup_T y = w$. For the strings that are shuffled along T_2 or T_0, we check the soundness using φ_{s1} or φ_{s2}, and the completeness using φ_{c1} or φ_{c2}. The choice of the Boolean formulae depends on the control characters $\gamma[|x|+1]$ and $\gamma[|x|+2]$. We explain the intuition behind the Boolean formulae for the case when $\gamma[|x|+1] = \gamma[|x|+2] =$ ①. Note that the strings in $X_\gamma \cap \Sigma^{|x|+1}$ and $X_\gamma \cap \Sigma^{|x|+2}$ are from $L(A_{\texttt{odd}})$, and the strings in $Y_\gamma \cap \Sigma^{|y|}$ and $Y_\gamma \cap \Sigma^{|y|+1}$ are from $L(A_{\texttt{even}})$. If (X_γ, Y_γ) is sound, for any two characters σ_1 and σ_2 satisfying $x\sigma_1\sigma_2 \in X_\gamma$ and $y \in Y_\gamma$, the string $(x\sigma_1\sigma_2) \sqcup_T y = (x \sqcup_T y)\sigma_1\sigma_2$ should also appear in R. This can be tested using φ_{s1}. Likewise, φ_{s2} tests the soundness of (X_γ, Y_γ) when $x\sigma_1 \in X_\gamma$ and $y\sigma_2 \in Y_\gamma$. Finally, if (X_γ, Y_γ) is complete, then $(x \sqcup_T y)\sigma_1\sigma_2 \in R$ implies either $(x\sigma_1\sigma_2, y)$ or $(x\sigma_1, y\sigma_2)$ to be in $X_\gamma \times Y_\gamma$. This can be tested using a disjunction of the formulae φ_{c1} and φ_{c2}. We can apply a similar intuition to other choices of $\gamma[|x| + 1]$ and $\gamma[|x| + 2]$ as well.

We are now ready to construct the automaton accepting the valid control strings of a given regular language R. The *decomposition automaton* is an alternating Büchi automaton $\texttt{D}(R) = (Q_D, \Gamma, \delta_D, s_D, F_D)$, where $Q_D = \Gamma^{\leq 2} \cup (\sharp Q \times \Gamma^2) \cup \{\texttt{acc}, \texttt{snk}\}$ is the set of states, $s_D = \lambda \in \Gamma^{\leq 2}$ is the start state, and $F_D = Q_D \setminus \{\texttt{snk}\}$ is the set of accepting states. Finally, for an input control character $g \in \Gamma$ the transition function is defined as follows. For a state $z \in \Gamma^{\leq 1}$, we have $\delta_D(z, g) = zg$. For a state $z \in \Gamma^2$, we have $\delta_D(z, g) = ((s_R, s_{\texttt{odd}}, s_{\texttt{even}}, s_{T_0}, s_{T_2}), z[2]g)$ if $[\lambda \in R \Leftrightarrow \lambda \in X_z \sqcup_{T_{\leq 2}} Y_z]$ for the decomposition candidate (X_z, Y_z) of R with respect to z, otherwise we have $\delta_D(z, g) = \texttt{snk}$. For a pair $(\sharp q, z) \in \sharp Q \times \Gamma^2$, we have $\delta_D((\sharp q, z), g) = (\bigwedge_{\sigma_1\sigma_2 \in \Sigma^2} N_{\sigma_1\sigma_2}^{(\sharp q, z, g)}) \wedge (\bigwedge_{\sigma \in \Sigma} M_\sigma^{(\sharp q, z, g)})$, where $N_{\sigma_1\sigma_2}^{(\sharp q, z, g)} = (\sharp\delta(\sharp q, \sigma_1\sigma_2), z[2]g)$ if the soundness and completeness conditions are both true for each corresponding case in the following table:

z		Conditions	
$z[0]$	$z[1]$	Soundness	Completeness
①	①	$\varphi_{s1}(\sharp q, \texttt{odd}, \texttt{even}, \sigma_1\sigma_2)$ $\wedge\varphi_{s2}(\sharp q, \texttt{odd}, \texttt{even}, \sigma_1\sigma_2)$	$\varphi_{c1}(\sharp q, \texttt{odd}, \texttt{even}, \sigma_1\sigma_2)$ $\vee\varphi_{c2}(\sharp q, \texttt{odd}, \texttt{even}, \sigma_1\sigma_2)$
	②	$\varphi_{s1}(\sharp q, T_0, \texttt{even}, \sigma_1\sigma_2)$	$\varphi_{c1}(\sharp q, T_0, \texttt{even}, \sigma_1\sigma_2)$
	③	$\varphi_{s2}(\sharp q, \texttt{odd}, T_2, \sigma_1\sigma_2)$	$\varphi_{c2}(\sharp q, \texttt{odd}, T_2, \sigma_1\sigma_2)$
②	①	$\varphi_{s2}(\sharp q, T_0, \texttt{even}, \sigma_1\sigma_2)$	$\varphi_{c2}(\sharp q, T_0, \texttt{even}, \sigma_1\sigma_2)$
	②	True	$[\delta_R(\sharp q[R], \sigma_1\sigma_2) \notin F_R]$
	③	$\varphi_{s2}(\sharp q, T_0, T_2, \sigma_1\sigma_2)$	$\varphi_{c2}(\sharp q, T_0, T_2, \sigma_1\sigma_2)$
③	①	$\varphi_{s1}(\sharp q, \texttt{odd}, T_2, \sigma_1\sigma_2)$	$\varphi_{c1}(\sharp q, \texttt{odd}, T_2, \sigma_1\sigma_2)$
	②	$\varphi_{s1}(\sharp q, T_0, T_2, \sigma_1\sigma_2)$	$\varphi_{c1}(\sharp q, T_0, T_2, \sigma_1\sigma_2)$
	③	True	$[\delta_R(\sharp q[R], \sigma_1\sigma_2) \notin F_R]$

otherwise we have $N_{\sigma_1\sigma_2}^{(\sharp q,z,g)} = \texttt{snk}$, and $M_{\sigma}^{(\sharp q,z,g)}$ is defined as $\texttt{acc}$ if $\varphi_o(\sharp q, z[1], \sigma)$ is True, otherwise $M_{\sigma}^{(\sharp q,z,g)} = \texttt{snk}$. Finally, we have $\delta_D(\texttt{acc}, g) = \texttt{acc}$ and $\delta_D(\texttt{snk}, g) = \texttt{snk}$. We claim that the decomposition automaton accepts the valid control strings γ of R, which we prove in Lemma 9.

Lemma 9. *For a regular language R and a control string γ of R, we have $\gamma \in L(\mathtt{D}(R))$ if and only if the control string γ is valid.*

For our example language $R_C = \{(abba)^k u \mid k \geq 0, u \in \{\lambda, a, ab\}\}$, let $\mathtt{D}(R_C)$ be its decomposition automaton. Following the construction, we consider a valid infinite control string $\gamma_1 =$ ① (① ②)$^\omega$. Then, the resulting decomposition candidate $(X_{\gamma_1}, Y_{\gamma_1})$ with respect to γ_1 is the pair of infinite regular languages: $X_{\gamma_1} = \{(ab)^i(\lambda + a) \mid i \geq 0\}$, $Y_{\gamma_1} = \{(ba)^i \mid i \geq 0\}$. Likewise, if we select another valid infinite control string then $\gamma_2 =$ ① (① ③)$^\omega$, the corresponding decomposition candidate $(X_{\gamma_2}, Y_{\gamma_2})$ with respect to γ_2 is also the pair of infinite regular languages: $X_{\gamma_2} = \{(ab)^i a \mid i \geq 0\}$, $Y_{\gamma_2} = \{(ba)^i(\lambda + b) \mid i \geq 0\}$. In both cases, we observe that $X_{\gamma_1} ⧢_{T_{\leq 2}} Y_{\gamma_1} = X_{\gamma_2} ⧢_{T_{\leq 2}} Y_{\gamma_2} = R_C$.

Then, we can decide whether or not R has a non-trivial decomposition with respect to the 2-comet shuffle through an emptiness test on $\mathtt{D}(R)$. We now explain how to construct a pair of regular languages (X, Y) satisfying $X ⧢_{T_{\leq 2}} Y = R$ when R has a non-trivial decomposition. From Lemma 9, we have an alternating Büchi automaton that accepts all the valid control strings for R. We apply Proposition 2 on the alternating Büchi automaton $\mathtt{D}(R)$ and obtain an equivalent Büchi automaton B. If $L(\mathtt{D}(R))$ is non-empty, then there must exist a cycle including an accepting state in B that is reachable from the start state of B from Proposition 3; let us denote the concatenation of the labels of the shortest non-empty path as γ_p and the concatenation of the labels in the cycle starting at the state where the shortest non-empty path ends as γ_r. Clearly, γ_p and γ_r are finite strings over Γ and can be computed from B. Furthermore, the string $\gamma_p(\gamma_r)^\omega$ must be accepted by $L(\mathtt{D}(R))$. We consider the valid control string $\gamma = \gamma_p(\gamma_r)^\omega$ and construct a pair of automata each for X_γ and Y_γ.

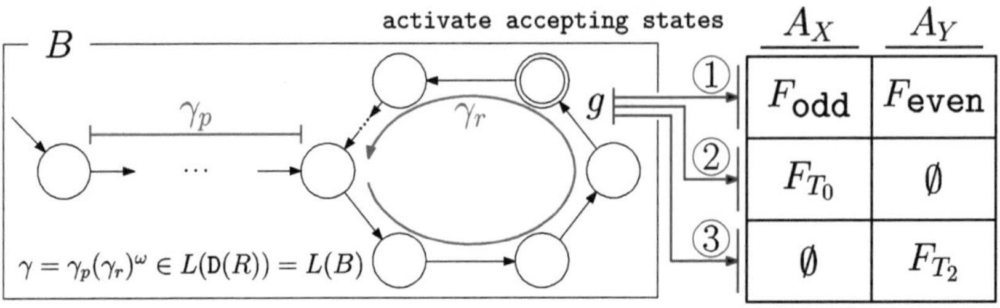

Fig. 4. An illustration of the construction for the shuffle decomposition of R. The automaton B is the Büchi automaton equivalent to $\mathtt{D}(R)$.

For the automaton for X_γ, the idea is to simulate the transition on the cycle and the path leading up to the cycle in B using a counter that grows up to $|\gamma_p\gamma_r|$. At each computation step, we use the counter to retrieve the current control character from $\gamma_p\gamma_r$, and then choose accepting states between F_{odd}, F_{T_0} and none. Formally, let $A_X = (Q_X, \Sigma, \delta_X, s_X, F_X)$ be a DFA defined as follows. The set Q_X of states is defined as $Q_{\mathsf{odd}} \times Q_{T_0} \times \{1, 2, \ldots, |\gamma_p\gamma_r|\}$, and the start state s_X is $(s_{\mathsf{odd}}, s_{T_0}, 1)$. Then, the transition function δ_X is defined as

$$\delta_X((q_{\mathsf{odd}}, q_{T_0}, i), \sigma) = (\delta_{\mathsf{odd}}(q_{\mathsf{odd}}, \sigma), \delta_{T_0}(q_{T_0}, \sigma), j),$$

where $j = i + 1$ if $i < |\gamma_p\gamma_r|$ and $j = |\gamma_p| + 1$ otherwise. In other words, we simultaneously simulate A_{odd} and A_{T_0} on the input string while maintaining a counter that increases by 1 for each consumed character. This counter starts at 1 and is reset to $|\gamma_p| + 1$ once it exceeds $|\gamma_p\gamma_r|$. Therefore, for any string $w \in \Sigma^*$, if the integer i satisfies $\delta_X^*(s_X, w) = (q_{\mathsf{odd}}, q_{T_0}, i)$, then we also have $\gamma[|w| + 1] = \gamma[i] = (\gamma_p\gamma_r)[i]$. Since γ_p is non-empty, all transitions out of a state with counter $|\gamma_p\gamma_r|$ will reset the counter to a value greater than 1; therefore the state $(q_{\mathsf{odd}}, q_{T_0}, 1)$ for any $q_{\mathsf{odd}} \in Q_{\mathsf{odd}}$ and $q_{T_0} \in Q_{T_0}$ does not have any in-transitions.

Let us now recall the definition of X_γ; the set of strings in X_γ of a particular length $k+1$ is one of $\mathsf{odd}(R \cap \Sigma^{2k+1})$, $(R \rightsquigarrow_{T_0} \Sigma^*) \cap \Sigma^{k+1}$ or $\emptyset$, where the choice depends solely on $\gamma[k+1]$. Then, the idea is to choose between the three cases using F_X as follows. Consider a string $x \in \Sigma^*$. If $x = \lambda$, then $x \in L(A_X)$ if and only if $(\gamma_p\gamma_r)[1] =$ ①. Recall that $\lambda \in X_\gamma$ if and only if $\gamma[1] =$ ①. Therefore, we have $L(A_X) \cap \Sigma^0 = X_\gamma \cap \Sigma^0$. For $x \neq \lambda$, since A_X is a DFA, there must be only one state $(q_{\mathsf{odd}}, q_{T_0}, i)$ that satisfies $\delta_X^*(s_X, x) = (q_{\mathsf{odd}}, q_{T_0}, i)$. By the definition of X_γ, we have $X_\gamma \cap \Sigma^{|x|} = L(A_{\mathsf{odd}}) \cap \Sigma^{|x|}$ if $\gamma[|x| + 1] =$ ①,

$X_\gamma \cap \Sigma^{|x|} = L(A_{T_0}) \cap \Sigma^{|x|}$ if $\gamma[|x|+1] = ②$, and $X_\gamma \cap \Sigma^{|x|} = \emptyset$ if $\gamma[|x|+1] = ③$. In other words, the set F_X of accepting states is the union of

$$\{(q_{\text{odd}}, q_{T_0}, i) \mid i \geq 2, (\gamma_p\gamma_r)[i] = ①, q_{\text{odd}} \in F_{\text{odd}}\},$$

$$\{(q_{\text{odd}}, q_{T_0}, i) \mid i \geq 2, (\gamma_p\gamma_r)[i] = ②, q_{T_0} \in F_{T_0}\}, \text{ and}$$

$$\{(s_{\text{odd}}, s_{T_0}, 1) \mid (\gamma_p\gamma_r)[1] = ①\}.$$

Note that for all states with counter $i \geq 2$, its membership in F_X is the same as the membership of the current state in A_{odd} if $\gamma[i] = ①$, or the membership of the current state in A_{T_0} if $\gamma[i] = ②$. No states are accepting if $\gamma[i] = ③$. Thus, $x \in X_\gamma$ if and only if $x \in L(A_X)$, and we have $L(A_X) = X_\gamma$. See Fig. 4 for an illustration of the construction.

We can construct a DFA A_Y for Y_γ in an analogous manner, and establish the following result.

Lemma 10. *If there exists a decomposition of R with respect to $T_{\leq 2}$, then there exists a decomposition (X, Y) of R with respect to $T_{\leq 2}$, where X and Y are both regular. Furthermore, X and Y are effectively constructable.*

Theorem 2. *Given a regular language R, we can decide whether or not there exists a decomposition (X, Y) of R with respect to $T_{\leq 2}$, and if there exists such a decomposition, compute a regular decomposition.*

5 Conclusions

We have solved the language decomposition problem under shuffle along trajectories for the trajectory sets $T_P = L((01)^* + (10)^*)$ and $T_{\leq 2} = L((01)^*0^{\leq 2})$. In both cases, we have obtained algorithms that decide whether or not there is a decomposition for a given regular language R. Furthermore, we have constructed finite-state automata for the decomposition of R in the case when R has a decomposition. This result is in-line with the conjecture that if a regular language has a decomposition under shuffle along regular trajectories, then it has a decomposition whose languages are both regular.

There remain some open problems related to our work. We note that we have considered trajectories that are 2-thin. In general, the decidability of the decomposition problem for 2-thin regular trajectories is still open [8]. Also, it is natural to consider Problem 3 for an arbitrary k. We remark that it is difficult to directly adapt the control string technique to k-comet trajectories with $k \geq 3$. The reason is that there is no easy bound on the number of different $X \cap \Sigma^i$'s and $Y \cap \Sigma^i$'s over all possible decompositions (X, Y) with respect to $T_{\leq k}$, and we can no longer express each case using a finite control alphabet. Pushing k to the limit, the shuffle decomposition problem under the unbounded comet trajectory $T = L((01)^*0^*)$ is also an interesting variant. This is a restricted case for the initial literal shuffle, which is also a trajectory set whose decomposition problem remains open [7].

Acknowledgements. This research was supported by the NRF grant (RS-2025-00562134) funded by the Korean government, and the Natural Sciences and Engineering Research Council of Canada (NSERC).

References

1. Andrasiu, M., Paun, G., Dassow, J., Salomaa, A.: Language-theoretic problems arising from Richelieu cryptosystems. Theoret. Comput. Sci. **116**(2), 339–357 (1993)
2. Björklund, H., Bojanczyk, M.: Shuffle expressions and words with nested data. In: Proceedings of the 32nd International Symposium on Mathematical Foundations of Computer Science, MFCS, pp. 750–761 (2007)
3. Brzozowski, J.A., Cohen, R.S.: On decompositions of regular events. J. ACM **16**(1), 132–144 (1969)
4. Câmpeanu, C., Salomaa, K., Yu, S.: Tight lower bound for the state complexity of shuffle of regular languages. J. Autom. Lang. Comb. **7**(3), 303–310 (2002)
5. Domaratzki, M.: Deletion along trajectories. Theoret. Comput. Sci. **320**(2–3), 293–313 (2004)
6. Domaratzki, M., Salomaa, K.: State complexity of shuffle on trajectories. J. Autom. Lang. Comb. **9**(2/3), 217–232 (2004)
7. Domaratzki, M., Salomaa, K.: Decidability of trajectory-based equations. Theoret. Comput. Sci. **345**(2–3), 304–330 (2005)
8. Domaratzki, M., Salomaa, K.: Restricted sets of trajectories and decidability of shuffle decompositions. Int. J. Found. Comput. Sci. **16**(5), 897–912 (2005)
9. Domaratzki, M., Salomaa, K.: On language decompositions and primality. In: Rainbow of Computer Science - Dedicated to Hermann Maurer on the Occasion of His 70th Birthday, pp. 63–75 (2011)
10. Emerson, E.A., Lei, C.: Temporal model checking under generalized fairness constraints. In: Proceedings of the 18th Annual Hawaii International Conference on System Sciences (1985)
11. Emerson, E.A., Lei, C.: Modalities for model checking: branching time logic strikes back. Sci. Comput. Program. **8**(3), 275–306 (1987)
12. Garg, V.K., Ragunath, M.T.: Concurrent regular expressions and their relationship to Petri nets. Theoret. Comput. Sci. **96**(2), 285–304 (1992)
13. Gelade, W., Martens, W., Neven, F.: Optimizing schema languages for XML: numerical constraints and interleaving. SIAM J. Comput. **38**(5), 2021–2043 (2009)
14. Ginsburg, S., Spanier, E.H.: Mappings of languages by two-tape devices. J. ACM **12**(3), 423–434 (1965)
15. Gischer, J.L.: Shuffle languages, Petri nets, and context-sensitive grammars. Commun. ACM **24**(9), 597–605 (1981)
16. Han, Y.S., Salomaa, A., Salomaa, K., Wood, D., Yu, S.: On the existence of prime decompositions. Theoret. Comput. Sci. **376**(1–2), 60–69 (2007)
17. Jedrzejowicz, J., Szepietowski, A.: Shuffle languages are in P. Theoret. Comput. Sci. **250**(1–2), 31–53 (2001)
18. Kuhlmann, M., Satta, G.: Treebank grammar techniques for non-projective dependency parsing. In: Proceedings of the 12th Conference of the European Chapter of the Association for Computational Linguistics, EACL, pp. 478–486. The Association for Computer Linguistics (2009)

19. Mateescu, A., Rozenberg, G., Salomaa, A.: Shuffle on trajectories: Syntactic constraints. Theoret. Comput. Sci. **197**(1–2), 1–56 (1998)
20. Miyano, S., Hayashi, T.: Alternating finite automata on ω-words. Theoret. Comput. Sci. **32**(3), 321–330 (1984)
21. Paun, G., Salomaa, A.: Thin and slender languages. Discret. Appl. Math. **61**(3), 257–270 (1995)

Pattern Matching Under $\mathcal{R}$-Congruence

Sungmin Kim, Hyundong Jin, and Yo-Sub Han(✉)

School of Computing, Yonsei University, 50 Yonsei-Ro, Seodaemun-Gu, Seoul 03722, Republic of Korea
{rena_rio,tuzi04,emmous}@yonsei.ac.kr

Abstract. For an integer parameter k, two strings are congruent under the Simon's congruence $\sim_k$ if they have the same set of subsequences of length at most k. The $\mathcal{R}$-congruence is a refinement of Simon's congruence; two strings are defined to be $\sim_k^{\mathcal{R}}$ -congruent if each prefix of one string is congruent to some prefix of the other string under $\sim_k$. This congruence relation is a crucial characterization of languages whose syntactic monoids are $\mathcal{R}$-trivial. We study the $\mathcal{R}$-congruence pattern matching problem that asks to find substrings of a given text $\mathtt{T}$ that are $\mathcal{R}$-congruent to a given pattern $\mathtt{P}$. We first present an optimal linear-time algorithm that tests the $\mathcal{R}$-congruence between two strings. Then, we propose an $O(|\mathtt{T}||\Sigma|k)$ time algorithm for a given parameter k, and an $O(|\mathtt{T}||\mathtt{P}|)$ time algorithm that considers all parameters $k \leq |\mathtt{P}|$.

Keywords: Pattern matching · Subsequences · $\mathcal{R}$-congruence · Simon's congruence

1 Introduction

A string u is a *subsequence* of a string w if u can be obtained by deleting characters from w. For example, the string abc is a subsequence of the string $aabac$. Then, for a given integer k, two strings are $\sim_k$ -congruent if they have the same set of subsequences of length at most k, where the length of a string is the number of characters in a string. This congruence is called the *Simon's congruence*, and has been studied from the language theoretical [13,15,16,21,22,24,27,28], algorithmic [1,4,5,8,9,14,17,23], and string-combinatorial perspectives [6,7]. Simon [27] introduced the Simon's congruence as a characterization of the $\mathcal{J}$-trivial monoids, where $\mathcal{J}$ is one of Green's relations [10]. He showed that a language is closed with respect to $\sim_k$ for some integer parameter k if and only if its syntactic monoid is $\mathcal{J}$-trivial [27].

For the remaining two relations $\mathcal{R}$ and $\mathcal{L}$ among the Green's relations [10], Brzozowski and Fich [2] introduced two refinements of $\sim_k$ that characterize exactly the languages whose syntactic monoids are $\mathcal{R}$- and $\mathcal{L}$-trivial, respectively. For an integer $k \geq 0$, the relation $\sim_k^{\mathcal{R}}$ is a congruence over Σ^*, where two strings u and v over Σ satisfy $u \sim_k^{\mathcal{R}} v$ when for each prefix u' of u, there exists a prefix v' of v that satisfies $u' \sim_k v'$, and for each prefix v' of v, there exists

T. J. Smith (Ed.): CIAA 2026, LNCS 16695, pp. 108–123, 2026.
https://doi.org/10.1007/978-3-032-31176-4_8

a prefix u' of u that satisfies $v' \sim_k u'$. This relation is called the $\mathcal{R}$-*congruence.* For example, for two string $x = abab$ and $y = baba$ and an integer $k = 2$, we have $x \sim_k y$ because every string of length 2 appears as subsequences of both x and y. However, x and y are not $\mathcal{R}$-congruent since there is no prefix of y that is $\sim_2$ -congruent to the prefix a of x. The $\mathcal{L}$-congruence is defined symmetrically, where the suffixes of u and v are $\sim_k$ -congruent instead of the prefixes of u and v. Then, similar to the Simon's congruence characterization of the $\mathcal{J}$-trivial languages, the syntactic monoid of a language is $\mathcal{R}$-trivial if and only if the language is closed under $\sim_k^{\mathcal{R}}$ for some integer k [2].

While there are several research articles on $\mathcal{R}$-trivial languages and their characterizations [3,12,20,21], there is little work on the algorithmic problems under $\sim_k^{\mathcal{R}}$. For example, after Brzozowski and Fich [2] proposed an exponential runtime algorithm that tests the $\sim_k^{\mathcal{R}}$ -congruence between two strings, there has been little improvement since. Since the congruence testing and pattern matching problems for the Simon's congruence are studied well [1,9,14,17], it is natural to consider similar problems for the $\mathcal{R}$-congruence as well.

Another motivation for examining the $\mathcal{R}$-congruence is related to the recent deep-learning architecture—Transformer [29]—the baseline architecture for large-language models such as ChatGPT and Gemini. Recently, Li and Cotterell [19] proved that the Transformer with fixed-precision and no positional encodings as a language acceptor recognizes exactly the languages whose syntactic monoid is $\mathcal{R}$-trivial. This characterization of the Transformer motivates us to design efficient algorithms for $\mathcal{R}$-trivial languages and the $\mathcal{R}$-congruence.

Given a string w, let $|w|$ be the length of w and $w[i]$ be the $(i+1)^{\text{th}}$ character for positions $0 \leq i \leq |w|-1$—then, $w = w[0]w[1] \cdots w[|w|-1]$. For two positions i and j with $i \leq j$, let $w[i:j]$ denote the substring $w[i]w[i+1] \cdots w[j]$. We consider the following three problems:

Problem 1 (Testing $\mathcal{R}$-congruence). Given two strings w and v and an integer k, decide whether or not $w \sim_k^{\mathcal{R}} v$.

Problem 2 (Pattern matching under $\mathcal{R}$-congruence). Given a text $\mathtt{T}$, a pattern $\mathtt{P}$ and an integer k, compute all position pairs (i, j) of $\mathtt{T}$ that satisfy $\mathtt{P} \sim_k^{\mathcal{R}} \mathtt{T}[i:j]$.

Problem 3 (Complete pattern matching under $\mathcal{R}$-congruence). Given a text $\mathtt{T}$ and a pattern $\mathtt{P}$, compute for each position pair (i, j) of $\mathtt{T}$ the maximum k allowing $\mathtt{P} \sim_k^{\mathcal{R}} \mathtt{T}[i:j]$.

Notice that while k is part of the input for Problems 1 and 2, we can assume without loss of generality that $k \leq |\mathtt{P}|$. This is because two strings with length less than k are $\sim_k^{\mathcal{R}}$ -congruent if and only if the two strings are exactly the same, just as in Simon's congruence. Furthermore, we remark that for Simon's congruence, the testing problem has an optimal linear-time algorithm [1], and the matching problem has an $O(|\Sigma||\mathtt{T}|(|\Sigma|^2 + k))$ time solution [17], where $|\Sigma|$ denotes the cardinality of Σ.

In Sect. 2, we define some notations and recall important tools that are used in our algorithms. We solve Problem 1 using an $O(|w| + |v|)$ time algorithm in

Sect. 3.1, where we combine the approaches of Brzozowski and Fich [2] and Fleischer and Kufleitner [5]. Then, we propose an $O(|\Sigma||\mathtt{P}| + |\Sigma||\mathtt{T}|k)$ time algorithm for Problem 2 using a filtering strategy in Sect. 3.2, where we split the pattern into a hierarchy of at most $|\Sigma|$ patterns. Finally, for Problem 3, we present an $O(|\mathtt{T}||\mathtt{P}|)$ time algorithm using an incremental matching strategy in Sect. 3.3. Finally, we recap our results in Sect. 4.

2 Preliminaries

Let $\chi_{\mathtt{True}}$ be an indicator function that maps `True` to 1 and `False` to 0. For an integer i, we denote the set $\{1, 2, \ldots, i\}$ as $[i]$, and the set $[i] \cup \{0\}$ as $[i]_0$. Without loss of generality, let us assume that the alphabet Σ is totally ordered. The *empty string* denoted as λ is a string of length 0. For two positions i and j with $i > j$, the string $w[i : j]$ corresponds to the empty string. A string u is a *prefix* or *suffix* of a string w if $u = w[0 : |u| - 1]$ or $u = w[|w| - |u| : |w| - 1]$, respectively. For a string w, let us denote the prefix of w obtained by deleting the final character of w as $\mathtt{pop}(w)$. For a string w over Σ, the *underlying alphabet* $\mathtt{alph}(w)$ of w is the set of characters in Σ that occur in w. We denote the set of subsequences of w as $\mathbb{S}(w)$, and the set of subsequences of length at most k as $\mathbb{S}_k(w)$.

Given a text $\mathtt{T}$, a position i of $\mathtt{T}$, and a character $\sigma \in \mathtt{alph}(\mathtt{T})$, we denote the smallest position $j > i$ that satisfies $\mathtt{T}[j - 1] = \sigma$ as $R_X(\mathtt{T}, i, \sigma)$. If such a position does not exist, we define $R_X(\mathtt{T}, i, \sigma) = \infty$. This function R_X is called an *X-ranker* [5,26,30]. For a given text $\mathtt{T}$, one can compute $R_X(\mathtt{T}, i, \sigma)$ for all positions i of $\mathtt{T}$ and characters $\sigma \in \mathtt{alph}(\mathtt{T})$ in overall $O(|\Sigma||\mathtt{T}|)$ time [5]. For two strings u and v where $u = v\sigma$ for a character σ, let us denote the sets $\mathbb{S}_k(v\sigma) \setminus \mathbb{S}_k(v)$ and $\mathbb{S}(v\sigma) \setminus \mathbb{S}(v)$ as $\mathbb{D}(u)$ and $\mathbb{D}(u)$, respectively. For a string w and a position i of w, the *X-coordinate* of w at position i is the length of the shortest string in $\mathbb{D}(w[0 : i])$. We denote the X-coordinate of w at position i as $X(w, i)$. For a given string w over Σ, Fleischer and Kufleitner [5] proposed an $O(|w||\Sigma|)$-time algorithm that computes the X-coordinates at each position of w, which was improved by Barker et al. [1] to run in optimal $O(|w|)$ time. We note that the X-coordinate of a position i of w is defined in the literature as the minimum number of recursive calls to the X-ranker to reach exactly i when starting from 0. However, we use the alternative definition of the X-coordinate that depends on $\mathbb{D}(w[0 : i])$ to emphasize the connection between the X-coordinates and the subsequences of prefixes of w.

The *alphabetical signature* of a string w is the subsequence of w containing at most one occurrence of each character in the order of their first appearances in w. For example, the alphabetical signature of *aabac* is *abc*. We denote the alphabetical signature of w as $\mathtt{sig}(w)$. The name of the alphabetical signature comes from the *subsequence universality signature* of a string defined by Schnoebelen and Veron [25], which uses $\mathtt{sig}(w)$ as a component. For a string w, the *arch factorization* of w is a unique decomposition of w defined by the following procedure. Starting from position 0, we repeatedly remove the length-minimal prefix of w that contains all characters in $\mathtt{alph}(w)$ until the remaining string does not contain all characters in $\texttt{alph}(w)$. The ith prefix removed in this procedure is the ith *arch* of w, which we denote as $\texttt{ar}_i(w)$. Then, the remaining string is called the *rest* of w, which we denote as $\texttt{rest}(w)$. The number of arches of a string w is called the *universality index* $\iota(w)$ of w, which is the maximum integer j that allows $\texttt{alph}(w)^j \subseteq \mathbb{S}(w)$. Furthermore, we denote the concatenation of the ith to j^{th} arches of w for $i \leq j \leq \iota(w)$ as $\texttt{ar}_{i..j}(w)$. Finally, we call the string of length $\iota(w)$ obtained by concatenating the last characters of each arch of w as the *modus* of w, which we denote as $\texttt{modus}(w)$. This factorization scheme has been introduced by Hébrard [11]. Note that we define the arch factorization not in terms of the alphabet Σ but the underlying alphabet $\texttt{alph}(w)$ of w. Then, Kim et al. [17] defined the *X-tree*, which is an extension of the arch factorization to the substrings of a given text. Formally, an X-tree $T_X(\texttt{T})$ of a text $\texttt{T}$ is a tree defined by the followings. First, the root node is labeled ∞. All other nodes are labeled with distinct integers in $[|\texttt{T}|]_0$. With slight abuse of notation, we refer to the node of an X-tree by its label. Second, there exists a node i if and only if there exists some integer $j < i$ where $\texttt{T}[j : i-1]$ is a single arch without a rest with respect to $\texttt{alph}(\texttt{T})$. Finally, there exists an edge between two nodes i and j, where $j < i$, if $\texttt{T}[j : i-1]$ is a single arch without a rest with respect to $\texttt{alph}(\texttt{T})$. We denote such an edge as $\langle j, i\rangle$, and call the higher and lower value between i and j as the parent and child, respectively. The authors prove that there are $|\texttt{alph}(\texttt{T})|\iota(\texttt{T})$ edges in an X-tree, which is in turn bounded by $|\texttt{T}|$. Furthermore, they proved that the X-tree can be computed in $O(|\texttt{T}||\Sigma|)$ time.

3 Pattern Matching Under $\mathcal{R}$-Congruence

Before we tackle the pattern matching problems, we briefly discuss an algorithm that tests the $\mathcal{R}$-congruence between two strings in Sect. 3.1. Then, we move on to the matching problems, where we consider the setting when k is given in Sect. 3.2, and the setting where we consider all possible $k \leq |\texttt{P}|$ in Sect. 3.3. Note that we can assume that Σ is the set of characters that occur both in $\texttt{P}$ and $\texttt{T}$, since for two strings to be $\sim_k^{\mathcal{R}}$ -congruent, the set of characters in each string must be the same. Especially, for Problems 2 and 3, we can first slice the text into length-maximal substrings that use exactly the characters of the pattern and then solve the problem for each of these substrings. For the sake of clarity, we reiterate some notations we frequently use throughout Sect. 3 in Table 1.

3.1 Testing $\mathcal{R}$-Congruence

We first recall a useful characterization of the congruence classes with respect to $\sim_k^{\mathcal{R}}$. Brzozowski and Fich [2] discovered that for any congruence class modulo $\sim_k^{\mathcal{R}}$, the length-minimal string is unique and can be obtained by deleting characters from any string in the congruence class.

Proposition 1 (Brzozowski and Fich [2]). *For a string w and a position i of w, we have $\mathbb{S}_k(w[0:i-1]) = \mathbb{S}_k(w[0:i])$ if and only if $w \sim_k^{\mathcal{R}} z$, where z is the string obtained by deleting $w[i]$ from w. If no position satisfies the above, then w is uniquely length-minimal in its congruence class with respect to $\sim_k^{\mathcal{R}}$.*

For example, consider the example string $w_e = abaab$ and $k = 2$; since $\mathbb{S}_2(w_e[0:2]) = \mathbb{S}_2(w_e[0:3]) = \{\lambda, a, b, aa, ab, ba\}$, we have $abaab \sim_2^{\mathcal{R}} abab$. For a string w, let us denote the unique shortest string among all strings that are $\sim_k^{\mathcal{R}}$ -congruent to w as $\texttt{Short}_k(w)$. Based on Proposition 1, the authors proposed an algorithm that tests $w \sim_k^{\mathcal{R}} v$ between two strings w and v by first computing $\texttt{Short}_k(w)$ and $\texttt{Short}_k(v)$, and then testing equivalence of the two strings. However, their method explicitly constructs $\mathbb{S}_k(w_p)$ for each prefix w_p of w, which incurs an exponential running time. We propose an alternative algorithm that constructs $\texttt{Short}_k(w)$ using the X-coordinates. Since the X-coordinate for each position i of w is the length of the shortest string in $\mathbb{S}(w[0:i]) \setminus \mathbb{S}(w[0:i-1])$, we observe that $\mathbb{S}_k(w[0:i]) = \mathbb{S}_k(w[0:i-1])$ whenever $X(w,i) > k$.

Corollary 1. *For a string w and a position i of w, we have $X(w,i) > k$ if and only if $w \sim_k^{\mathcal{R}} w[0:i-1]w[i+1:|w|-1]$.*

Therefore, by Corollary 1, we can delete $w[i]$ for positions with $X(w,i) > k$ to obtain a shorter string that is $\sim_k^{\mathcal{R}}$ -congruent to w. Indeed, in our example $w_e = abaab \sim_2^{\mathcal{R}} abab$, we confirm that $X(w_e, 3) = 3 > 2$. Furthermore, Lemma 1 allows us to re-use the X-coordinates after a single delete. We can then compute $\texttt{Short}_k(w)$ from w by first computing all the X-coordinates in $O(|w|)$ time, and then deleting all characters at the positions with X-coordinates greater than k without the need to re-compute the X-coordinates.

Lemma 1. *For a string w and a position i of w, let $z = w[0:i-1]w[i+1:|w|-1]$ be $\sim_k^{\mathcal{R}}$ -congruent to w. For any position $j \neq i$ of w, we have $X(w,j) > k$ if and only if $X(z, j - \chi_{\texttt{True}}) > k$.*

Proposition 2. *Given two strings w and v and an integer k, we can decide whether or not $w \sim_k^{\mathcal{R}} v$ in $O(|w| + |v|)$ time.*

3.2 Matching for a Given *k*

We now solve Problem 2, which is the pattern matching problem for a given integer k. Since we consider a single k, we assume without loss of generality that $\texttt{P} = \texttt{Short}_k(\texttt{P})$ throughout Sect. 3.2. Then, we either have $\iota(\texttt{P}) < k$, or $\texttt{P}$ has no rest with $\iota(\texttt{P}) = k$. Observe that two strings with different lengths can be $\sim_k^{\mathcal{R}}$ -congruent to each other. Therefore, a naive algorithm that tests the $\mathcal{R}$-congruence between the pattern and each substring of $\texttt{T}$ takes $O(|\texttt{T}|^3)$ time, where we spend linear time for a quadratic number of substrings of $\texttt{T}$. This algorithm can be improved into $O(|\texttt{T}|^2)$ by performing the X-coordinate computation only once per a suffix of $\texttt{T}$, and checking the $\sim_k^{\mathcal{R}}$ -congruence in an incremental manner. However, we still need to consider all starting positions of the string, which is highly inefficient. Our strategy is to match the pattern starting from the middle of the pattern so that we do not need to consider every suffix of $\texttt{T}$. Therefore, we characterize the $\sim_k^{\mathcal{R}}$ -congruence between two strings with respect to the congruence between their substrings in the following lemmas.

Lemma 2. *For two strings u and v and an integer $k \geq 2$, we have $u \sim_k^{\mathcal{R}} v$ if and only if $\mathtt{ar}_1(u) \sim_k^{\mathcal{R}} \mathtt{ar}_1(v)$ and $u[i : |u|-1] \sim_{k-1}^{\mathcal{R}} v[j : |v|-1]$, where $i = |\mathtt{ar}_1(u)|$ and $j = |\mathtt{ar}_1(v)|$.*

Lemma 3. *For a pattern $\mathtt{P}$, a string w and an integer k, we have $\mathtt{P} \sim_k^{\mathcal{R}} w$ if and only if the following four conditions are true: $\mathtt{pop}(\mathtt{ar}_1(\mathtt{P})) \sim_k^{\mathcal{R}} \mathtt{pop}(\mathtt{ar}_1(w))$, $\mathtt{modus}(\mathtt{P}) = \mathtt{modus}(w)$, $\mathtt{ar}_{2..\iota(\mathtt{P})}(\mathtt{P}) \sim_{k-1}^{\mathcal{R}} \mathtt{ar}_{2..\iota(w)}(w)$, and $\mathtt{rest}(\mathtt{P}) \sim_{k-\iota(\mathtt{P})}^{\mathcal{R}} \mathtt{rest}(w)$.*

Consider the example strings

$$\mathtt{P}_e = abacaabcc \text{ and } w_e = abacaaabccc$$

and the integer $k = 3$. Since $\mathtt{Short}_k(\mathtt{P}_e) = \mathtt{Short}_k(w_e) = abacaabcc$, we have $\mathtt{P}_e \sim_2^{\mathcal{R}} w_e$. We confirm that each condition in Lemma 3 holds; we have

$$\mathtt{pop}(\mathtt{ar}_1(\mathtt{P}_e)) = aba \sim_3^{\mathcal{R}} aba = \mathtt{pop}(\mathtt{ar}_1(w_e)),\ \mathtt{modus}(\mathtt{P}_e) = \mathtt{modus}(w_e) = cc,$$

$$\mathtt{ar}_2(\mathtt{P}_e) = aabc \sim_2^{\mathcal{R}} aaabc = \mathtt{ar}_2(w_e) \text{ and } \mathtt{rest}(\mathtt{P}_e) = c \sim_1^{\mathcal{R}} cc = \mathtt{rest}(w_e).$$

Note that the conditions $\mathtt{pop}(\mathtt{ar}_1(\mathtt{P})) \sim_k^{\mathcal{R}} \mathtt{pop}(\mathtt{ar}_1(w))$ and $\mathtt{rest}(\mathtt{P}) \sim_{k-\iota(\mathtt{P})}^{\mathcal{R}} \mathtt{rest}(w)$ check the congruence between strings whose underlying alphabet is a strict subset of the underlying alphabet of $\mathtt{P}$. Therefore, we further introduce some notations and extend the definition of the X-tree with respect to a given basis alphabet. For a position ℓ of $\mathtt{T}$ with $\mathtt{T}[\ell] \in \Gamma$, we define the *arch end* of $\mathtt{T}$ starting at ℓ with respect to the basis Γ as the position $\mathtt{jmp}^{\Gamma}(\mathtt{T}, \ell) = m$ that satisfies $\mathtt{alph}(\mathtt{T}[\ell : m-2]) \subsetneq \mathtt{alph}(\mathtt{T}[\ell : m-1]) = \Gamma$, or ∞ if such a position does not exist. Then, letting $\mathtt{jmp}_1^{\Gamma}(\mathtt{T}, \ell) = \mathtt{jmp}^{\Gamma}(\mathtt{T}, \ell)$, we define $\mathtt{jmp}_i^{\Gamma}(\mathtt{T}, \ell) = \mathtt{jmp}_1^{\Gamma}(\mathtt{T}, \mathtt{jmp}_{i-1}^{\Gamma}(\mathtt{T}, \ell))$. Finally, we define the X-tree $T_X^{\Gamma}(\mathtt{T})$ of $\mathtt{T}$ *for the basis* Γ as the rooted tree with the following nodes and edges. First, the set of nodes of $T_X^{\Gamma}(\mathtt{T})$ is exactly $\{\mathtt{jmp}_1^{\Gamma}(\mathtt{T}, \ell) \mid \ell \in [|\mathtt{T}|]_0\}$. Second, there exists an edge $\langle c, p \rangle$ between a child c and a parent p if $\mathtt{jmp}_1^{\Gamma}(\mathtt{T}, c) = p$. For a text $\mathtt{T}$ and a string w, we denote the X-tree of $\mathtt{T}$ for the basis $\mathtt{alph}(w)$ with the shorthand $T_X^w(\mathtt{T})$. We illustrate an example of an X-tree for some basis in Fig. 1.

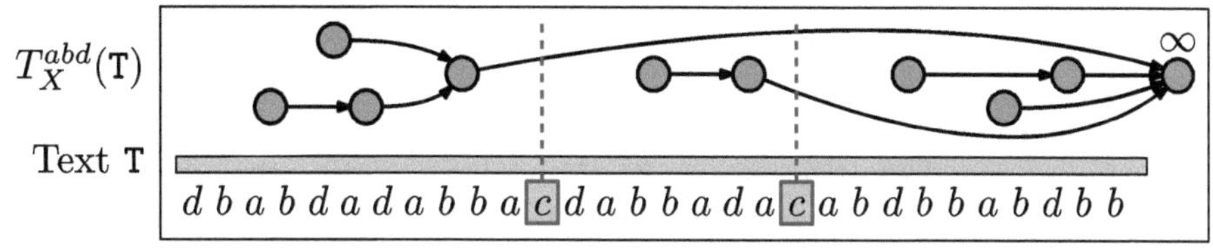

Fig. 1. An illustration of the X-tree of a text $\mathtt{T} = dbabdadabbacdabbadacabdbbabdbb$ for the basis $\Gamma = \{a, b, d\}$. Note that $c \notin \Gamma$, and therefore any occurrence of c before the completion of an arch results in an edge towards the root ∞.

Lemma 4. *For a finite set S and two integers m and n, let f be a mapping from $S \times [n]_0$ to $[m]_0$. We can sort the elements s of S with respect to the increasing order of $f(s,i)$ for each $i \in [n]_0$ in overall $O(|S|n+m)$ time.*

Lemma 5. *For a given text $\mathtt{T}$ over Σ and a string w of length n where each character in Σ appears at most once, we can compute $T_X^{w[0:i]}(\mathtt{T})$ for all $i \in [n-1]_0$ in overall $O(|\Sigma||\mathtt{T}|)$ time.*

Recall that the set of nodes in an X-tree of the text is either labeled ∞ or the arch endpoints of the substrings of the text. Then, for any match $\mathtt{T}[i:j-1]$ of $\mathtt{P}$ with respect to $\sim_k^{\mathcal{R}}$, the first arch of $\mathtt{T}[i:j-1]$ will end at some node of the X-tree of $\mathtt{T}$ for the basis $\mathtt{alph}(\mathtt{P})$. Using this observation, our strategy is to repeatedly filter out the nodes of the X-tree whenever the match candidates whose first arch end at the node do not satisfy the four conditions in Lemma 3 (Fig. 2).

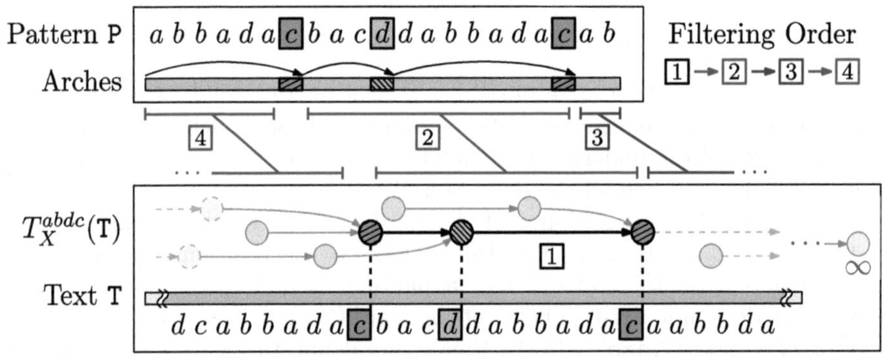

Fig. 2. An illustration of the four filtering steps from Lemma 3. Since $\mathtt{modus}(\mathtt{P}) = cdc$, the first step is to find the match candidates whose modus is cdc. Then, we further filter the match candidates using the sub-patterns marked 2, 3 and 4, in that order.

Our first filtering step is to check the condition $\mathtt{modus}(\mathtt{P}) = \mathtt{modus}(w)$. In other words, our goal is to compute the set of nodes i of $T_X^{\mathtt{P}}(\mathtt{T})$ whose character $\mathtt{T}[i-1]$ is exactly $\mathtt{modus}(\mathtt{P})[0]$, and the modus of the following $\iota(\mathtt{P})-1$ arches with respect to the basis $\mathtt{alph}(\mathtt{P})$ are equal to $\mathtt{modus}(\mathtt{P})[1:\iota(\mathtt{P})-1]$. The main idea is to construct the KMP automaton [18] on the reversal of $\mathtt{modus}(\mathtt{P})$. Then, we perform a top-down traversal on $T_X^{\mathtt{P}}(\mathtt{T})$ while simulating the automaton by one step on the character associated with the current node, starting from the state associated with the parent node. If the automaton arrives at a final state, we report that node and continue matching. We denote the set of nodes obtained from the first filtration step as $\mathtt{F1}(\mathtt{T},\mathtt{P})$.

Lemma 6. *Given a text $\mathtt{T}$, a pattern $\mathtt{P}$ and its modus $\mathtt{modus}(\mathtt{P})$ and the X-tree $T_X^{\mathtt{P}}(\mathtt{T})$, we can compute all the nodes i in $T_X^{\mathtt{P}}(\mathtt{T})$ that satisfy $\mathtt{T}[i-1] =$*

$\texttt{modus}(\texttt{P})[0]$ *and* $\texttt{modus}(\texttt{T}[i : \texttt{jmp}^{\texttt{alph(P)}}_{\iota(\texttt{P})-1}(\texttt{T}, i) - 1]) = \texttt{modus}(\texttt{P})[1 : \iota(\texttt{P}) - 1]$ *in* $O(\iota(\texttt{P}) + |T^{\texttt{P}}_X(\texttt{T})|)$ *time.*

Our second filtering step is to check the condition $\texttt{ar}_{2..\iota(\texttt{P})}(\texttt{P}) \sim^{\mathcal{R}}_{k-1} \texttt{ar}_{2..\iota(w)}(w)$ in Lemma 3. For each node i in $\texttt{F1}(\texttt{T},\texttt{P})$, we directly test the $\sim^{\mathcal{R}}_{k-1}$ - congruence between $\texttt{ar}_{2..\iota(\texttt{P})}(\texttt{P})$ and $\texttt{T}[i : \texttt{jmp}^{\texttt{alph(P)}}_{\iota(\texttt{P})-1}(\texttt{T}, i) - 1]$ using Proposition 2. Let us denote the subset of $\texttt{F1}(\texttt{T},\texttt{P})$ that pass the second filtration step as $\texttt{F2}_k(\texttt{T},\texttt{P})$. Since the number of nodes in $T^{\texttt{P}}_X(\texttt{T})$ can grow up to $|\texttt{T}|$, a naive analysis on the running time for the second filtering step gives an $O(|\texttt{T}|^2)$ bound. However, we can give a tighter $O(|\texttt{T}|\iota(\texttt{P}))$ bound using the following lemmas.

Lemma 7. *Let i be a position of* $\texttt{T}$*, and let σ_1, σ_2 be characters in* $\texttt{alph}(\texttt{T})$*, which are not necessarily distinct. Then, there is at most one edge $\langle c, p\rangle$ of $T^{\texttt{P}}_X(\texttt{T})$ with $c \leq i < p \neq \infty$, $\texttt{T}[c-1] = \sigma_1$ and $\texttt{T}[p-1] = \sigma_2$.*

Lemma 8. *For a text* $\texttt{T}$*, a pattern* $\texttt{P}$ *and a position j of* $\texttt{T}$*, there are at most $\iota(\texttt{P}) - 1$ nodes i in $\texttt{F1}(\texttt{T},\texttt{P})$ that satisfy $i \leq j < \texttt{jmp}^{\texttt{alph(P)}}_{\iota(\texttt{P})-1}(\texttt{T}, i)$.*

Recall that each congruence test between $\texttt{P}$ and $\texttt{T}[i : \texttt{jmp}^{\texttt{alph(P)}}_{\iota(\texttt{P})-1}(\texttt{T}, i) - 1]$ takes time linear in the length of $\texttt{T}[i : \texttt{jmp}^{\texttt{alph(P)}}_{\iota(\texttt{P})-1}(\texttt{T}, i) - 1]$. We charge $O(1)$ time of the testing procedure to each position ℓ between i and $\texttt{jmp}^{\texttt{alph(P)}}_{\iota(\texttt{P})-1}(\texttt{T}, i)$, and by Lemma 8, the overall running time of the second filtration step becomes $O(|\texttt{T}|\iota(\texttt{P}))$.

Corollary 2. *Given a text* $\texttt{T}$*, a pattern* $\texttt{P}$*, the X-tree $T^{\texttt{P}}_X(\texttt{T})$, an integer k and the set $\texttt{F1}(\texttt{T},\texttt{P})$, we can compute the set $\texttt{F2}_k(\texttt{T},\texttt{P})$ in overall $O(|\texttt{T}|\iota(\texttt{P}))$ time.*

The third filtration step checks whether or not the rest of the pattern and the rest of the match candidate are congruent under $\sim^{\mathcal{R}}_{k-\iota(\texttt{P})}$. Before we explain this filtration step, however, we observe in Lemma 9 that we can extend matches of $\texttt{P}$ in $\texttt{T}$ towards the right until some forbidden character appears. We remark that a similar result was obtained by Kim et al. [17] for Simon's congruence. However, unlike Simon's congruence, the same argument cannot be applied in extending matches under $\sim^{\mathcal{R}}_{k}$ to the left of the text.

Lemma 9. *For a pattern* $\texttt{P}$ *and an integer k, let $A = \{\sigma \in \Sigma \mid \texttt{P} \sim^{\mathcal{R}}_{k} \texttt{P}\sigma\}$. Then, for any string $w \in A^*$ and any character $b \in \Sigma \setminus A$, we have $\texttt{P} \sim^{\mathcal{R}}_{k} \texttt{P}w$ and $\texttt{P} \not\sim^{\mathcal{R}}_{k} \texttt{P}wb$.*

In other words, the right endpoints of any match of $\texttt{P}$ in $\texttt{T}$ will form a contiguous range. Therefore, for the third filtration step, let us define an array $\texttt{F3}_k(\texttt{T},\texttt{P})$ of length $|\texttt{T}|$. Each cell holds either a range of positions or $\texttt{NULL}$, where $\texttt{F3}_k(\texttt{T},\texttt{P})[i] \neq \texttt{NULL}$ if and only if $i \in \texttt{F2}_k(\texttt{T},\texttt{P})$, and there exists a position x of $\texttt{T}$ where $\texttt{T}[\texttt{jmp}^{\texttt{alph(P)}}_{\iota(\texttt{P})-1}(\texttt{T}, i) : x-1]$ is $\sim^{\mathcal{R}}_{k-\iota(\texttt{P})}$ -congruent to $\texttt{rest}(\texttt{P})$. Then, whenever there exists such a position x, the ith cell holds the set of positions in the form of an inclusion-maximal range.

We now explain how to compute $\texttt{F3}_k(\texttt{T},\texttt{P})$. First, we initialize $\texttt{F3}_k(\texttt{T},\texttt{P})$ with every value initialized as NULL. Then, we iterate over all $i \in \texttt{F2}_k(\texttt{T},\texttt{P})$ and repeat the following procedure for each i. We compute the value $\ell = \texttt{jmp}^{\texttt{alph(P)}}_{\iota(\texttt{P})-1}(\texttt{T}, i)$ by starting at node i and moving up the X-tree $T^{\texttt{P}}_X(\texttt{T})$ using exactly $\iota(\texttt{P}) - 1$ edges. Let r be the smallest position among $R_X(\texttt{T}, \ell, a)$ over all $a \in \texttt{alph}(\texttt{T}) \setminus \texttt{alph}(\texttt{rest}(\texttt{P}))$, and let $r = |\texttt{T}|$ if no such position exists. Now, we compute the range of positions by the following method. We compute the X-coordinates for the string $\texttt{T}[\ell : r-1]$. Then, scanning the string from left to right, we incrementally compute $\texttt{Short}_{k-\iota(\texttt{P})}(\texttt{T}[\ell : r-1])$ by appending characters at positions whose X-coordinates are no greater than $k - \iota(\texttt{P})$, and skipping the positions with X-coordinates that exceed $k - \iota(\texttt{P})$. Meanwhile, we also scan the string $\texttt{rest}(\texttt{P})$ from left to right as follows. Whenever we append a character in the incremental computation of $\texttt{Short}_{k-\iota(\texttt{P})}(\texttt{T}[\ell : r-1])$, we compare the character to the current character of $\texttt{rest}(\texttt{P})$. If they differ, then we discard the computation for node $i \in \texttt{F2}_k(\texttt{T},\texttt{P})$ and continue to the next node. Otherwise, we move on to the next character of $\texttt{rest}(\texttt{P})$ until there are no more characters to read from $\texttt{rest}(\texttt{P})$. If no more characters can be read from $\texttt{rest}(\texttt{P})$, then we have obtained a minimal match that ends at the current position s in the text. Then, letting $A = \{\sigma \in \Sigma \mid \texttt{P} \sim^{\mathcal{R}}_k \texttt{P}\sigma\}$, we compute the minimum position e among the values $R_X(\texttt{T}, s, b)$ over all $b \in \Sigma \setminus A$, and if all values are ∞, we let $e = |\texttt{T}| + 1$. Finally, we assign $(s, e-1)$ to the cell $\texttt{F3}_k(\texttt{T},\texttt{P})[i]$ and move on to the next node in $\texttt{F2}_k(\texttt{T},\texttt{P})$.

Lemma 10. *Given a text* T*, a pattern* P*, the X-tree* $T^{\texttt{P}}_X(\texttt{T})$*, an integer* k *and the set* $\texttt{F2}_k(\texttt{T},\texttt{P})$*, we can compute the set* $\texttt{F3}_k(\texttt{T},\texttt{P})$ *in overall* $O(|\texttt{T}|\iota(\texttt{P}))$ *time.*

It remains to apply the remaining condition from Lemma 3 to the match candidates w, which is the $\sim^{\mathcal{R}}_k$-congruence between $\texttt{pop}(\texttt{ar}_1(\texttt{P}))$ and $\texttt{pop}(\texttt{ar}_1(w))$. Recall that testing $\sim^{\mathcal{R}}_k$-congruence between two strings required scanning the string from left to right. However, after the third filtration step, we only have the candidates of the right endpoints of the matches of $\texttt{pop}(\texttt{ar}_1(\texttt{P}))$. Indeed, the ith cell of the array $\texttt{F3}_k(\texttt{T},\texttt{P})$ is not NULL if and only if a string starting at $\texttt{T}[i]$ matches $\texttt{P}[|\texttt{ar}_1(\texttt{P})| : |\texttt{P}|-1]$ under $\sim^{\mathcal{R}}_{k-1}$-congruence. In other words, we still have a linear number of positions of T which can be the starting position of a match of P. If we apply the testing algorithm from Proposition 2 for each of these starting positions, then the overall time for testing the congruence between the first arch alone takes time quadratic in the length of the text. Therefore, we need another method for the fourth and final filtration step.

We note that the underlying alphabet of $\texttt{pop}(\texttt{ar}_1(\texttt{P}))$ is exactly the set of characters obtained by removing $\texttt{modus}(\texttt{P})[0]$ from the underlying alphabet of P. In other words, we can re-apply the filtration methods for the smaller pattern $\texttt{pop}(\texttt{ar}_1(\texttt{P}))$ under a strictly smaller underlying alphabet. However, we slightly alter the third filtration step for the re-application of Lemma 10 as follows. After computing the array $\texttt{F3}_k(\texttt{T}, \texttt{pop}(\texttt{ar}_1(\texttt{P})))$, we iterate over all positions i of T. If $\texttt{F3}_k(\texttt{T}, \texttt{pop}(\texttt{ar}_1(\texttt{P})))[i]$ is not NULL, we retrieve the pair $(s, e-1)$ in the cell. Now, if $\texttt{T}[e-1] = \texttt{modus}(\texttt{P})[0]$, then we copy the cell $\texttt{F3}_k(\texttt{T},\texttt{P})[e]$

to $\mathtt{F3}_k(\mathtt{T}, \mathtt{pop}(\mathtt{ar}_1(\mathtt{P})))[i]$, and otherwise set the cell $\mathtt{F3}_k(\mathtt{T}, \mathtt{pop}(\mathtt{ar}_1(\mathtt{P})))[i]$ to NULL. This additional step is to merge the filtration results for P with the results for $\mathtt{pop}(\mathtt{ar}_1(\mathtt{P}))$. Note that we can implement the additional computation in $O(|\mathtt{T}|)$ time, since we only spend constant time for accessing arrays for each position of T. Furthermore, $\mathtt{sig}(\mathtt{P})$ is exactly the sequence of characters that closes the first arches of each pattern in the recursive step of the filtration. Finally, the base case of the recursion is when the alphabet is unary, which we can test $\sim_k^{\mathcal{R}}$-congruence by the following lemma.

Lemma 11. *For a unary string w and an integer k, a string u satisfies $u \sim_k^{\mathcal{R}} w$ if and only if $\mathtt{alph}(u) = \mathtt{alph}(w)$, and either $|u|, |w| \geq k$ or $|u| = |w|$.*

We are ready to describe our main algorithm. For the preprocessing, we first compute the values $R_X(\mathtt{T}, i, \sigma)$ and $R_X(\mathtt{P}, j, \sigma)$ for all positions i of T, positions j of P and characters $\sigma \in \Sigma$. Then, we compute $\mathtt{sig}(\mathtt{P})$ in $O(|\Sigma| \log |\Sigma|)$ time by sorting the characters in Σ in increasing order of $R_X(\mathtt{P}, 0, \sigma)$. Furthermore, letting $\ell_m = R_X(\mathtt{P}, 0, \mathtt{sig}(\mathtt{P})[m-1])$ for each $m \leq |\mathtt{alph}(\mathtt{P})|$ and $\ell_{|\mathtt{alph}(\mathtt{P})|+1} = |\mathtt{P}| + 1$, we perform the arch factorization for the strings $\mathtt{P}[\ell_m : \ell_{m+1} - 2]$ in overall $O(|\Sigma||\mathtt{P}|)$ time. Finally, we compute the X-trees $T_X^{\mathtt{sig}(\mathtt{P})[0:m]}(\mathtt{T})$ for all $m \leq |\mathtt{alph}(\mathtt{P})| - 1$ in $O(|\Sigma||\mathtt{T}|)$ time by Lemma 5.

We now apply Lemma 6, Corollary 2 and Lemma 10 in this order with respect to each pattern $\mathtt{P}_m = \mathtt{P}[0 : \ell_m - 1]$ in decreasing order of m to filter out the match candidates in T. Since we already have the X-trees for each pattern, we spend at most $O(|\mathtt{T}|k)$ time in computing $\mathtt{F1}(\mathtt{T}, \mathtt{P}_m)$, $\mathtt{F2}_k(\mathtt{T}, \mathtt{P}_m)$ and $\mathtt{F3}_k(\mathtt{T}, \mathtt{P}_m)$. Finally, after the iteration ends for $m = 1$, we identify the ranges of the endpoints for each starting position i in T by returning the contents of each cell $\mathtt{F3}_k(\mathtt{T}, \mathtt{P}_1)[i]$.

Theorem 1. *Given a text T, a pattern P and an integer k, we can find all matches of P in T under $\sim_k^{\mathcal{R}}$ in $O(|\Sigma||\mathtt{P}| + |\Sigma||\mathtt{T}|k)$ time.*

3.3 Matching for All Possible k's

We now consider Problem 3, where we are given a text T and a pattern P and the goal is to compute for all position pairs (i, j) of T the maximum number k that allows $\mathtt{T}[i : j] \sim_k^{\mathcal{R}} \mathtt{P}$. Since we consider multiple values of k in this problem, we can no longer assume that $\mathtt{P} = \mathtt{Short}_k(\mathtt{P})$ from now on. On the other hand, we now assume that $\Sigma = \mathtt{alph}(\mathtt{T}) = \mathtt{alph}(\mathtt{P})$, since the $\mathcal{R}$-congruence between the pattern and a substring of the text requires that they have the same set of characters. A naive solution for Problem 3 is to simply iterate over all $k \leq |\mathtt{P}|$ and apply the $O(|\Sigma||\mathtt{T}|k)$ algorithm from Theorem 1. However, this solution admits an algorithm that runs in $O(|\Sigma||\mathtt{T}||\mathtt{P}|^2)$ time, since k grows up to $|\mathtt{P}|$ in Problem 3.

We present a faster solution for Problem 3. Before we present our algorithm, however, we first reformulate Problem 3 as follows. For a text T, a position i of T and a pattern P, let $\mathtt{cmpMatch}_{\mathtt{T}}^{\mathtt{P}}$ be a $|\mathtt{T}|$-by-$|\mathtt{P}|$ array, where $\mathtt{cmpMatch}_{\mathtt{T}}^{\mathtt{P}}[i][k]$ holds an inclusion-maximal range of positions in T. Here, a position j of T is in the range $\texttt{cmpMatch}_{\texttt{T}}^{\texttt{P}}[i][k]$ if and only if k is the maximum integer satisfying $\texttt{P} \sim_k^{\mathcal{R}} \texttt{T}[i : j-1]$. If no such position exists, then the range $\texttt{cmpMatch}_{\texttt{T}}^{\texttt{P}}[i][k]$ is empty. Then, the goal of Problem 3 is to compute the array $\texttt{cmpMatch}_{\texttt{T}}^{\texttt{P}}$.

We now explain the main idea for computing the table $\texttt{cmpMatch}_{\texttt{T}}^{\texttt{P}}$. The strategy is to incrementally match the prefixes of P to the prefixes of $\texttt{T}[i : |\texttt{T}|-1]$, where we start with $k = |\texttt{P}|$ and decrement the value of k whenever the prefix of the pattern does not match the prefix of $\texttt{T}[i : |\texttt{T}|-1]$. We initialize the table $\texttt{cmpMatch}_{\texttt{T}}^{\texttt{P}}$ with every cell initialized as the empty range. Then, for each starting position i of T, we set $j = i$ and $k = |\texttt{P}|$, and iterate over positions ℓ of P. At each position ℓ, we check whether or not $\texttt{P}[0 : \ell-1] \sim_k^{\mathcal{R}} \texttt{P}[0 : \ell]$. If so, we move on to the next position for ℓ. Otherwise, we increment j by one until $\texttt{P}[0 : \ell-1] \not\sim_k^{\mathcal{R}} \texttt{T}[i : j]$. Then, if $\texttt{P}[\ell] = \texttt{T}[j]$, we move on to the next position for ℓ, except we increment j by one at the start of the computation for the next ℓ. If not, then we decrease the value of k by one and start over the iteration for the same ℓ. Finally, if k reaches 0 or j reaches $|\texttt{T}|$ before ℓ reaches $|\texttt{P}|$, then no match starts at position i and we move on to the next value for i. Otherwise, when ℓ reaches $|\texttt{P}|$, we assign a half-open range $[j, \infty)$ to the cell $\texttt{cmpMatch}_{\texttt{T}}^{\texttt{P}}[i][k]$ to mark a minimal match of P starting at T under $\sim_k^{\mathcal{R}}$. We continue the same strategy of increasing j until $\texttt{P} \not\sim_k^{\mathcal{R}} \texttt{T}[i : j]$ and decreasing k by one until k reaches 0 or j reaches $|\texttt{T}|$, except now we do not increment ℓ. Here, each time just before k is decreased or when j reaches $|\texttt{T}|$, we close the half-open interval in the cell $\texttt{cmpMatch}_{\texttt{T}}^{\texttt{P}}[i][k]$ with the current value of j, and if $k \geq 2$, assign a half-open range $[j+1, \infty)$ to the cell $\texttt{cmpMatch}_{\texttt{T}}^{\texttt{P}}[i][k-1]$. This ends the computation of the array $\texttt{cmpMatch}_{\texttt{T}}^{\texttt{P}}$. See Fig 3 for an illustration of the algorithm.

Lemma 12. *Right after we increase j or decrease k in any point in the computation of $\texttt{cmpMatch}_{\texttt{T}}^{\texttt{P}}$, we have $\texttt{P}[0 : \ell-1] \sim_k^{\mathcal{R}} \texttt{T}[i : j-1]$.*

If we naively compare each character of T with each character of P, we obtain an algorithm that is at best quadratic in the length of the text. Since usually the length of the pattern is much smaller than the length of the text, our goal is to optimize the algorithm to run in $O(|\texttt{T}||\texttt{P}|)$ time. We achieve the speedup by avoiding incrementing the value of j one-by-one. Rather, we update j to the maximum value that achieves $\texttt{P}[0 : \ell-1] \sim_k^{\mathcal{R}} \texttt{T}[i : j-1]$ and $\texttt{P}[0 : \ell-1] \not\sim_k^{\mathcal{R}} \texttt{T}[i : j]$ in constant time. If we can update the value of j in each iteration in constant time, then note that the overall time to compute $\texttt{cmpMatch}_{\texttt{T}}^{\texttt{P}}$ is already $O(|\texttt{T}||\texttt{P}|)$. Indeed, for a fixed position i of T, we either increment ℓ or decrement k once in each iteration, and for each iteration we spend constant time. Especially, if we spend $O(|\texttt{P}|)$ preprocessing time for computing the X-coordinates of the pattern, we can check whether or not $\texttt{P}[0 : \ell-1] \sim_k^{\mathcal{R}} \texttt{P}[0 : \ell]$ by testing $X(\texttt{P}, \ell) > k$ in constant time using Corollary 1. Since ℓ grows from 0 up to $|\texttt{P}|$ and k is decreased starting from $|\texttt{P}|$ until 1, we spend $O(|\texttt{P}|)$ time for each position of T. This results in an overall running time of $O(|\texttt{T}||\texttt{P}|)$. The following lemma describes the key idea for the speedup.

Lemma 13. *For a nonempty string w, a position j of w and an integer k, let*

$$p_j = \max(\{p \leq j \mid X(w, p_j - 1) < k\} \cup \{0\}), \text{ and}$$

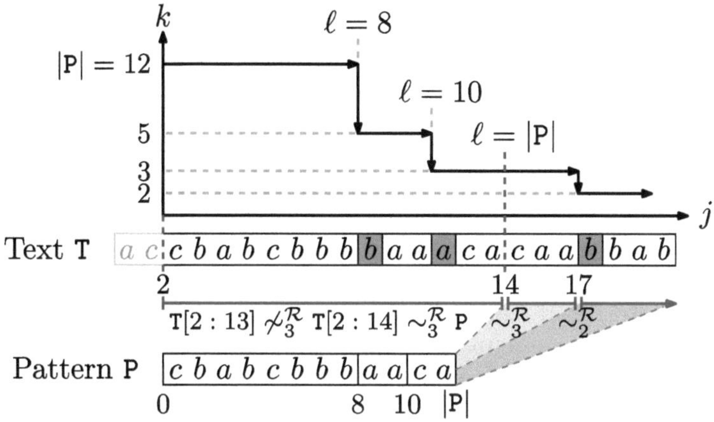

Fig. 3. The evolution of the values for j, k and ℓ over an execution of the complete pattern matching algorithm for the starting position $i = 2$ of T. Note that P is $\sim_3^{\mathcal{R}}$-congruent to the string T$[2:13]$, but is not $\sim_3^{\mathcal{R}}$-congruent to T$[2:12]$ nor $\sim_4^{\mathcal{R}}$-congruent to T$[2:13]$.

$$j' = \min(\{R_X(w, p_j, \sigma) > j \mid \sigma \in \Sigma\} \cup \{\infty\}).$$

If $j' = \infty$, *then* $w[0:j-1] \sim_k^{\mathcal{R}} w$. *Otherwise, we have* $w[0:j-1] \not\sim_k^{\mathcal{R}} w[0:j'-1]$, *and for all* $x \in [j, j'-1]$, *we have* $w[0:j-1] \sim_k^{\mathcal{R}} w[0:x-1]$.

For each iteration for ℓ in the computation of $\texttt{cmpMatch}_{\texttt{T}}^{\texttt{P}}$, the idea is to maintain an array S of $|\texttt{P}|$ integer pairs, where for all $t \leq k$, the cell $S[t-1]$ holds a tuple (p_j, x) defined as follows. The integer p_j is the maximum position of T that satisfies $p_j \leq j$, and either $p_j = i$ or $X(\texttt{T}[i : |\texttt{T}|-1], p_j - i - 1) < t$. The integer x is the number of different characters in the string $\texttt{T}[p_j : j-1]$. Then, in updating j, we first access the cell $S[k-1] = (p_j, x)$. Afterwards, we invoke Lemma 13 and update the value of j to the $(x+1)^{\text{th}}$ smallest value of $R_X(\texttt{T}, p_j, \sigma)$ over all $\sigma \in \Sigma$, which we can retrieve in constant time if we spend $O(|\Sigma||\texttt{T}|)$ time for preprocessing as follows. First, we compute the values for the X-rankers of T in $O(|\Sigma||\texttt{T}|)$ time. Then, we sort all pairs $(\sigma, i) \in \Sigma \times [|\texttt{T}|]_0$ with the same i in increasing order with respect to $R_X(\texttt{T}, i, \sigma)$ in $O(|\Sigma||\texttt{T}|)$ time using Lemma 4. Note that we can now access the character σ with the $(x+1)^{\text{th}}$ smallest value of $R_X(\texttt{T}, p_j, \sigma)$, and thus we can update j in constant time.

Finally, we explain how we can maintain the array S in amortized constant time in each iteration. We spend an additional $O(|\Sigma||\texttt{T}|)$ preprocessing time to compute the number of occurrences of each character $\sigma \in \Sigma$ in each prefix of T by a single sweep of the text. This allows us to test whether or not a given character σ occurs between two positions of T in constant time, which is done by checking whether or not the number of occurrences of σ in the prefixes ending at the two positions are the same. Now, at the beginning of the iteration for i, we construct the array S with $S[0] = (i, 0)$ and all other cells initialized as the wildcard $*$; we treat the wildcard as the tuple $(j, 0)$. Furthermore, we initialize a variable $t = 0$ to keep track of the maximum cell of S that is not the wildcard. Then, in each iteration for ℓ, consider the moment right after we test $\mathtt{P}[\ell] = \mathtt{T}[j]$, which is the step right after j is incremented using S. If $\mathtt{P}[\ell] \neq \mathtt{T}[j]$, before decreasing k by one, we update $S[k] = *$ and if $k = t$, we decrease t by one as well. Otherwise, we have $\mathtt{P}[\ell] = \mathtt{T}[j]$, and before we increment j by one, we compute the maximum integer $t' \leq t$ where for the first item p of $S[t']$, the character $\mathtt{T}[j]$ occurs in the string $\mathtt{T}[p : j-1]$. This is done by starting from $t' = t$ and decreasing t' until the condition is satisfied. If no integer satisfies the above, we let $t' = -1$. Then, by Lemma 13, t' is the maximum integer allowing $\mathtt{P}[0 : \ell - 1] \sim_{t'}^{\mathcal{R}} \mathtt{T}[i : j]$. If $t = t' < k$, then we assign $(j+1, 0)$ to the cell $S[t'+1]$. Otherwise, if $t' < t$, we update all cells $S[d]$ to $*$ for all integers $d \in [t'+2, t]$, and increment the second component of the pair in the cell $S[t'+1]$ by one. Finally, we update $t = t' + 1$.

We analyze the time to maintain S as follows. At each value of ℓ, we write at most one tuple (j, x) to S and modify at most one tuple in S. We charge the constant time needed to overwrite the tuple with a wildcard to the time that the tuple is written to S, including the constant time associated with checking whether or not a character occurs in a substring of $\mathtt{T}$. Since any tuple in S

the total time for maintaining S in the iteration for i is $O(|\mathtt{P}|)$. Summing up, we can augment the $O(|\mathtt{T}|^2)$ algorithm for Problem 3 to run in $O(|\mathtt{T}||\mathtt{P}|)$ time.

Theorem 2. *Given a text* $\mathtt{T}$ *and a pattern* $\mathtt{P}$*, we can compute for each position* i *of* $\mathtt{T}$ *and for each integer* $k \leq |\mathtt{P}|$ *the set of all positions* j *that satisfy* $\mathtt{P} \sim_k^{\mathcal{R}} \mathtt{T}[i : j-1]$ *and* $\mathtt{P} \not\sim_{k+1}^{\mathcal{R}} \mathtt{T}[i : j-1]$ *in overall* $O(|\mathtt{T}||\mathtt{P}|)$ *time.*

Remark 1. We can solve Problem 2 in $O(|\mathtt{T}||\mathtt{P}|)$ time using $\mathtt{cmpMatch}_{\mathtt{T}}^{\mathtt{P}}$. This is an improvement on Theorem 1 for the special case when $|\mathtt{Short}_k(\mathtt{P})| < |\Sigma|k$. Note that for $\mathtt{P}$ over a given Σ, $|\mathtt{Short}_k(\mathtt{P})|$ can grow up to

$$\left(\frac{k}{|\Sigma|}\right)^{|\Sigma|-1} \log_{|\Sigma|} \frac{k}{|\Sigma|}.$$

Table 1. Notations we frequently use in Sect. 3.

Notation	Meaning
$\mathtt{pop}(w)$	$w[0 : \lvert w\rvert - 2]$; string obtained by removing the last character of w
$\mathtt{sig}(w)$	the subsequence of w obtained by choosing the first occurrences of each character in $\mathtt{alph}(w)$
$\mathtt{Short}_k(\mathtt{P})$	the shortest string that is $\sim_k^{\mathcal{R}}$ -congruent to $\mathtt{P}$
$R_X(\mathtt{T}, i, \sigma)$	$\min(\{j > i \mid \mathtt{T}[j-1] = \sigma\} \cup \{\infty\})$
$X(w, i)$	the length of the shortest string in $\mathbb{D}(w[0 : i])$
$\mathtt{jmp}_i^{\Gamma}(\mathtt{T}, \ell)$	right endpoint of the ith arch with respect to Γ, starting at $\mathtt{T}[\ell]$
$T_X^w(\mathtt{T})$	The X-tree of $\mathtt{T}$ for the basis $\mathtt{alph}(w)$

The bound on the length of $|\mathtt{Short}_k(\mathtt{P})|$ in Remark 1 is obtained by the following idea. Consider a total ordering of strings in Σ^*, where the shorter strings precede the longer strings. Let us map each congruence class modulo $\sim_k$ to strings in Σ^*, where the ith congruence class is mapped to the ith string with respect to the total order on Σ^*. Then, $\mathtt{Short}_k(\mathtt{P})$ is longer than the first string in Σ^* without any congruence class mapped to it. The number $(\frac{k}{|\Sigma|})^{|\Sigma|-1} \log_{|\Sigma|} \frac{k}{|\Sigma|}$ follows from a lower bound on number of congruence classes modulo $\sim_k$ [13].

4 Conclusions

We have solved pattern matching problems for $\sim_k^{\mathcal{R}}$. We have first presented a linear time solution for testing $\sim_k^{\mathcal{R}}$-congruence between two strings. Then, we have presented an $O(|\Sigma||\mathtt{P}| + |\Sigma||\mathtt{T}|k)$ time solution for the case when k is given, which is based on a filtering approach that parses the pattern into multiple sub-patterns. For the case when we consider all values of k, we have designed an $O(|\mathtt{T}||\mathtt{P}|)$ time solution that incrementally matches $\mathtt{P}$ under a decreasing k. Interestingly, we observe that our algorithm for the pattern matching problem under $\sim_k^{\mathcal{R}}$-congruence asymptotically runs faster than the best known algorithm for pattern matching under Simon's congruence [17]. In this direction, we think that it is interesting to investigate the relation between the $\mathcal{R}$-congruence and the Simon's congruence in terms of the computational complexity of problems under each congruence. Furthermore, we expect that the running time of both matching algorithms can be improved; therefore we intend to investigate faster algorithms and matching lower bounds in future work.

Acknowledgements. This research was supported by the NRF grant (RS-2025-00562134) funded by the Korean government.

References

1. Barker, L., Fleischmann, P., Harwardt, K., Manea, F., Nowotka, D.: Scattered factor-universality of words. In: Proceedings of the 24th International Conference on Developments in Language Theory, DLT, pp. 14–28 (2020)
2. Brzozowski, J., Fich, F.E.: Languages of $\mathcal{R}$-trivial monoids. J. Comput. Syst. Sci. **20**(1), 32–49 (1980)
3. Brzozowski, J., Li, B.: Syntactic complexity of $\mathcal{R}$- and $\mathcal{J}$-trivial languages. Int. J. Found. Comput. Sci. **25**(07), 807–821 (2014)
4. Day, J.D., Fleischmann, P., Kosche, M., Koß, T., Manea, F., Siemer, S.: The edit distance to k-subsequence universality. J. Comput. Syst. Sci. **154**, 103681 (2025)
5. Fleischer, L., Kufleitner, M.: Testing Simon's congruence. In: Proceedings of the 43rd International Symposium on Mathematical Foundations of Computer Science, MFCS, pp. 62:1–62:13 (2018)
6. Fleischmann, P., Haschke, L., Höfer, J., Huch, A., Mayrock, A., Nowotka, D.: Nearly k-universal words - investigating a part of Simon's congruence. Theoret. Comput. Sci. **974**, 114113 (2023)

7. Fleischmann, P., Höfer, J., Huch, A., Nowotka, D.: α-β-factorization and the binary case of Simon's congruence. In: Proceedings of the 24th International Symposium on Fundamentals of Computation Theory, FCT, pp. 190–204 (2023)
8. Fleischmann, P., et al.: Matching patterns with variables under Simon's congruence. In: Proceedings of the 17th International Conference on Reachability Problems, RP, pp. 155–170 (2023)
9. Gawrychowski, P., Kosche, M., Koß, T., Manea, F., Siemer, S.: Efficiently testing Simon's congruence. In: Proceedings of the 38th International Symposium on Theoretical Aspects of Computer Science, STACS, pp. 34:1–34:18 (2021)
10. Green, J.A.: On the structure of semigroups. Ann. Math. **54**(1), 163–172 (1951)
11. Hébrard, J.: An algorithm for distinguishing efficiently bit-strings by their subsequences. Theoret. Comput. Sci. **82**(1), 35–49 (1991)
12. Jirásková, G., Masopust, T.: On the state complexity of the reverse of $\mathcal{R}$- and $\mathcal{J}$-trivial regular languages. In: Proceedings of the International Workshop on Descriptional Complexity of Formal Systems, pp. 136–147 (2013)
13. Karandikar, P., Kufleitner, M., Schnoebelen, P.: On the index of Simon's congruence for piecewise testability. Inf. Process. Lett. **115**(4), 515–519 (2015)
14. Kim, S., Han, Y.S.: Pattern mining under Simon's congruence. In: Proceedings of the 29th International Conference on Developments in Language Theory, DLT, pp. 182–196 (2025)
15. Kim, S., Han, Y.S., Ko, S.K., Salomaa, K.: On Simon's congruence closure of a string. Theoret. Comput. Sci. **972**, 114078 (2023)
16. Kim, S., Han, Y.S., Ko, S.K., Salomaa, K.: On the Simon's congruence neighborhood of languages. In: Proceedings of the 27th International Conference on Developments in Language Theory, DLT, pp. 168–181 (2023)
17. Kim, S., Ko, S.K., Han, Y.S.: Simon's congruence pattern matching. Theoret. Comput. Sci. **994**, 114478 (2024)
18. Knuth, D.E., Morris, J.H., Jr., Pratt, V.R.: Fast pattern matching in strings. SIAM J. Comput. **6**(2), 323–350 (1977)
19. Li, J., Cotterell, R.: Characterizing the expressivity of fixed-precision transformer language models. In: Proceedings of the 39th Annual Conference on Neural Information Processing Systems, NeurIPS (2025)
20. Masopust, T., Krötzsch, M.: Deciding universality of ptNFAs is PSPACE-complete. In: Tjoa, A.M., Bellatreche, L., Biffl, S., van Leeuwen, J., Wiedermann, J. (eds.) SOFSEM 2018. LNCS, vol. 10706, pp. 413–427. Springer, Cham (2018). https://doi.org/10.1007/978-3-319-73117-9_29
21. Masopust, T., Krötzsch, M.: Partially ordered automata and piecewise testability. Logical Methods Comput. Sci. **17**(2) (2021)
22. Masopust, T., Thomazo, M.: On the complexity of k-piecewise testability and the depth of automata. In: Proceedings of the 19th International Conference on Developments in Language Theory, DLT, pp. 364–376 (2015)
23. Pach, P.P.: Normal forms under Simon's congruence. In: Semigroup Forum, vol. 97, pp. 251–267 (2018)
24. Pin, J.E.: Varieties of Formal Languages. North Oxford Academic (1989), translated by A. Howie
25. Schnoebelen, P., Veron, J.: On arch factorization and subword universality for words and compressed words. In: Proceedings of the 14th International Conference on Combinatorics on Words, WORDS, pp. 274–287 (2023)
26. Schwentick, T., Thérien, D., Vollmer, H.: Partially-ordered two-way automata: A new characterization of DA. In: Proceedings of the 5th International Conference on Developments in Language Theory, DLT, pp. 239–250 (2001)

27. Simon, I.: Piecewise testable events. In: Brakhage, H. (ed.) GI-Fachtagung 1975. LNCS, vol. 33, pp. 214–222. Springer, Heidelberg (1975). https://doi.org/10.1007/3-540-07407-4_23
28. Straubing, H., Thérien, D.: Partially ordered finite monoids and a theorem of I. Simon. J. Algebra **119**(2), 393–399 (1988)
29. Vaswani, A., et al.: Attention is all you need. In: Proceedings of the 30th Annual Conference on Neural Information Processing Systems, pp. 5998–6008. NeurIPS (2017)
30. Weis, P., Immerman, N.: Structure theorem and strict alternation hierarchy for FO^2 on words. Logical Methods in Computer Science 5 (2009)

Hyper-Minimization for Deterministic Register Automata

Yong Li[1], Qiyi Tang[2], and Di-De Yen[2(✉)]

[1] Key Laboratory of System Software (Chinese Academy of Sciences), Institute of Software Chinese Academy of Sciences, Beijing, China
[2] School of Computer Science and Informatics, University of Liverpool, Liverpool, UK
d.d.yen@liverpool.ac.uk

Abstract. We investigate hyper-minimization for deterministic register automata (DRAs). We begin by introducing DRA counterparts of classical notions from deterministic finite automata. Building on these foundations, we present an algorithm for hyper-minimizing well-typed DRAs, where each state is associated with a unique register type. The resulting automata are minimal with respect to both the number of states and registers among all well-typed DRAs. We prove the correctness of the proposed algorithm, thereby establishing the decidability of hyper-minimization for well-typed DRAs.

Keywords: Register Automata · Hyper-Minimization · Dense Domains

1 Introduction

The minimization problem is a fundamental topic in automata theory, dating back to [14], where an automaton is minimal if it has no smaller equivalent counterpart. For finite automata (FAs), minimality corresponds to having the smallest number of states among all equivalent FAs. While every FA can be minimized, the general problem is PSPACE-complete [10]; in contrast, deterministic finite automata (DFAs) admit an $\mathcal{O}(n \log n)$ minimization algorithm [8], where n is the number of states.

For register automata (RAs), which extend FAs with registers for storing and comparing data values, minimality requires minimizing both the number of states and registers. The minimization problem for deterministic register automata (DRAs) over dense ordered and unordered domains is decidable [4].

The notion of DFA minimization was generalized to *hyper-minimization* in [2]. A DFA $\mathcal{A}'$ is hyper-minimal for a DFA $\mathcal{A}$ if the symmetric difference between their languages is finite. For example, let $\mathcal{A}$ recognize $L_{even+x^{49}}$ over $\Sigma = \{a, b\}$, consisting of all even-length words together with a^{49} and b^{49}. Any

A full version of this paper is available at http://arxiv.org/abs/2605.03535.

T. J. Smith (Ed.): CIAA 2026, LNCS 16695, pp. 124–137, 2026.
https://doi.org/10.1007/978-3-032-31176-4_9

DFA for this language requires at least 101 states: one initial state, 49 states each to count a^{49} and b^{49}, and two states to track parity. In contrast, a hyper-minimal DFA needs only two states, recognizing L_{even}, which consists of all even-length words. This illustrates that allowing finitely many errors can yield significantly more succinct automata. An $\mathcal{O}(n \log n)$ algorithm for DFA hyper-minimization was later given in [7], matching the complexity of standard minimization.

Hyper-minimization has been studied for several extensions of DFAs, including deterministic tree automata [9] and deterministic weighted automata [12]. In contrast, it has not been systematically investigated for automata over infinite alphabets, such as DRAs or symbolic finite automata. One reason is that hyper-minimization is often considered less meaningful in this setting, since over infinite alphabets any non-empty symmetric difference is typically infinite [6].

For example, consider extending $L_{even+x^{49}}$ to the alphabet $\mathbb{Q}$, yielding the language of all even-length words together with all words of the form c^{49} for $c \in \mathbb{Q}$. For any RA $\mathcal{A}$, if $c^{49} \in L(\mathcal{A})$ for some $c \in \mathbb{Q}$, then $d^{49} \in L(\mathcal{A})$ for all $d \in \mathbb{Q}$. Hence, if $\mathcal{A}$ does not recognize $L_{even+x^{49}}$ exactly, the symmetric difference $L_{even+x^{49}} \ominus L(\mathcal{A})$ is necessarily infinite.

Nevertheless, if $L(\mathcal{A})$ consists of all even-length words over $\mathbb{Q}$, then all words in $L_{even+x^{49}} \ominus L(\mathcal{A})$ share the same *word type*, namely words of the form c^{49} for $c \in \mathbb{Q}$. Intuitively, two words have the same word type if they induce the same order relations among their positions. For example, the words $u = 3 \cdot 1 \cdot 5$ and $v = \frac{5}{2} \cdot 2 \cdot 9$ have the same word type over $\mathbb{Q}$ under the standard order $<$, as the relative ordering between any pair of positions is identical in both words.

Motivated by this observation, we study hyper-minimization of DRAs modulo word types. A DRA $\mathcal{A}'$ is hyper-minimal for a given DRA $\mathcal{A}$ if $L(\mathcal{A}) \ominus L(\mathcal{A}')$ contains only *finitely many word types*. Under this notion, the hyper-minimal DRA for $L_{even+x^{49}}$ over $\mathbb{Q}$ recognizes L_{even} and requires only two states.

In general, a minimal DRA need not exist due to a trade-off between the number of states and registers; the same issue arises for hyper-minimization. However, for *well-typed* DRAs—where each state has a unique register type—a minimal DRA is guaranteed to exist [4]. We show that an analogous result holds for hyper-minimization, and present an algorithm for hyper-minimizing well-typed DRAs, thereby establishing its decidability.

2 Preliminaries

We use $\mathbb{R}$ (resp. $\mathbb{Q}$, $\mathbb{Z}$, $\mathbb{N}$) to denote the sets of real (resp. rational, integer, and non-negative) numbers. A set $S \subseteq \mathbb{R}$ is *dense* in $\mathbb{R}$ if, for every $s < t$, there exists $r \in S$ with $s < r < t$ [16]. Accordingly, $\mathbb{Q}$ is dense in $\mathbb{R}$, whereas $\mathbb{Z}$ is not.

Let Σ be an alphabet and R a binary relation on Σ, where Σ may be finite or infinite. A *(data) word* is a finite sequence over Σ. For words $u = a_1 \dots a_m$, $v = b_1 \dots b_n$ in Σ^* and a symbol $d \in \Sigma$, we write $d \in u$ if $d = a_i$ for some $1 \leq i \leq m$, $u \cup v$ as the set $\{a_1, \cdots, a_m, b_1, \cdots, b_n\}$ and $u \cdot v$ as the concatenation $a_1 \dots a_m b_1 \dots b_n$. Given two symbols a and b, we use $u[a/b]$ to denote the word obtained by replacing each a in u with b. The length of u is denoted by $|u|$.

The relation R induces an equivalence relation $\sim_R$ on Σ^*: for $u = a_1 \dots a_m$ and $v = b_1 \dots b_n$, we say u and v are in the same equivalence τ iff (1) $m = n$, and (2) $(a_i, a_j) \in R$ if and only if $(b_i, b_j) \in R$ for all $1 \le i, j \le n$. Each equivalence class τ of $\sim_R$ is called a $\sim_R$-*word type*. τ has length n if $|\tau| = n$. A language $L \subseteq \Sigma^*$ is a *data language* over (Σ, R) if, for all words u, v of the same word type, $u \in L \Leftrightarrow v \in L$. In this paper, we consider only two kinds of domains: dense ordered domains or R is the identity relation.

Given two data languages L and L', we say that L and L' are *almost-equal* if there exists a *finite* set T of word types such that every word in the symmetric difference $L \ominus L'$ has a word type $\tau \in T$. In other words, two almost-equal data languages differ only on finitely many word types.

Register automata (RAs) extend finite-state automata to infinite alphabets and recognize data languages. There are several equivalent definitions in the literature [3,4,11,15]. We adopt the definition of RAs from [4] in this paper.

Definition 1. *Given $k \in \mathbb{N}$, an alphabet Σ, and a binary relation R on Σ, a k-register automaton (k-RA) $\mathcal{A}$ over (Σ, R) is a tuple (Q, q_0, F, Δ) where: Q is a set of states, partitioned into disjoint subsets $Q_0, \dots, Q_k$, $q_0 \in Q_0$ is the* initial state, *$F \subseteq Q$ is the set of* final states, *and Δ is a finite set of* transitions *of the form (p, τ, E, q), where $p \in Q_i$ and $q \in Q_j$ for some $0 \le i, j \le k$, τ is a $\sim_R$-word type of length $i+1$, and $E \subseteq \{1, \dots, i+1\}$.*

An RA $\mathcal{A}$ is *deterministic* (DRA) if, for any two transitions (p, τ, E, q) and (p', τ', E', q'), we have $(E, q) = (E', q')$ whenever $(p, \tau) = (p', \tau')$.

A *configuration* of $\mathcal{A}$ is a pair (q, v), where $q \in Q_i$ and v is a word over Σ of length i, for some $0 \le i \le k$. We say a configuration (q, v) is *accepting* if $q \in F$. For configurations (p, u) and (q, v) and a symbol $a \in \Sigma$, we write $(p, u) \xrightarrow{a}_{\mathcal{A}} (q, v)$ or $(p, u) \xrightarrow{\tau:E}_{\mathcal{A}} (q, v)$ if there exists a transition $\delta = (p, \tau, E, q)$ such that $u \cdot a = a_1 \dots a_n$ is of type τ and v is obtained from $u \cdot a$ by removing all a_i with $i \in E$, where (1) $E \neq \emptyset$ if $|u| = k$, preventing memory overflow, and (2) $j \in E$ whenever $1 \le j < n$ and $a_j = a_n (= a)$, ensuring no data duplication. The subscript $\mathcal{A}$ is omitted when clear from context.

Let $w = b_1 \dots b_m \in \Sigma^*$. A sequence of configurations $\pi = (p_0, u_0) \dots (p_m, u_m)$ is a *run* on w, written $(p_0, u_0) \xrightarrow{w} (p_m, u_m)$, if $(p_i, u_i) \xrightarrow{b_i} (p_{i+1}, u_{i+1})$ for all $0 \le i < m$. It is *accepting* if $(p_0, u_0) = (q_0, \varepsilon)$ and (p_m, u_m) is accepting. We say $\mathcal{A}$ is *complete* if every word $u \in \Sigma^*$ has a run in $\mathcal{A}$ from (q_0, ϵ). Any DRA can be made complete by introducing a *sink state*. The language recognized by $\mathcal{A}$ is $L(\mathcal{A}) = \{w \in \Sigma^* \mid (q_0, \epsilon) \xrightarrow{w} (q_f, u), q_f \in F, u \in \Sigma^*\}$, which is a data language over (Σ, R). Two DRAs $\mathcal{A}$ and $\mathcal{A}'$ are *equivalent* if $L(\mathcal{A}) = L(\mathcal{A}')$. Given a configuration c of $\mathcal{A}$, we use $\mathcal{A}_c$ to denote the RA obtained from $\mathcal{A}$ such that all accepting runs start from configuration c. Note that if $\mathcal{A}$ is a finite automaton, then a configuration is simply a state.

A state q of $\mathcal{A}$ is *useful* if there exist words w, w' such that $(q_0, \epsilon) \xrightarrow{w}_{\mathcal{A}} (q, u)$ and $(q, v) \xrightarrow{w'}_{\mathcal{A}} (q_f, x)$ for some u, v, x and an accepting state q_f. $\mathcal{A}$ is *trimmed* if all its states, excluding the sink state, are useful. In this paper, we primarily consider DRAs over dense or non-ordered domains (i.e., with R being identity test). Since the reachability problem for these models is decidable [4,5], we assume all DRAs in the remainder of this paper are trimmed.

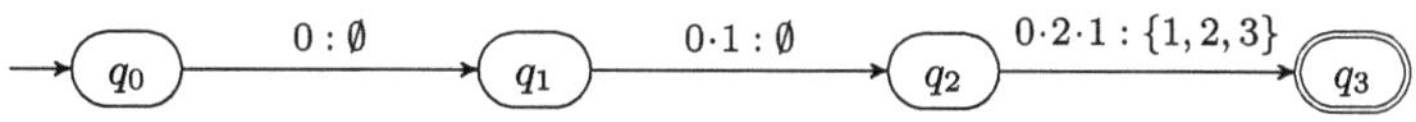

Fig. 1. An example DRA $\mathcal{A}$.

Example 2. Let $L_{mid} = \{a_1a_2a_3 \mid a_1 < a_3 < a_2\}$ over $(\mathbb{Q}, <)$. The DRA in Fig. 1, with initial state q_0 and accepting state q_3, recognizes L_{mid}. Transitions are labeled by $\tau : E$. For instance, the transition from q_2 to q_3 is labeled $0 \cdot 2 \cdot 1 : \{1, 2, 3\}$. On input $w = 3 \cdot 9 \cdot 7$, the automaton has the accepting run $(q_0, \varepsilon) \xrightarrow{3} (q_1, 3) \xrightarrow{9} (q_2, 3 \cdot 9) \xrightarrow{7} (q_3, \varepsilon)$, showing that $w \in L_{mid}$. □

Definition 3. *Let $\mathcal{A}$ be an RA over (Σ, R). We call $\mathcal{A}$* well-typed *if for every two configuration transitions $(p, u) \xrightarrow{a}_{\mathcal{A}} (q, v)$ and $(p', u') \xrightarrow{a'}_{\mathcal{A}} (q, v')$ ending in the same state q, we have v and v' are of the same word type τ.*

Intuitively, an RA is well-typed if the word type τ of its registers is uniquely determined by the current state p and we say p is of *τ-register-type.*

Let L be a data language over the data domain D, and let $\mathcal{A}$ and $\mathcal{B}$ be two RAs over D. We say that $\mathcal{A}$ and $\mathcal{B}$ are *almost-equivalent* (resp., *equivalent*) if the data languages $L(\mathcal{A})$ and $L(\mathcal{B})$ are almost-equal (resp., equal); furthermore, $\mathcal{A}$ *almost-recognizes* L if $L(\mathcal{A})$ is almost-equal to L. Suppose that $\mathcal{A}$ has n states and k registers. We say that $\mathcal{A}$ is *hyper-data-minimal* (resp., *data-minimal*) for L if every RA that almost-recognizes (resp., recognizes) L has at least k registers, and it is *hyper-state-minimal* (resp., *state-minimal*) for L if every RA that almost-recognizes (resp., recognizes) L has at least n states. Finally, $\mathcal{A}$ is *hyper-minimal* (resp., *minimal*) if it is both hyper-data-minimal and hyper-state-minimal (resp., both data-minimal and state-minimal).

Given a directed graph G with an initial node v_0, a node v of G is a *preamble* if there are finitely many paths from v_0 to v, otherwise it is a *kernel.* Intuitively, v is a preamble if no path from v_0 to v contains a loop. A state of an automaton $\mathcal{A}$ is a *preamble state* if it is a preamble node in the underlying graph with the initial state as the initial node. The *kernel states* of $\mathcal{A}$ are defined analogously.

3 Characteristics of Well-Typed DRAs

An important property of DRAs over dense or non-ordered domains is the decidability of the minimization problem [4]. The derivation of this result in [4] is analogous to that of DFA minimization, which is closely related to the Nerode congruence.

For regular languages, two words are Nerode-equivalent if no distinguishing extension exists. For data languages, however, equivalence additionally depends on *memorability* [4], which captures the symbols that influence future acceptance. Intuitively, a symbol is L-memorable in a word u if modifying it (while preserving type) can change the membership of some extension of u. Consider the language L_{mid} in Example 2. For any DRA $\mathcal{A}$ recognizing L_{mid}, after reading $u = 2 \cdot 6$, the values 2 and 6 must be stored in registers to check whether the next input lies between them. Hence, both 2 and 6 are *L-memorable* in u. The formal definition follows:

Definition 4. *Let L be a language over (Σ, R). A symbol $a \in \Sigma$ is* L-memorable *in a word $u \in \Sigma^*$ if there is a symbol $b \in \Sigma$ and a word $w \in \Sigma^*$ such that (1) $u \cdot w \sim_R (u \cdot w)[a/b]$; and (2) $uw \in L \Leftrightarrow u \cdot w[a/b] \notin L$.*

Intuitively, a letter a in u is memorable if it can distinguish between continuations w and $w[a/b]$ such that $uw \sim_R (uw)[a/b]$, yet replacing a with a nearby symbol b in w changes the acceptance of uw versus $uw[a/b]$.

We write $mem_L(u)$ for the sequence of L-memorable symbols in u. With the notion of memorability, we can then define the Nerode congruence relation $\cong_L$ and the canonical automaton $\mathcal{C}_L$ for a data language L. For details, see [4] or the full version of this paper. As a result, we have the Myhill-Nerode theorem for DRAs:

Theorem 5. *[4] A data language L is recognized by a DRA iff $\cong_L$ has finite index. Moreover, every DRA-recognizable language L has a canonical automaton $\mathcal{C}_L$ which is the unique minimal well-typed DRA up to isomorphism*[1].

The well-typedness condition in Theorem 5 is necessary and sufficient for the existence and uniqueness of the minimal DRA. Without it, a trade-off between the number of states and registers may arise. In the remainder of this paper, we assume all DRAs are well-typed unless stated otherwise.

Lastly, we present a pumping property for DRAs that is employed in our hyper-minimization algorithms.

For a finite automaton $\mathcal{A}$ with n states, any run π of length greater than n contains a non-empty subrun that can be pumped or depumped. This is the classical Pumping Lemma [8], which can be extended to multiple runs. In particular, given two runs π_1 and π_2 of $\mathcal{A}$ over the same word w with $|w| > n^2$, there exist non-empty subruns of π_1 and π_2 that can be pumped or depumped simultaneously. We establish a similar property for DRAs:

Lemma 6. *Let $\mathcal{A}$ be a DRA over (Σ, R) with n states and k registers, and let $len^2_{\mathcal{A}} = n^2 \cdot (2k)!$. Given a word w and two runs π and ρ from configurations (p, u) and (q, v) over w, ending in states r and s, respectively, there exists a word w' with $|w'| \leq len^2_{\mathcal{A}}$ and corresponding runs π' and ρ' from (p, u) and (q, v) that also end in r and s. Moreover, if $|w| > len^2_{\mathcal{A}}$, then for every $\ell \in \mathbb{N}$, there exists a word w'' with $|w''| > \ell$ and corresponding runs π'' and ρ'' from (p, u) and (q, v) that end in r and s.*

[1] This claim does not hold when the well-typeness constraint is dropped.

4 Hyper-Minimization

Before introducing our hyper-minimization algorithm for DRAs, we briefly recall hyper-minimization for DFAs, which are a special class of DRAs over a finite alphabet and without registers.

We begin with the *merge* operation from DFA hyper-minimization. For a DFA $\mathcal{A}$, a state p can be merged into a distinct state q if:

- p is a *preamble* state, i.e., no loops in any path from the initial state to p;
- p and q are *almost-equivalent*, i.e., $L(\mathcal{A}_p) \ominus L(\mathcal{A}_q)$ is finite.

We denote by $\texttt{merge}(p \to q, \mathcal{A})$ the DFA obtained by performing the merge operation, i.e., redirecting all incoming transitions of p to q. Let $\mathcal{B} = \texttt{merge}(p \to q, \mathcal{A})$. It is not difficult to see that $L(\mathcal{A})$ is almost-equivalent to $L(\mathcal{B})$. In particular, $L(\mathcal{A}) \ominus L(\mathcal{B}) = \{u \cdot w \mid q_0 \xrightarrow{u} p, w \in L(\mathcal{A}_p) \ominus L(\mathcal{A}_q)\}$. This symmetric difference set is finite since p is a preamble state—hence only finitely many prefixes u lead to p—and $L(\mathcal{A}_p) \ominus L(\mathcal{A}_q)$ is finite by almost-equivalence definition. By recursively applying the merge operation, we eventually obtain a minimal DFA that is almost-equivalent to $\mathcal{A}$.

The basic DFA hyper-minimization algorithm proceeds as follows: it first minimizes the input DFA and then recursively merges a preamble state to its almost-equivalent state, where $p \approx q$ denotes that $L(\mathcal{A}_p)$ and $L(\mathcal{A}_q)$ are almost-equivalent.

The most interesting part, which is also missed so far, is the computation of the almost-equivalence relation on the states in a DFA $\mathcal{A}$. Existing algorithms compute the almost-equivalence relation based on the following result:

Lemma 7. ([2]) *Let $\mathcal{A}$ be a minimal DFA with n states. Then, two distinct states p, q are almost-equivalent, i.e., $p \approx q$ if and only if there is an integer $\ell > 0$ such that $\mathcal{A}_p(w) = \mathcal{A}_q(w)$ for all $w \in \Sigma^*$ with $|w| \geq \ell$, where $\mathcal{A}_r(w)$ is the state reached over w from the initial state for $r = p, q$.*

Let n be the number of states in $\mathcal{A}$. We can restrict ℓ in Lemma 7 to be at most n^2. To see this, suppose that for every $k \geq n^2$ there exists a word w with $|w| \geq k$ such that $\mathcal{A}_p(w) \neq \mathcal{A}_q(w)$; that is, $p' = \mathcal{A}_p(w) \neq \mathcal{A}_q(w) = q'$. Since $\mathcal{A}$ is minimal, this implies that $L(\mathcal{A}_{p'}) \neq L(\mathcal{A}_{q'})$. Hence, there exists a word $w' \in L(\mathcal{A}_{p'}) \ominus L(\mathcal{A}_{q'})$. Because $|w| \geq n^2$, the word w can be decomposed as $w = xuy$ such that $(\mathcal{A}_p(x), \mathcal{A}_q(x)) = (\mathcal{A}_p(xu), \mathcal{A}_q(xu))$. It then follows that $xu^iyw' \in L(\mathcal{A}_p) \ominus L(\mathcal{A}_q)$ for all $i \geq 1$, yielding infinitely many distinguishing words. This contradicts the assumption that $p \approx q$, so the lemma follows.

Moreover, this lemma yields a brute-force algorithm for checking whether two states p and q are almost equivalent: enumerate all words of length up to n^2. By Lemma 7, we have $p \approx q$ if and only if p and q differ only on words of length smaller than n^2. We briefly mention this algorithm for comparison with the almost-equivalence computation for DRAs. In fact, more efficient algorithms exist for computing the almost-equivalence relation over all states of DFAs, running in $\mathcal{O}(n^2)$ [1] and even $\mathcal{O}(n \log n)$ [7].

We fix a **well-typed** DRA $\mathcal{A} = (Q, q_0, F, \Delta)$ with n states and k registers, and we want to hyper-minimize it in the remainder of the paper.

Our hyper-minimization algorithm for DRAs follows the general approach used for DFAs. However, extending hyper-minimization to DRAs is considerably more involved, as identifying suitable analogues of the key notions is non-trivial. We identify the two main challenges as follows, and address them in Sects. 4.1 and 4.2, respectively.

C1 How to define and compute the almostspsequivalence relation over states given that a run on a word of a DRA is not a sequence of states but configurations?

C2 How to design and apply merge operations, and ensure that the final resulting DRA remains both hyper-data-minimal and hyper-state-minimal?

4.1 Almost-Equivalence Relation over States

In this section, we introduce the almost-equivalence relation over states and then provide the means to compute them over dense domains or non-ordered domains. Firstly, we define the *almost-equivalent* relation over configurations.

Definition 8. *Let* (p, u) *and* (q, v) *be two configurations of* $\mathcal{A}$. *We say they are* almost-equivalent, *denoted* $(p, u) \approx (q, v)$, *if there exist only finitely many word types* $w \in \Sigma^*$ *s.t.* $(p, u) \xrightarrow{w} (p', u')$ *and* $(q, v) \xrightarrow{w} (q', v')$ *with* $p' \in F \not\Leftrightarrow q' \in F$.

Now, we are ready to define the DRA analogue of the almost-equivalence relation over states.

Definition 9. *Let* p *and* q *be two states in* $\mathcal{A}$. p *and* q *are* almost-equivalent, *denoted by* $p \approx q$, *if for each register word* u *of* p, *there exists a register word* v *of* q *such that* $(p, u) \approx (q, v)$, *and vice versa.*

Two configurations (p, u) and (q, v) are *equivalent*, denoted by $(p, u) \equiv (q, v)$, if $L(\mathcal{A}_{(p,u)}) = L(\mathcal{A}_{(q,v)})$. This equivalence relation $\equiv$ can be extended to states, analogously to almost-equivalence (cf. Definition 9). Moreover, both almost-equivalence and equivalence relations naturally extend to configurations or states of *different* DRAs over the same data domain (Σ, R). For example, for configurations (p, u) of $\mathcal{A}$ and (q, v) of $\mathcal{B}$, we have $(p, u) \approx (q, v)$ if there are finitely many word types $w \in \Sigma^*$ such that $(p, u) \xrightarrow{w} (p', u')$ and $(q, v) \xrightarrow{w} (q', v')$ with $p' \in F_{\mathcal{A}} \not\Leftrightarrow q' \in F_{\mathcal{B}}$, where $F_{\mathcal{A}}$ and $F_{\mathcal{B}}$ denote the sets of accepting states of $\mathcal{A}$ and $\mathcal{B}$, respectively. All properties of almost-equivalence extend to this setting.

To compute the almost-equivalence relation over states, we develop the following DRA analogue of Lemma 7.

Lemma 10. *Let* $(p, u), (q, v)$ *be two configurations of a DRA* $\mathcal{A}$. $(p, u) \approx (q, v)$, *iff there exists* $\ell \in \mathbb{N}$ *s.t. for all* $w \in \Sigma^*$ *with* $|w| > \ell$, *the configurations reached from* (p, u) *and* (q, v) *after reading* w *either both accepting or both rejecting.*

Using Lemma 10, one can prove that the successors of almost-equivalent configurations over a given letter are also almost-equivalent.

Lemma 11. *Let $\mathcal{A}$ be a DRA over (Σ, R), and let c and d be two configurations with $c \approx d$. For every $a \in \Sigma$, if $c \xrightarrow{a} f$ and $d \xrightarrow{a} g$, then $f \approx g$.*

Further, by combining Lemma 6 and Lemma 10, we immediately derive:

Corollary 12. *Let $\mathcal{A}$ be a DRA and $(p, u), (q, v)$ be two configurations of $\mathcal{A}$. $(p, u) \approx (q, v)$, if and only if for all $w \in \Sigma^{>len^2_{\mathcal{A}}}$, the configurations reached from (p, u) and (q, v) after reading w either both accepting or both rejecting.*

Note that if (p, u) and (q, v) in the lemma are from different DRAs $\mathcal{A}$ and $\mathcal{B}$, where $\mathcal{A}$ has m states and k registers and $\mathcal{B}$ has n states and ℓ registers, then the bound $len^2_{\mathcal{A}}$ should be replaced with $m \cdot n \cdot (k + \ell)!$.

The equivalence problem for DRAs is decidable because it can be reduced to the emptiness problem using the product automaton technique; emptiness is itself decidable for DRAs [4,11]. Note that if an automaton recognizes the empty language, its minimal form consists of a single rejecting state. The minimization problem for DRAs is also decidable [4]. So by Corollary 12, the configuration almost-equivalence problem is also decidable if we can compute DRAs recognizing $L(\mathcal{A}_{(p,u)}) \cap \Sigma^{\geq \ell}$ and $L(\mathcal{A}_{(q,v)}) \cap \Sigma^{\geq \ell}$. Let $\mathcal{A}_\ell$ be a DFA that recognizes the language $\Sigma^{\geq \ell}$. Consequently, the product automaton of $\mathcal{A}_{(p,u)}$ and $\mathcal{A}_\ell$ is a DRA that recognizes $L(\mathcal{A}_{(p,u)}) \cap \Sigma^{\geq \ell}$. Analogously, the product automaton of $\mathcal{A}_{(q,v)}$ and $\mathcal{A}_\ell$ is a DRA recognizing $L(\mathcal{A}_{(q,v)}) \cap \Sigma^{\geq \ell}$. As a result, we have:

Lemma 13. *Let $\mathcal{A}$ be a DRA, and let c, d be two configurations of $\mathcal{A}$. It is decidable to determine whether $c \approx d$.*

By Definition 9, when checking whether two states are almost-equivalent, it seems that we need to enumerate all possible register words of the two states (which are infinite), and check the almost-equivalence relation over configurations using Lemma 10. Nevertheless, with Lemma 14, we only need to consider one configuration for a state.

Lemma 14. *lemAlmostEqConfigurationImpAlmostEqStates Let $\mathcal{A}$ be a DRA, and let (p, u) and (q, v) be two configurations of $\mathcal{A}$. If $(p, u) \approx (q, v)$ (resp., $(p, u) \equiv (q, v)$), then $p \approx q$ (resp., $p \equiv q$).*

Let p and q be two states of $\mathcal{A}$ with register types τ and η, respectively, and let u be a word of type τ. By Lemma 14, we have $p \approx q$ if and only if there exists a word v of type η such that $(p, u) \approx (q, v)$. Moreover, by Lemma 13, almost-equivalence of configurations is decidable. Hence, to decide whether $p \approx q$, it suffices to determine whether there exists such a word v.

The main difficulty lies in the fact that there are infinitely many candidates for v. Our key observation is that it is unnecessary to consider all such words. Instead, one can effectively compute a *finite* set $W_\eta \subset \Sigma$ such that there exists a word v with $(p, u) \approx (q, v)$ if and only if there exists v' over W_η satisfying $(p, u) \approx (q, v')$. The following lemma helps define the finite set W_η.

Lemma 15. *Let $\mathcal{A}$ be a DRA over a dense domain and $(p, u), (q, v)$ be two configurations of $\mathcal{A}$. Then, if $(p, u) \approx (q, v)$, then for every word $v' \in \Sigma^*$ with $uv \sim_R uv'$, we have $(p, u) \approx (q, v')$.*

To examine whether there exists a register assignment v for a given configuration (p, u) and a state q such that $(p, u) \approx (q, v)$, Lemma 15 suggests that we only need to consider the word type of the concatenation $u \cdot v$; the exact values in v do not affect the result. Therefore, given a word u and v of η-type with $|\eta| = k$, to compute the alphabet W_η of v' that cover all possible types of $u \cdot v$ (and thus uv'), i.e., $uv \sim_R uv'$, we do the following:

1. Let $a_1, a_2, \cdots, a_m$ be the increasing sequence of the letters in u. Let $a_0 = -\infty$ and $a_{m+1} = +\infty$.
2. Let $b_{i,1}, b_{i,2}, \cdots, b_{i,k}$ be k fresh letters in increasing order such that $a_i < b_{i,j} < a_{i+1}$ for all $0 \leq i \leq m$ and $1 \leq j \leq k$.
3. Then, we define $W_\eta = u \cup \bigcup_{i=0}^{m} \{b_{i,1}, \cdots, b_{i,k}\}$.

Note that if the binary relation R is identity test, then we only need to add all letters in u together with k fresh letters into W_η.

Hence, according to Lemma 15, it is not hard to see:

Corollary 16. *Let p has register type u and q has register type η. Let W_η be the finite set computed above. Then, there exists a word v such that $(p, u) \approx (q, v)$ if, and only if, there exists a word v' over W_η such that $(p, u) \approx (q, v')$.*

We now give an example to explain how to compute W_η. Consider a DRA $\mathcal{A}$ over $(\mathbb{Q}, <)$: let p and q be two states where p has a register type $\tau = 1 \cdot 2$ while q has the register type $\eta = 2 \cdot 1$. Recall that τ and η represent the relative order of registers. Let (p, u) be a configuration of p where $u = 3 \cdot 7$. To cover all possible configurations (q, v') associated with q, it suffices to consider valuations v over the finite alphabet $W_\eta = \{3, 7\} \cup \{1, 2\} \cup \{4, 5\} \cup \{9, 11\}$, constructed as follows. Let $a_1 = 3$ and $a_2 = 7$ in u, and define $a_0 = -\infty$ and $a_3 = +\infty$.

- Between $a_0 = -\infty$ and $a_1 = 3$, we pick $b_{0,1} = 1$ and $b_{0,2} = 2$.
- Between $a_1 = 3$ and $a_2 = 7$, we select $b_{1,1} = 4$ and $b_{1,2} = 5$.
- Between $a_2 = 7$ and $a_3 = +\infty$, we choose $b_{2,1} = 9$ and $b_{2,2} = 11$.

Let p and q be two states in $\mathcal{A}$ where τ is the register type of p and η is the register type of q. To check whether $p \approx q$ holds, we perform the following steps:

1. Find a word u of type τ.
2. Compute the finite set W_η based on u and η.
3. Enumerate all v' of type η over W_η and check whether $(p, u) \approx (q, v')$.
4. If there is a word v' with $(p, u) \approx (q, v')$, then we conclude that $p \approx q$; otherwise, we know that $p \not\approx q$.

Since words of a given type over the finite set W_η can be enumerated and, by Lemma 13, the configuration almost-equivalence problem for DRAs is decidable, we immediately obtain the following result:

Theorem 17. *The state almost-equivalence problem for DRAs is decidable.*

Algorithm 1: Hyper-minimization of a DRA $\mathcal{A}$.

```
A = minimize(A);
A = hyper-data-minimize(A);
while ∃p ≠ q with the same register type.(p ≈ q and p is a preamble) ∨ (p ≡ q) do
    A = merge(p → q, A);
return A;
```

Algorithm 2: The `hyper-data-minimize` step of a minimal DRA $\mathcal{A}$.

```
for preamble state q ∈ Q and w is a shortest word leading A to q do
    Let u be such that (q_0, ε) --w--> (q, u) in the original well-typed DRA;
    for symbol a ∈ u do
        if a is not len²_A-memorable then
            remove a from u;
    make q u-register-type;
```

4.2 The Hyper-Minimization Process for DRAs

With the check for the almost-equivalence relation between two states established, we now present Algorithm 1 to hyper-minimize a given DRA $\mathcal{A}$. For ease of explanation, we assume that every preamble state can reach an accepting state via infinitely many distinct paths in the underlying graph. Otherwise, such a state can simply be removed from the automaton, as the resulting language remains almost-equivalent to the original.

Similar to the process for DFAs, we first minimize the DRA. However, before entering the loop to merge almost-equivalent states, we perform a `hyper-data-minimize` step to ensure that only those symbols that can affect the acceptance of infinitely many extensions are kept in the registers. The details of this `hyper-data-minimize` step are provided in Algorithm 2. We then check whether there exist almost-equivalent states p and q such that p is a preamble state and p, q share the *same* register type; if so, they can be merged.

First, consider the process of hyper-data-minimization. Recall that a symbol in a word x is memorable if its appearance in some word extension w is relevant to whether $x \cdot w$ is accepted. Furthermore, in a DRA recognizing the language, every memorable symbol must be kept in the registers. However, there are no restrictions on the length of w. On the contrary, by Corollary 12, we have that two configurations (p, u) and (q, v) of a given DRA $\mathcal{A}$ are almost-equivalent if and only if for every word w of length greater than $len^2_{\mathcal{A}}$, the runs from these configurations on w are either both accepting or both rejecting. This suggests that if a symbol $a \in u$ is not relevant to the acceptance of any word w with length greater than $len^2_{\mathcal{A}}$, then a hyper-data-minimal DRA for $\mathcal{A}$ does not need to keep a in its registers. To formalize this, we introduce the following definition:

Definition 18 (ℓ-Memorable Letters). *Let $\mathcal{A}$ be a DRA, (p, u) be a configuration, and $\ell \in \mathbb{N}$. A symbol $a \in u$ is called ℓ-memorable if there exist a word $w \in \Sigma^{\geq \ell}$ and a symbol b such that $w \sim_R w[a/b]$, and the configurations reached from (p, u) after reading w and $w[a/b]$, respectively, are such that exactly one of them is accepting.*

The notion of ℓ-memorability identifies whether a specific data value stored in a register is essential for the future acceptance of word extensions of length at least ℓ. Since the hyper-data-minimization process relies on removing symbols that are not $len^2_{\mathcal{A}}$-memorable, it is necessary to ensure that this property can be effectively checked. Note the distinction from the notion of L-memorability (cf. Definition 4): here, l is a parameter in $\mathbb{N}$.

Lemma 19. *Let $\mathcal{A}$ be a DRA, (p, u) a configuration of $\mathcal{A}$, and $\ell \in \mathbb{N}$. It is decidable whether a symbol in u is ℓ-memorable.*

By Corollary 12, two configurations are almost-equivalent if and only if their runs on every word extension of length greater than $len^2_{\mathcal{A}}$ either both accept or both reject. One might therefore assume that the hyper-minimization process aims to produce an automaton that does not retain any non-$len^2_{\mathcal{A}}$-memorable symbols. However, given a configuration (p, u), if p is a kernel state and a symbol $a \in u$ is memorable, then a must be retained regardless of whether it is $len^2_{\mathcal{A}}$-memorable. Otherwise, by the definitions of kernel states and memorability, the resulting automaton would not be almost-equivalent to the original. Accordingly, we have the following lemma:

Lemma 20. *Let $\mathcal{A}$ be a minimal DRA. Then, $\mathcal{A}$ is hyper-data-minimal if and only if for every configuration (p, u) of $\mathcal{A}$ where p is a preamble state, every symbol $a \in u$ is $len^2_{\mathcal{A}}$-memorable.*

Recall our assumption that all preamble states can reach an accepting state via infinitely many distinct paths. Without this assumption, there could exist a configuration (p, u) in a hyper-data-minimal $\mathcal{A}$ where p is a preamble state and $a \in u$. In such a case, the symbol a would not be $len^2_{\mathcal{A}}$-memorable simply because the language recognized from p contains only finitely many word types; consequently, none of the values in u would be $len^2_{\mathcal{A}}$-memorable.

Intuitively, our hyper-data-minimization process ensures that the automaton does not retain any non-$len^2_{\mathcal{A}}$-memorable symbols for preamble states. However, when a non-$len^2_{\mathcal{A}}$-memorable symbol is removed from a state's register type, the automaton may become non-deterministic. For instance, suppose state p has register type $1 \cdot 3 \cdot 5$, and there are two transitions $(p, 1 \cdot 3 \cdot 5 \cdot 2, E, q)$ and $(p, 1 \cdot 3 \cdot 5 \cdot 4, E', q')$. If we remove the value at index 3 from the types $1 \cdot 3 \cdot 5$, $1 \cdot 3 \cdot 5 \cdot 2$, and $1 \cdot 3 \cdot 5 \cdot 4$, then p may have non-deterministic outgoing transitions. To resolve this problem, we require the following lemma:

Lemma 21. *Let p be a preamble state, (p, u) a configuration of a DRA $\mathcal{A}$ over (Σ, R) with $a \in u$, and let v be the register assignment obtained by removing a from u. Let $\delta = (p, \tau, E, q)$ and $\delta' = (p, \tau', E', q')$ be two outgoing transitions* *from* p, *where* $\tau = u \cdot b$ *and* $\tau' = u \cdot b'$ *such that* $v \cdot b \sim_R v \cdot b'$. *If* a *is not* $len^2_{\mathcal{A}}$-*memorable in* u, *then the automaton obtained by removing* δ *or* δ' *and setting* p *to be* v-*typed is almost-equivalent to* $\mathcal{A}$.

Using Lemma 20, we can derive the hyper-data-minimization process for DRAs, as shown in Algorithm 2. For each preamble state q:

- Obtain a representative word w such that $(q_0, \epsilon) \xrightarrow{w} (q, u)$ for some configuration u in the original input DRA.
- Check every symbol $a \in u$ to determine whether a is $len^2_{\mathcal{A}}$-memorable, which is decidable by Lemma 19.
- For every incoming transition (p, τ, E, q) to q, modify the set E s.t. any symbol that is not $len^2_{\mathcal{A}}$-memorable is no longer kept in the registers upon entering q.
- Update the word type η of every outgoing transition (q, η, G, r) from q. By Lemma 21, if there is more than one such transition, we can arbitrarily retain one of them.

Next, we consider the merging of almost-equivalent states p and q in Algorithm 1, starting from Line 3. There are two cases in which we cannot merge p and q: (i) Both p and q are kernel. (ii) p and q have different register types.

Case (i) is also considered in DFA hyper-minimization. Let $L_{q_0 \to p}$ be the set of words for which the run from the initial configuration ends in p. The set $L_{q_0 \to p}$ is finite if and only if p is a preamble state. Since $p \approx q$, the set $L_{q_0 \to p} \cdot (L(\mathcal{A}_p) \ominus L(\mathcal{A}_q))$ contains only finitely many word types if and only if $L_{q_0 \to p}$ contains only finitely many word types. Accordingly, we can merge p into q only if p is a preamble state; otherwise, the resulting automaton would not be almost-equivalent to the original.

Consider Case (ii). If $(p, u) \approx (q, v)$, Lemma 20 implies that for every symbol a, $a \in u \Leftrightarrow a \in v$. However, if p and q have different register types, then $u \not\sim_R v$. Suppose w, x, y are three words such that $|w| > len^2_{\mathcal{A}}$, $(q_0, \epsilon) \xrightarrow{x} (p, u) \xrightarrow{w} c$, and $(q_0, \epsilon) \xrightarrow{y} (q, v) \xrightarrow{w} d$ for some configurations c and d. Since $(p, u) \approx (q, v)$, c and d are either both accepting or both rejecting. Because all values in u and v are $len^2_{\mathcal{A}}$-memorable, we can assume w.l.o.g. that modifying these values would cause c and d to switch their acceptance status (from accepting to rejecting, or vice versa). Furthermore, since $u \not\sim_R v$, we cannot redirect transitions ending in p to q without eliminating values in u and v. Accordingly, any such elimination would result in an automaton that is no longer almost-equivalent to the original.

Lemma 22. *Let* $\mathcal{A}$ *be a DRA, and* p, q *be two almost-equivalent states of* $\mathcal{A}$ *with the same register type. If* p *is preamble, then the DRA* $\mathcal{B}$ *obtained by performing the merging operation* $\texttt{merge}(p \to q, \mathcal{A})$ *is almost-equivalent to* $\mathcal{A}$.

So, according to Lemma 22, at the end of Line 4 of Algorithm 1, the resulting DRA is almost-equivalent to the input DRA.

Lemma 23. *For every two almost-equivalent hyper-data-minimal DRAs* $\mathcal{A}$ *and* $\mathcal{B}$ *with sets* $Q_{\mathcal{A}}$ *and* $Q_{\mathcal{B}}$ *of states, respectively, there exists a function* $h : Q_{\mathcal{A}} \to Q_{\mathcal{B}}$ *such that for all* $q \in Q_{\mathcal{A}}$:

(a) $q \approx h(q)$ *and* $q, h(q)$ *have the same register type.*
(a) $h(q)$ *is a kernel state if* q *is a kernel state.*

Suppose that $\mathcal{A}$ is the DRA obtained at the end of Line 4 of Algorithm 1, $\mathcal{B}$ is a hyper-state-minimal DRA, and h is the mapping satisfying Lemma 23. For every $p, q \in Q_{\mathcal{A}}$, if $h(p) = h(q)$, then by Item (a), we have $p \approx q$, and they have the same register type. If at least one of p or q is a preamble state, then they must be the same state (i.e., $p = q$); otherwise, they would have been merged during the execution of Algorithm 1. If both are kernel states, then by Lemma 14 and Item (a), it follows that $p \equiv q$. In this case, we must also have $p = q$, as they otherwise would have been merged by the algorithm. Accordingly, h is an injective mapping, and the DRA resulting from Line 4 of Algorithm 1 is a hyper-state-minimal DRA for the given input. Consequently, we obtain the following result:

Theorem 24. *For every input DRA* $\mathcal{A}$*, the output DRA of Algorithm 1 is a hyper-minimal DRA for the language* $L(\mathcal{A})$*.*

5 Conclusion

We introduced a notion of hyper-minimization for DRAs and presented an algorithm to compute hyper-minimal DRAs. To the best of our knowledge, this is the first work addressing hyper-minimization in this setting.

Our algorithm follows the general structure of classical hyper-minimization for DFAs: it checks almost-equivalence between states and performs merge operations. Extending this approach to DRAs, however, is technically challenging. In particular, it requires an additional hyper-data-minimization step, and all components—hyper-data-minimization, almost-equivalence checking, and state merging—must be carefully designed to ensure that the resulting automaton is both hyper-data-minimal and hyper-state-minimal. Finally, we note that, as in the DFA setting, hyper-minimal DRAs are *not* unique.

An interesting direction for future work is to investigate the precise complexity of hyper-minimization for DRAs. In particular, it remains open whether hyper-minimization can be performed in polynomial time for DRAs with identity tests, for which language equivalence checking is known to be polynomial [13].

Acknowledgements. We would like to thank the anonymous reviewers for their suggestions that helped improve the paper. This work was supported in part by the National Natural Science Foundation of China (Grant No. 62102407), CAS Project for Young Scientists in Basic Research (Grant No. YSBR-040), ISCAS New Cultivation Project ISCAS-PYFX-202201, and the Engineering and Physical Sciences Research Council (EPSRC) through grant EP/X042596/1. This project is part of the European Union's Horizon Europe programme under the Marie Skłodowska-Curie grant agreement No. 101208673.

References

1. Badr, A.: Hyper-minimization in $O(n^2)$. In: Ibarra, O.H., Ravikumar, B. (eds.) Implementation and Applications of Automata (CIAA 2008). Lecture Notes in Computer Science, vol. 5148, pp. 223–231. Springer, Heidelberg (2008)
2. Badr, A., Geffert, V., Shipman, I.: Hyper-minimizing minimized deterministic finite state automata. RAIRO Theor. Inform. Appl. **43**(1), 69–94 (2009)
3. Balachander, M., Filiot, E., Gentilini, R., Tzevelekos, N.: Register automata with permutations. In: Gawrychowski, P., Mazowiecki, F., Skrzypczak, M. (eds.) 50th International Symposium on Mathematical Foundations of Computer Science, MFCS 2025, August 25-29, 2025, Warsaw, Poland, vol. 345. LIPIcs, pp. 14:1–14:18. Schloss Dagstuhl - Leibniz-Zentrum für Informatik, 2025
4. Benedikt, M., Ley, C., Puppis, G.: Long version of What You Must Remember When Processing Data Words (2010). Technical report Minimal memory automata
5. Chen, Y.-F., Wang, B.-Y., Yen, D.-D.: A finite exact representation of register automata configurations. In: Holik, L., Clemente, L. (eds.) Proceedings 15th International Workshop on Verification of Infinite-State Systems, Hanoi, Vietnam, 14th October 2013, volume 140 of Electronic Proceedings in Theoretical Computer Science, pp. 16–34. Open Publishing Association (2014)
6. D'Antoni, L., Veanes, M.: Minimization of symbolic automata. In: Proceedings of the 41st ACM SIGPLAN-SIGACT Symposium on Principles of Programming Languages, POPL '14, pp. 541–553. Association for Computing Machinery, New York (2014)
7. Holzer, M., Maletti, A.: An nlogn algorithm for hyper-minimizing a (minimized) deterministic automaton. Theoret. Comput. Sci. **411**(38), 3404–3413 (2010). Implementation and Application of Automata (CIAA 2009)
8. Hopcroft, J.E.: Introduction to Automata Theory, Languages, and Computation, Pearson Addison Wesley (2007). 3rd edition
9. Jeż, A., Maletti, A.: Hyper-minimization for deterministic tree automata. In: Proceedings of the 17th International Conference on Implementation and Application of Automata, CIAA'12, pp. 217–228. Springer, Berlin (2012)
10. Jiang, T., Ravikumar, B.: Minimal nfa problems are hard. SIAM J. Comput. **22**(6), 1117–1141 (1993)
11. Kaminski, M., Francez, N.: Finite-memory automata. Theoret. Comput. Sci. **134**(2), 329–363 (1994)
12. Maletti, A., Quernheim, D.: Hyper-minimisation of deterministic weighted finite automata over semifields. In: International Conference on Automata and Formal Languages (2011)
13. Murawski, A.S., Ramsay, S.J., Tzevelekos, N.: Polynomial-time equivalence testing for deterministic fresh-register automata. In: Potapov, I., Spirakis, P.G., Worrell, J. (eds.) 43rd International Symposium on Mathematical Foundations of Computer Science, MFCS 2018, August 27-31, 2018, Liverpool, UK, vol. 117. LIPIcs, pp. 72:1–72:14. Schloss Dagstuhl - Leibniz-Zentrum für Informatik (2018)
14. Nerode, A., Saeks, R.: Fundamental concepts in the theory of systems. WADC Technical Report 57-624, Wright Air Development Center, November 1957
15. Neven, F., Schwentick, T., Vianu, V.: Finite state machines for strings over infinite alphabets. ACM Trans. Comput. Logic **5**(3), 403–435 (2004)
16. Royden, H., Fitzpatrick, P.: Real Analysis. Prentice Hall (2010)

On the State Complexity of Two-Way Reversible Finite Automata

Daniil Lyubaev, Maria Radionova(✉), and Alexander Okhotin

Department of Mathematics and Computer Science, St. Petersburg State University, 7/9 Universitetskaya nab., Saint Petersburg 199034, Russia
st084881@student.spbu.ru, alexander.okhotin@spbu.ru

Abstract. It is proved that every two-way reversible finite automaton (2RFA) with n states can be simulated by a partial one-way deterministic finite automaton (1DFA) with $\varphi(n)$ states, where φ is a certain integer function of growth rate $\varphi(n) = O\big((\frac{n}{2})^{n/2+3/2}\big)$. Furthermore, it is shown that exactly $\varphi(n)$ states are necessary in the worst case, with witness languages defined by sweeping reversible automata over a 4-symbol alphabet. The second result of the paper is a new lower bound $2n + \frac{2n}{\ln n + o(1)}$ on the complexity of transforming a deterministic two-way finite automaton (2DFA) to a 2RFA, cf. the known upper bound of $4n+3$ states.

Keywords: Finite automata · Two-way automata · Reversible automata · Sweeping automata · Reversibility · State complexity

1 Introduction

Different kinds of two-way finite automata are known: deterministic (2DFA), nondeterministic (2NFA), unambiguous (2UFA), alternating (2AFA) and reversible (2RFA); all of them recognize the same family of languages: the regular languages. However, very little is known about the number of states needed to transform one kind of two-way automata to another. In particular, whether every n-state 2NFA can be transformed to a 2DFA with polynomially many states in n is a major open problem notable for its close connection to the famous problem in the complexity theory of whether L (the deterministic logarithmic space) equals NL (the nondeterministic logarithmic space), which was discovered by Berman and Lingas [1]. Later, Kapoutsis [9] established a related characterization of the NL *vs.* L/poly problem in terms of 2NFA and 2DFA. Still, all that is known for certain is that transforming an n-state 2NFA to a 2DFA requires at least $\Omega(n^2)$ states. Among the few known results on the size of two-way automata, there are also constructions for complementing 2DFA and unary 2NFA by Geffert, Mereghetti and Pighizzini [5], and for complementing 2AFA by Geffert, Kapoutsis and Zakzok [3].

This work was supported by the Russian Science Foundation, project 23-11-00133.

T. J. Smith (Ed.): CIAA 2026, LNCS 16695, pp. 138–150, 2026.
https://doi.org/10.1007/978-3-032-31176-4_10

What is known much better about two-way automata is the complexity of transforming them to various classes of one-way automata. The classical transformation of 2DFA to one-way deterministic automata (1DFA) by Shepherdson [18] was later refined by Kapoutsis [7], who proved that $n(n^n + (n-1)^n) + 1$ states are sufficient and in the worst case necessary. Kapoutsis [7] has also determined the precise number of states for other transformations between the main models: from n-state 2NFA or 2DFA to one-way nondeterministic automata (1NFA) with $\binom{2n}{n+1}$ states, and from n-state 2NFA to 1DFA with $2^{n^2-O(n)}$ states. Whereas the lower bounds by Kapoutsis [7] rely on an alphabet of exponential size, Geffert and Okhotin [6] showed that this transformation can be improved in the case of a subexponential-sized alphabet. In the case of a unary alphabet, Chrobak [2] showed that the transformation yields much fewer states, that is, of the order of Landau's function—the largest least common multiple of any partition of n, known to be $e^{(1+o(1))\sqrt{n \ln n}}$. Chrobak's bounds were later made more precise by Kunc and Okhotin [11]. The complexity of transforming unary two-way automata to 1DFA and 1NFA was studied by Mereghetti and Pighizzini [14] and by Geffert et al. [4]. Vardi [20] studied the complexity of transforming two-way automata to 1NFA recognizing the complement of the language. Petrov and Okhotin [15] obtained some bounds on the transformation of two-way automata to one-way unambiguous finite automata (1UFA). Recently, Radionova and Okhotin [17] established the exact number of states needed to transform sweeping permutation automata to one-way permutation automata.

Can these precise results on the transformations from two-way automata to one-way help in estimating any state complexity tradeoffs between different kinds of two-way automata? In theory, yes: for instance, since the transformation of 2NFA to 1DFA requires more states than the transformation of 2DFA to 1DFA, this implies a certain lower bound on the transformation of 2NFA to 2DFA. However, since the growth rates of the two functions are not that much different (ca. 2^{n^2} vs. ca. n^n), the resulting lower bound will be very weak, something like $\Omega(\frac{n^2}{\log n})$, and this will not contribute anything new.

The goal of this paper is to investigate the transformation of two-way reversible finite automata (2RFA) to 2DFA, and approach it by first estimating the complexity of transforming 2RFA to partial 1DFA. The model of 2RFA was introduced by Kondacs and Watrous [10] in their study of quantum finite automata, and they proved that 2RFA recognize all regular languages, which makes them more powerful than one-way reversible finite automata (1RFA). Later Kunc and Okhotin [13] proved that an n-state 2DFA can be transformed to a 2RFA with $4n + 3$ states, and, in the case of a unary alphabet, to a 2RFA with $2n + 5$ states. Also, Kunc and Okhotin [12] established a lower bound of $2n - 2$ states on the complexity of this transformation for a unary alphabet. In the case of an arbitrary alphabet, there is a gap between the lower bound $2n-2$ and the upper bound $4n + 3$. This paper aims to reduce this gap.

The first result of this paper, established in Sect. 3, is a transformation of an n-state 2RFA to a partial 1DFA with $\varphi(n)$ states, where φ is an integer function defined by a combinatorial condition. In the next Sect. 4, it is proved that exactly $\varphi(n)$ states are necessary in the worst case already for a transformation of sweeping reversible finite automata (sRFA) to partial 1DFA. Section 5 simplifies the definition of φ down to a combinatorial sum. The growth rate of $\varphi(n)$ is bounded as $\varphi(n) = O\big((\frac{n}{2})^{n/2+3/2}\big)$ in Sect. 6, confirming that this transformation is substantially more efficient than the transformation of 2DFA to 1DFA.

In the last Sect. 7, the notable difference in complexity between the transformations of 2RFA and 2DFA to 1DFA is used to infer a new lower bound $2n + \frac{2n}{\ln n + o(1)}$ on the number of states in a 2RFA needed to simulate an n-state 2DFA.

2 Definitions

Definition 1. *A* two-way deterministic finite automaton *(2DFA) is a quintuple $\mathcal{A} = (\Sigma, Q, q_0, \delta, F)$, in which*

- *Σ is a finite alphabet, which does not contain two special symbols: the left end-marker ($\vdash$) and the right end-marker ($\dashv$);*
- *Q is a finite set of states;*
- *$q_0 \in Q$ is the initial state;*
- *$\delta \colon Q \times (\Sigma \cup \{\vdash, \dashv\}) \to Q \times \{-1, +1\}$ is a partially defined transition function;*
- *$F \subseteq Q$ is the set of accepting states, effective at the right end-marker ($\dashv$).*

Given an input string $w \in \Sigma^*$, the 2DFA $\mathcal{A}$ operates on a read-only tape containing the string $\vdash w \dashv$. It begins its computation in the state q_0, with the head observing the left end-marker $\vdash$. At every step of the computation, when $\mathcal{A}$ is in a state $q \in Q$ and observes a square of the tape containing a symbol $a \in \Sigma \cup \{\vdash, \dashv\}$, the value $\delta(q, a)$ defines the next state and the direction of motion. The computation of $\mathcal{A}$ on w is defined uniquely; if $\mathcal{A}$ eventually reaches an accepting state in F while at the right end-marker ($\dashv$), this computation is accepting; otherwise, the automaton either encounters an undefined transition or gets into an infinite loop. The set of strings, on which the computation is accepting, is the language recognized by the 2DFA, denoted by $L(\mathcal{A})$.

Two-way automata of a special kind, which remember the direction of their motion and can turn only at the end-markers, are known as *sweeping automata* [19]. There is also a larger class of *direction-determinate automata* that remember the direction, in which they came to the current state, but may turn at any point. The condition of direction-determinacy is an essential element of the definition of two-way reversible finite automata, as given by Kondacs and Watrous [10]; it can also be defined as a separate notion as follows.

Definition 2. (Kunc and Okhotin [13]). *A 2DFA $\mathcal{A} = (\Sigma, Q, q_0, \delta, F)$ is said to be* direction-determinate, *if there is a partial mapping $d \colon Q \to \{-1, +1\}$, so that every transition $\delta(p, a) = (q, d)$ must satisfy $d = d(q)$. For each $d \in \{-1, +1\}$, denote $Q_d = \{\, q \mid d(q) = d \,\}$.*

Definition 3. *A direction-determinate 2DFA* $\mathcal{A} = (\Sigma, Q, q_0, \delta, F)$ *is called* sweeping *(sDFA), if* $\delta(p, a) = (q, d)$ *implies* $p, q \in Q_d$*, as long as* $a \notin \{\vdash, \dashv\}$.

Definition 4. (Kondacs and Watrous [10]**).** *A two-way reversible finite automaton (2RFA) is a direction-determinate 2DFA, in which no transitions from two different states on the same symbol could lead to a single state, that is, whenever* $\delta(p, a) = \delta(p', a) = (q, d(q))$*, the states* p *and* p' *must be the same.*

Notably, reversible automata cannot loop, unless the loop goes through the initial configuration [13, Lemma 2]: indeed, if a cyclic computation does not return to the initial configuration, then there is irreversibility at the point of entry to the cycle.

A *sweeping reversible automaton* (sRFA) is a 2DFA that is both reversible and sweeping. The class of languages recognized by sRFA is known to be smaller than the class of all regular languages [16], whereas 2RFA of the general form recognize all regular languages [10].

Partial one-way automata (partial 1DFA) are a special case of 2DFA, which move to the right after every transition. This makes the end-markers unnecessary, hence they are not used.

3 Transformation of 2RFA to 1DFA

The proposed new transformation of 2RFA to 1DFA is yet another refinement of the classical transformation of 2DFA to 1DFA by Shepherdson [18]. It is based on the same idea of having a 1DFA remember the outcome of all computations of the 2DFA starting in different states at the last symbol of the prefix read so far, as well as the first state in which the two-way automaton moves to the right of this last symbol. In the case of 2RFA, the number of different possible states of this kind turns out to be smaller than in the general case of 2DFA.

Lemma 1. *Let* $\mathcal{A} = (\Sigma, Q, q_0, \delta, F)$ *be a 2RFA, let* $Q = Q_{+1} \cup Q_{-1}$ *be the partition of its set of states into states reachable by transitions to the right and to the left. Then there exists a partial 1DFA recognizing the same language, with the following set of states.*

$$\{\, (\widehat{q}, f) \mid \widehat{q} \in Q_{+1}, f : Q_{-1} \to Q_{+1} \textit{ is an injective partial function}, \widehat{q} \notin \operatorname{Im} f \,\}$$

Proof. A 1DFA simulating $\mathcal{A}$ is first constructed using the adaptation of Shepherdson's [18] method for direction-determinate automata given by Geffert and Okhotin [6, Cor. 1]. Next, it is shown that, for a reversible 2DFA, the resulting 1DFA shall always contain some unreachable states. As the last step, those unreachable states will be removed to obtain the final automaton.

Denote the first version of the 1DFA by $\mathcal{B} = (\Sigma, Q', q_0', \delta', F')$. It has the following set of states.

$$Q' = \{\, (\widehat{q}, f) \mid \widehat{q} \in Q_{+1}, f : Q_{-1} \to Q_{+1} \text{ is a partial function} \,\}$$

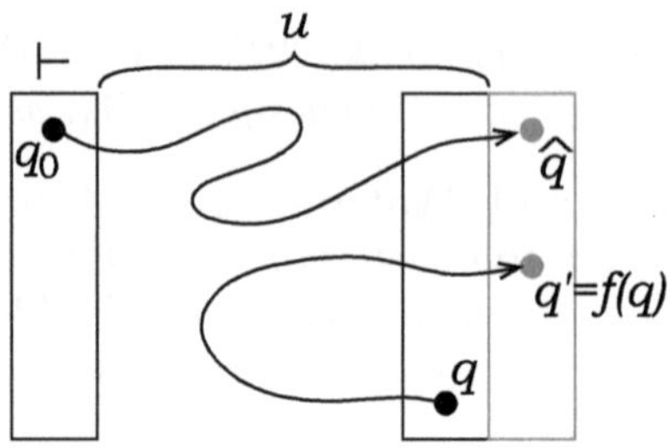

Fig. 1. How a state $(\widehat{q}, f)$ of the constructed 1DFA encodes the behaviour of the 2RFA on the prefix.

The idea is that $\mathcal{B}$, upon reading a prefix u, enters a state $(\widehat{q}, f)$, where $\widehat{q}$ is the first state in which $\mathcal{A}$ moves to the right of u, while f is a function that encodes the outcomes of all possible computations of $\mathcal{A}$ on the prefix u.

To be precise, the following correctness statement for the construction should hold.

Claim 1. *Upon reading a prefix $u \in \Sigma^*$, the 1DFA $\mathcal{B}$ reaches a state $(\widehat{q}, f)$, where:*

1. *$\widehat{q}$ is the state in which $\mathcal{A}$, having started in its initial configuration, first moves from the last symbol of u to the right, and*
2. *if $\mathcal{A}$, having started at the last symbol of u in a state $q \in Q_{-1}$, eventually moves from the last symbol of u to the right in a state $q' \in Q_{+1}$, and does so without having visited the initial configuration, then $f(q) = q'$; otherwise, $f(q)$ is undefined. These conditions are illustrated in (Fig. 1)*

The initial state is $q'_0 = (\widehat{q}_0, f_0)$, where $\delta(q_0, \vdash) = (\widehat{q}_0, +1)$, and f_0 is defined by $f_0(q) = q'$ if $q \neq q_0$ and $\delta(q, \vdash) = (q', +1)$. A small difference from the original construction is in having $f_0(q_0)$ undefined: this does not affect its correctness, because any computation that reuses the initial transition accordingly loops, and can be safely ignored.

The transition of the new automaton $\mathcal{B}$ in a state $(\widehat{q}, f) \in Q'$ upon reading the next symbol a determines the results of $\mathcal{A}$'s computations on a longer prefix extended with a. It is defined as $\delta'\big((\widehat{q}, f), a\big) = (\widehat{r}, g)$, where $\widehat{r}$ and g are defined as follows.

1. The state in which $\mathcal{A}$ first moves to the right of the symbol a is determined by first considering the transition $\delta(\widehat{q}, a)$. If $\delta(\widehat{q}, a) = (\widehat{r}, +1)$, then the desired state is $\widehat{r}$. Otherwise, if $\delta(\widehat{q}, a) = (p, -1)$, then the result of the computation beginning at the previous symbol is given by $f(p)$, and, next, the transition $\delta(f(p), a)$ is being similarly considered.

Overall, a sequence of the form $q_1, p_1, \ldots, q_k, p_k, \ldots$ is obtained, where $q_1 = \widehat{q}$, $\delta(q_i, a) = (p_i, -1)$ and $q_{i+1} = f(p_i)$ for all $i \geqslant 1$. This sequence may end in two ways: it may either finally have a transition to the right, $\delta(q_{k+1}, a) = (\widehat{r}, +1)$, so that $\widehat{r}$ is the desired state, or the sequence may bump into an undefined value of f or δ, in which case the entire transition $\delta'(\widehat{q}, f)$ is undefined.

Note that the sequence $\{q_k\}_{k \geqslant 1}$ cannot be infinite. If it were so, this would mean that the 2RFA loops, but since a reversible automaton can only loop through its initial configuration, whereas f only records computations that avoid the initial configuration, this is not possible.

2. The function $g\colon Q_{-1} \to Q_{+1}$, which describes the outcomes of all computations of $\mathcal{A}$ on a prefix extended by a, is defined separately for every argument $q \in Q_{-1}$. Consider $\delta(q, a)$. If $\delta(q, a) = (r, +1)$, then $g(q) = r$. Otherwise, let $\delta(q, a) = (p, -1)$; as in the previous case, a sequence $q_1, p_1, \ldots, q_k, p_k, \ldots$ is defined, with $q_1 = q$, $\delta(q_i, a) = (p_i, -1)$ and $q_{i+1} = f(p_i)$ for all $i \geqslant 1$. This time, there are three possibilities. First, the sequence may be infinite, that is, the 2RFA loops; in this case, $g(q)$ is undefined (note that looping does not contradict reversibility, as a self-contained loop with no incoming transitions is perfectly reversible). Secondly, the sequence may end in an undefined value of f or an undefined transition $\delta(q_i, a)$; then, $g(q)$ is undefined as well. Thirdly, the automaton may move from a to the right by a transition $\delta(q_{k+1}, a) = (r, +1)$, and the result is defined as $g(q) = r$.

The set of accepting states is defined by analyzing, for each state $(\widehat{q}, f)$, what happens if the right end-marker is read next. The automaton starts in the state $\widehat{q}$ at the end-marker, and the ensuing computation passes through a sequence of states $q_1, p_1, \ldots, q_k, p_k, \ldots$, constructed as in the definition of $\widehat{r}$ above. This time, no transitions to the right are possible, and the sequence can either reach an undefined transition, or enter an accepting state $q_k \in F$. In the latter case, the state $(\widehat{q}, f)$ is marked as accepting.

To see that the resulting 1DFA $\mathcal{B}$ recognizes the same language as the reversible 2DFA $\mathcal{A}$, one can prove the correctness statement by induction on the length of the prefix, and then consider the computations at the end-marker—just like in the transformations by Shepherdson [18] and Kapoutsis [8].

Let $Q'_{\text{reach}} \subseteq Q'$ be the set of reachable states in the 1DFA $\mathcal{B}$ constructed above.

Claim 2. *If $(\widehat{r}, g) \in Q'_{\text{reach}}$, then g is injective.*
Claim 3. *If $(\widehat{r}, g) \in Q'_{\text{reach}}$, then $\widehat{r} \notin \operatorname{Im} g$.*

Once all states found to be unreachable are removed from Q', the desired set of states given in the statement of the lemma is obtained. □

For each k and ℓ, denote by $\varphi(k, \ell)$ the number of pairs (j, f), where $j \in \{1, \ldots, k\}$ and $f\colon \{1, \ldots, \ell\} \to \{1, \ldots, k\}$ is an injective partial function, with $j \notin \operatorname{Im} f$. Then, Lemma 1 constructs a partial 1DFA with $\varphi(|Q_{+1}|, |Q_{-1}|)$ states.

4 Lower Bound on the 2RFA to 1DFA Tradeoff

In this section, it will be proved that the transformation of a two-way reversible automaton to a 1DFA given in Lemma 1 is actually optimal in terms of the number of states.

Lemma 2. *For every $\ell \geqslant 2$ and $k \geqslant \ell + 1$, there exists a sweeping reversible automaton (sRFA) defined over an alphabet $\Sigma = \{a, b, c, d\}$, with k right-bound states and ℓ left-bound states, such that every 1DFA recognizing the same language must have at least $\varphi(k, \ell)$ states.*

Proof. Let $Q_{+1} = \{r_0, \dots, r_{k-1}\}$ be the right-bound states of the sRFA, and let $Q_{-1} = \{p_0, \dots, p_{\ell-1}\}$ be its left-bound states. Let r_0 be the initial state (this, in particular, ensures that the initial configuration is not reenterable). The transitions are illustrated in Fig. 2 and are defined below. At the left end-marker, the automaton begins its computation by moving forward in its initial state, and in each state p_i it turns back in the state r_{i+1}.

$$\begin{aligned} \delta(r_0, \vdash) &= (r_0, +1) & \\ \delta(p_i, \vdash) &= (r_{i+1}, +1) & (0 \leqslant i \leqslant \ell - 1) \end{aligned}$$

Transitions on a implement a cycle on the states in Q_{+1}, and a acts as identity on Q_{-1}.

$$\begin{aligned} \delta(r_j, a) &= (r_{j+1 \bmod k}, +1) & (0 \leqslant j \leqslant k - 1) \\ \delta(p_i, a) &= (p_i, -1) & (0 \leqslant i \leqslant \ell - 1) \end{aligned}$$

The symbol b exchanges the first two states in Q_{+1}, and proceeds without changing the state in all other states.

$$\begin{aligned} \delta(r_j, b) &= (r_{1-j}, +1) & (j \in \{0, 1\}) \\ \delta(r_j, b) &= (r_j, +1) & (2 \leqslant j \leqslant k - 1) \\ \delta(p_i, b) &= (p_i, -1) & (0 \leqslant i \leqslant \ell - 1) \end{aligned}$$

On c, there is a cycle on Q_{-1} and the identity on Q_{+1}.

$$\begin{aligned} \delta(r_j, c) &= (r_j, +1) & (0 \leqslant j \leqslant k - 1) \\ \delta(p_i, c) &= (p_{i+1 \bmod \ell}, -1) & (0 \leqslant i \leqslant \ell - 1) \end{aligned}$$

The symbol d undefines the transition in r_0, and acts as identity on the rest of the states.

$$\begin{aligned} &\delta(r_0, d) \text{ is undefined} & \\ &\delta(r_j, d) = (r_j, +1) & (1 \leqslant j \leqslant k - 1) \\ &\delta(p_i, d) = (p_i, -1) & (0 \leqslant i \leqslant \ell - 1) \end{aligned}$$

At the right end-marker, the automaton turns back in the state r_0, accepts in the state r_{k-1}, and rejects in all remaining states.

$$\begin{aligned} &\delta(r_0, \dashv) = (p_0, -1) & \\ &\delta(r_j, \dashv) \text{ is undefined} & (1 \leqslant j \leqslant k - 1) \\ &\delta(p_i, \dashv) \text{ is undefined} & (0 \leqslant i \leqslant \ell - 1) \\ &F = \{r_{k-1}\} & \end{aligned}$$

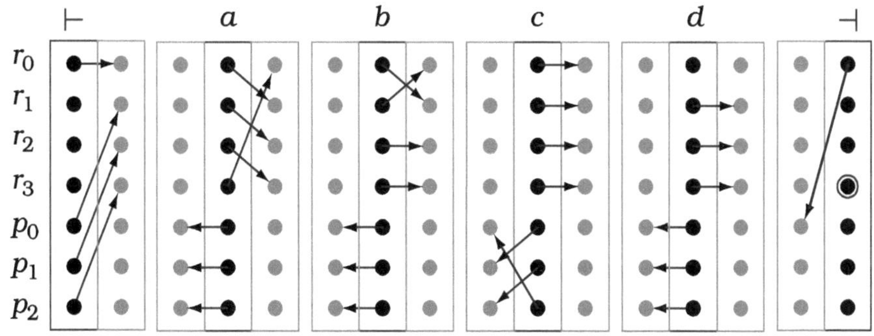

Fig. 2. Transitions of the sRFA with $k \geqslant \ell + 1$ in Lemma 2.

The first claim is essentially that, in the partial 1DFA constructed for the above sRFA using Lemma 1, every state is reachable.

Claim 4. *For every pair* $(\widehat{q}, f)$*, where* $\widehat{q} \in Q_{+1}$*,* $f : Q_{-1} \to Q_{+1}$ *is an injective partial function, and* $\widehat{q}$ *is not in the image of* f*, there exists a string* $u = u_{\widehat{q},f}$*, such that:*

1. *the sRFA first moves from the last symbol of* $\vdash u$ *to the right in the state* $\widehat{q}$*, and*
2. *if the sRFA begins at the last symbol of* $\vdash u$ *in a state* $p \in Q_{-1}$ *and eventually moves from the last symbol of* $\vdash u$ *to the right in a state* r*, then* $f(p) = r$*; otherwise,* $f(p)$ *is undefined.*

The proof is standard: first, the symbols a, b and d are used as generators to produce every partial injective function on Q_{+1}, and then it is sufficient to express the desired behaviour by such a function.

The second claim is that all the states reached in Claim 4 must be pairwise distinguishable.

Claim 5. *For every two strings* $u_{r_j,f}$ *and* $u_{r_{j'},f'}$*, constructed in Claim 4 for two distinct pairs* (r_j, f) *and* $(r_{j'}, f')$*, there exists a string* w*, such that one of the strings* $u_{r_j,f}w$ *and* $u_{r_{j'},f'}w$ *is accepted by the sRFA and the other is rejected.*

This is another standard construction, carried out separately for the cases of $r_j \neq r_{j'}$ and $r_j = r_{j'}$.

With Claims 4–5 established, one can see that every partial 1DFA recognizing the same language as the sRFA constructed above must have at least as many states as there are strings $u_{\widehat{q},f}$, where $\widehat{q} \in Q_{+1}$, $f : Q_{-1} \to Q_{+1}$ is an injective partial function, and $\widehat{q} \notin \operatorname{Im} f$—that is, at least $\varphi(k, \ell)$ states, as claimed. □

In Lemma 2, the desired lower bound has been established for the case of $k > \ell$. The proof in the case of $k \leqslant \ell$ is generally similar, with a few changes in the construction; it is omitted due to space constraints.

Lemma 3. *For every $k \geqslant 3$ and $\ell \geqslant k$, there exists a sweeping reversible automaton (sRFA) over an alphabet $\Sigma = \{a, b, c, d\}$, with k right-bound and ℓ left-bound states, such that every partial 1DFA recognizing the same language must have at least $\varphi(k, \ell)$ states.*

It has thus been proved that the precise state complexity of transforming 2RFA and sRFA to partial 1DFA is given by the function $\varphi(k, \ell)$, defined as the number of pairs (j, f), where $j \in \{1, \ldots, k\}$ and $f \colon \{1, \ldots, \ell\} \to \{1, \ldots, k\}$ is an injective partial function, with $j \notin \operatorname{Im} f$. The number of such pairs can be calculated by multiplying k ways to choose j by the number of partial injective functions from an ℓ-element set to a $(k-1)$-element set. If $i = |\operatorname{dom} f|$ is the number of values defined in f, which is within the bounds $0 \leqslant i \leqslant \min(k-1, \ell)$, then, for every fixed i, there are $\binom{\ell}{i}$ ways to choose the domain and $\binom{k-1}{i}$ ways to choose the image, and finally, $i!$ ways to define a bijection between this domain and this image. Overall, the number of pairs (j, f) is given by the following sum.

$$\varphi(k, \ell) = k \cdot \sum_{i=0}^{\min(k-1,\ell)} i! \binom{\ell}{i} \binom{k-1}{i}$$

Theorem 1. *For every two-way reversible finite automaton (2RFA) with k states reachable by right-moving transitions and ℓ states reachable by left-moving transitions, there exists a partial deterministic finite automaton (1DFA) with $\varphi(k, \ell) = k \cdot \sum_{i=0}^{\min(k-1,\ell)} i! \binom{\ell}{i} \binom{k-1}{i}$ states recognizing the same language. This number of states is necessary in the worst case, for all $k \geqslant 3$ and $\ell \geqslant 2$, already for the transformation of a sweeping reversible finite automaton (sRFA) to a partial 1DFA, and with witness languages defined over a 4-symbol alphabet.*

5 The Worst-Case Partition Into Right-Bound and Left-Bound States

In Theorem 1, the number of states in a partial 1DFA simulating a 2RFA of a given size was determined for any fixed values of k (the number of states reachable by moving to the right) and ℓ (the number of states reachable by left-moving transitions). The question is, what would be the required number of states if all that is known is that the 2RFA has n states in total, and its partition $n = k + \ell$ is not fixed? This is the standard notion of a succinctness tradeoff between two models, and the number of states that is sufficient and in the worst case necessary to do this transformation is then expressed by the following function.

$$\varphi(n) = \max_{n=k+\ell} \varphi(k, \ell)$$

In this section, a simpler expression for $\varphi(n)$ shall be obtained by determining the partition $n = k+\ell$ on which this maximum is reached. This optimal partition turns out to be $k = \lceil \frac{n+1}{2} \rceil$, and $\ell = \lfloor \frac{n-1}{2} \rfloor$.

Lemma 4. *$\varphi(n) = \varphi(\frac{n}{2}+1, \frac{n}{2}-1)$ for even n, and $\varphi(n) = \varphi(\frac{n+1}{2}, \frac{n-1}{2})$ for odd n.*

Proof. (a sketch) The proof is by showing that if k and ℓ are sufficiently different, then a higher value of φ could be obtained by incrementing the lesser of the two numbers and decrementing the greater of them. To be precise, it is proved that if $\ell \geqslant k$, then $\varphi(k, \ell) < \varphi(k+1, \ell-1)$, and if $\ell+3 \leqslant k$, then $\varphi(k, \ell) \leqslant \varphi(k-1, \ell+1)$.

Then, for every n, there is a unique partition that cannot be improved by such a redistribution: for even n, it is the one with $k = \ell+2$, which is $k = \frac{n}{2}+1$ and $\ell = \frac{n}{2}-1$; and for odd n, it satisfies $k = \ell+1$ and is therefore $k = \frac{n+1}{2}$, $\ell = \frac{n-1}{2}$. □

It remains to substitute the worst-case partition into $\varphi(k, \ell)$ to obtain the following consequence of Theorem 1.

Theorem 2. *For every two-way reversible finite automaton (2RFA) with n states, there exists a partial deterministic finite automaton (1DFA) recognizing the same language, with the following number of states.*

$$\varphi(n) = \begin{cases} \displaystyle\sum_{i=0}^{\frac{n}{2}-1} \binom{\frac{n}{2}-1}{i} \frac{(\frac{n}{2}+1)!}{(i+1)!}, & \text{if } n \text{ is even,} \\ \displaystyle\sum_{i=0}^{\frac{n-1}{2}} \binom{\frac{n-1}{2}}{i} \frac{(\frac{n+1}{2})!}{i!}, & \text{if } n \text{ is odd.} \end{cases}$$

This number of states is necessary in the worst case, for all $n \geqslant 1$, already for the transformation of sweeping reversible finite automata (sRFA) to partial 1DFA, and with witness languages defined over a 4-symbol alphabet.

6 Growth Rate of the 2RFA to 1DFA Tradeoff

The growth rate of the function $\varphi(n)$ representing the number of states in a partial 1DFA necessary to simulate every n-state 2RFA is not only interesting *per se*, but will also serve as a necessary tool for the ultimate result of this paper: the lower bound on the 2DFA to 2RFA tradeoff.

Theorem 3. *Let $\varphi(n)$ be the complexity of transforming 2RFA to partial 1DFA. Then, for every $n \geqslant 1$,*

$$\varphi(n) \leqslant \sqrt{\frac{2}{\pi}} \left(\frac{n}{2}\right)^{\frac{n}{2}+\frac{3}{2}} \exp\left(2\sqrt{\frac{n}{2}} - \frac{n}{2} + \frac{32}{9}\right).$$

Proof. (a sketch) For even n, the desired sum is $\sum_{i=0}^{\ell} \binom{\ell}{i} \frac{(\ell+2)!}{(i+1)!}$, where $\ell = \frac{n}{2}-1$, and it is estimated by first determining the largest summand. To this end, it is shown that the ratio of any two consecutive summands is greater than 1 for i up to $\sqrt{\ell+2}-2$, and is less than 1 afterwards. Hence, the largest summand is

$\binom{\ell}{i_*}\frac{(\ell+2)!}{(i_*+1)!}$, where $i_* = \lfloor\sqrt{\ell+2}\rfloor - 1$. Next, the whole sum is bounded by this summand times $\ell+1$, which is $(\ell+1)\binom{\ell}{i_*}\frac{\ell!\cdot(\ell+2)!}{i_*!\cdot(\ell-i_*)!\cdot(i_*+1)!}$. The latter expression is then estimated by using Stirling's approximation five times, and then doing some calculations on the resulting inequalities to obtain the following bound.

$$\varphi(n) = \sum_{i=0}^{\frac{n}{2}-1} \binom{\frac{n}{2}-1}{i} \frac{(\frac{n}{2}+1)!}{(i+1)!} \leqslant \sqrt{\frac{2}{\pi}} \left(\frac{n}{2}\right)^{\frac{n}{2}+\frac{3}{2}} \cdot \exp\left(2\sqrt{\frac{n}{2}} - \frac{n}{2} + \frac{32}{9}\right)$$

For n odd, the sum is $\sum_{i=0}^{\ell}\binom{\ell}{i}\frac{(\ell+1)!}{i!}$, where $\ell = \frac{n-1}{2}$, and it is estimated by the same method, resulting in the same upper bound.

$$\varphi(n) = \sum_{i=0}^{\frac{n-1}{2}} \binom{\frac{n-1}{2}}{i} \frac{(\frac{n-1}{2}+1)!}{i!} \leqslant \sqrt{\frac{2}{\pi}} \cdot \left(\frac{n}{2}\right)^{\frac{n}{2}+\frac{3}{2}} \exp\left(2\sqrt{\frac{n}{2}} - \frac{n}{2} + \frac{32}{9}\right)$$

□

7 Lower Bound on the 2DFA to 2RFA Transformation

So far we have established that the state complexity of transforming n-state 2RFA to equivalent partial 1DFA, represented by a function $\varphi(n) = O((\frac{n}{2})^{n/2+3/2})$, grows substantially slower than the state complexity of transforming 2DFA of the general form to partial 1DFA, which is $\psi(n) = n(n^n - (n-1)^n)$ [8]. It turns out that the difference in the growth rate between these two functions implies a lower bound on the complexity of transforming 2DFA to 2RFA, which is higher than the known lower bound by Kunc and Okhotin [12].

Example 1. There exists an 8-state 2DFA, for which every equivalent 2RFA must have at least 19 states.

Proof. Suppose the contrary, that every 8-state 2DFA can be transformed to a 2RFA with 18 states. Then the latter 2RFA can be further transformed to a partial 1DFA with $\varphi(18) = 45965530$ states. However, it is known from Kapoutsis [8] that there exist 2DFA with 8 states that require at least $\psi(8) = 88099320$ states in every equivalent partial 1DFA. The contradiction obtained proves the statement. □

Applying this argument to every n, using the bound on $\varphi(n)$ established in Theorem 3, one can obtain the following new lower bound on the complexity of transforming a 2DFA to a 2RFA.

Theorem 4. *For every $n \geqslant 3$, there exists a language defined over an alphabet of size $O(n^n)$, which is recognized by an n-state 2DFA, whereas every 2RFA recognizing this language must have at least $2n + \frac{2n}{\ln n + \frac{9}{\ln n}}$ states.*

Proof. (a sketch). Let the function $f(n)$ be the desired 2DFA to 2RFA tradeoff. It is known that $\psi(n) \leqslant \varphi(f(n))$, for otherwise one could transform a 2DFA to a smaller 1DFA than theoretically possible by transforming it first to an intermediate 2RFA, and then transforming the latter to a 1DFA. As closed-form expressions for $\varphi(n)$ and $\psi(n)$ are known, it remains to solve the equation in f. First, the precise expression for ψ is replaced with the following simpler function.

$$\psi(n) \geqslant n^{n+1}(1 - e^{-1})$$

For φ, the estimation from Theorem 3 is used.

$$\varphi(n) \leqslant \sqrt{\frac{2}{\pi}}\left(\frac{n}{2}\right)^{\frac{n}{2}+\frac{3}{2}} \exp\left(2\sqrt{\frac{n}{2}} - \frac{n}{2} + \frac{32}{9}\right)$$

Let $f(n) = 2n + 2g(n)$. Then the following chain of inequalities has been obtained.

$$n^{n+1}(1 - e^{-1}) \leqslant \psi(n) \leqslant \varphi(2n + 2g(n)) \leqslant$$
$$\leqslant \sqrt{\frac{2}{\pi}}\big(n + g(n)\big)^{n+g(n)+\frac{3}{2}} \exp\left(2\sqrt{n + g(n)} - n - g(n) + \frac{32}{9}\right)$$

Then, in order to obtain the desired lower bound on $f(n)$, it is proved that, under the substitution $g(n) = \frac{n}{\ln(n)+\frac{9}{\ln(n)}}$, the value of the leftmost expression is actually greater than the value of the rightmost expression for all $n > e^6$, that is, for all $n \geqslant 404$. For n between 3 and 403, the lower bound has been verified by calculating $\varphi(n)$ for n up to 923, and $\psi(n)$ for n up to 403. □

This new lower bound of $2n + \frac{2n}{\ln n+\frac{9}{\ln n}}$ states, obtained for an alphabet of size around n^n, is slightly greater than the existing lower bound of $2n - 2$ states established by Kunc and Okhotin [12] for a unary alphabet. However, the new lower bound is still far from the known upper bound of $4n + 3$ states [13]. Bringing the bounds on the complexity of the 2DFA to 2RFA transformation closer together is a question suggested for future research.

References

1. Berman, P., Lingas, A.: On complexity of regular languages in terms of finite automata. Report 304. Institute of Computer Science, Polish Academy of Sciences, Warsaw (1977)
2. Chrobak, M.: Finite automata and unary languages. Theoret. Comput. Sci. **47**, 149–158 (1986). https://doi.org/10.1016/0304-3975(86)90142-8
3. Geffert, V., Kapoutsis, C.A., Zakzok, M.: Complement for two-way alternating automata. Acta Informatica **58**(5), 463–495 (2020). https://doi.org/10.1007/s00236-020-00373-8
4. Geffert, V., Mereghetti, C., Pighizzini, G.: Converting two-way nondeterministic unary automata into simpler automata. Theoret. Comput. Sci. **295**(1–3), 189–203 (2003). https://doi.org/10.1016/S0304-3975(02)00403-6

5. Geffert, V., Mereghetti, C., Pighizzini, G.: Complementing two-way finite automata. Inf. Comput. **205**(8), 1173–1187 (2007). https://doi.org/10.1016/j.ic.2007.01.008
6. Geffert, V., Okhotin, A.: Deterministic one-way simulation of two-way deterministic finite automata over small alphabets. Int. J. Found. Comput. Sci., to appear, https://doi.org/10.1142/S0129054124420024
7. Kapoutsis, C.A.: Removing bidirectionality from nondeterministic finite automata. Mathematical Foundations of Computer Science (MFCS 2005: Gdańsk, Poland, 29 August-2 September 2005). LNCS **3618**, 544–555 (2005). https://doi.org/10.1007/11549345_47
8. Kapoutsis, C.A.: Algorithms and Lower Bounds in Finite Automata Size Complexity, Massachusetts Institute of Technology (2006). Ph. D. thesis
9. Kapoutsis, C.A.: Two-way automata versus logarithmic space. Theory Comput. Syst. **55**(2), 421–447 (2013). https://doi.org/10.1007/s00224-013-9465-0
10. Kondacs, A., Watrous, J.: On the power of quantum finite state automata. In: 38th Annual Symposium on Foundations of Computer Science (FOCS 1997: Miami Beach, Florida, USA, 19–22 October 1997). IEEE, pp. 66–75 (1997). https://doi.org/10.1109/SFCS.1997.646094
11. Kunc, M., Okhotin, A.: Describing periodicity in two-way deterministic finite automata using transformation semigroups. Developments in Language Theory (DLT 2011: Milan, Italy, 19–22 July 2011). LNCS, vol. 6795, pp. 324–336 (2011). https://doi.org/10.1007/978-3-642-22321-1_28
12. Kunc, M., Okhotin, A. https://tucs.fi/publications/view/?pub_id=tOkKu11a "Reversible two-way finite automata over a unary alphabet", TUCS Technical Report 1024, Turku Centre for Computer Science, December 2011
13. Kunc, M., Okhotin, A.: Reversibility of computations in graph-walking automata. Inf. Comput. **275** (2020). article 104631. https://doi.org/10.1016/j.ic.2020.104631
14. Mereghetti, C., Pighizzini, G.: Optimal simulations between unary automata. SIAM J. Comput. **30**(6), 1976–1992 (2001). https://doi.org/10.1137/S009753979935431X
15. Petrov, S., Okhotin, A.: On the transformation of two-way finite automata to unambiguous finite automata. Inf. Comput. **295A** (2023), article 104956. https://doi.org/10.1016/j.ic.2022.104956
16. Radionova, M., Okhotin, A.: A hierarchy of reversible finite automata. Implementation and Application of Automata (CIAA 2025: Palermo, Italy, 22–25 September 2025). LNCS, vol. 15981, pp. 316–329 (2025). https://doi.org/10.1007/978-3-032-02602-6_22
17. Radionova, M., Okhotin, A.: Sweeping permutation automata. Acta Informatica **63**(3) (2026), article 3. https://doi.org/10.1007/s00236-025-00519-6
18. Shepherdson, J.C.: The reduction of two-way automata to one-way automata. IBM J. Res. Dev. **3**, 198–200 (1959). https://doi.org/10.1147/rd.32.0198
19. Sipser, M.: Lower bounds on the size of sweeping automata. In: STOC 1979, 360–364. https://doi.org/10.1145/800135.804429
20. Vardi, M.: A note on the reduction of two-way automata to one-way automata. Inf. Process. Lett. **30**(5), 261–264 (1989). https://doi.org/10.1016/0020-0190(89)90205-6

Efficiently Finding Minimal Absent Subsequences in a String

Florin Manea(✉), Tina Ringleb, Stefan Siemer, and Maximilian Winkler

Department of Computer Science, University of Göttingen, Göttingen, Germany
{florin.manea,tina.ringleb,stefan.siemer, maximilian.winkler}@cs.uni-goettingen.de

Abstract. Given a string w, another string v is said to be a subsequence of w if v can be obtained from w by removing some of its letters; on the other hand, v is called an absent subsequence of w if v is not a subsequence of w. Of particular interest is the class of minimal absent subsequences, consisting of the absent subsequences whose every proper subsequence is not absent from w. The minimal absent subsequences define the border between the set of subsequences and that of absent subsequences, as each absent subsequence has, as subsequence, a minimal absent subsequence. Thus, the minimal absent subsequences seem fundamental to the understanding of both the set of subsequences and the set of absent subsequences of a string.

Our contributions to this area of research are related to the efficient design and implementation of automata, which can be used to index and enumerate minimal absent subsequences.

- We revisit the existing approaches related to the construction of subsequence automata and extend them by providing an optimal implementation of this automaton for strings over polylogarithmic alphabets – this is the first such optimal implementation of the respective automaton for non-constant alphabets.
- We show how to use the subsequence automaton for indexing minimal absent subsequences as efficiently as indexing subsequences or absent subsequences.
- We provide an efficient algorithm for determining the longest minimal absent subsequence, which significantly improves the state-of-the art solution for this problem.
- We show the first optimal algorithm (with linear time preprocessing and constant delay) for the enumeration of the set of minimal absent subsequences of a given string, based on an efficient implementation of an automaton accepting the respective set.

1 Introduction

A string v of length ℓ is a *subsequence* of a string w if there exist (possibly empty) strings $x_1, \ldots, x_{\ell+1}$ and letters $v_1, \ldots, v_\ell$ such that $v = v_1 \cdots v_\ell$ and $w = x_1 v_1 \cdots x_\ell v_\ell x_{\ell+1}$, i.e., v can be obtained from w by removing some of its

T. J. Smith (Ed.): CIAA 2026, LNCS 16695, pp. 151–170, 2026.
https://doi.org/10.1007/978-3-032-31176-4_11

letters. Any string that is not a subsequence of w is called an *absent subsequence* of w.

Subsequences (scattered factors, subwords) frequently appear in many different areas of computer science, e.g., in combinatorics on strings [23,43,46,48], formal languages [2,20,31,50,51] and automata theory [28,56,57], as well as in more applied settings: for modeling concurrency [45,49], in database theory, in settings related to event stream processing [6,27,58], in data mining [39,40], or in problems related to bioinformatics [14]. They also prominently appear in the area of algorithm-design. Classical problems on subsequences range from string-algorithmic problems, e.g., computing the longest common subsequences or shortest common supersequences (see [10] and the references therein), subsequence indexing [12,13,17] (see also [54] for indexing of *absent* subsequences), testing of the Simon's congruence of strings (see [26,52] and the references therein), to the combinatorial and algorithmic study of subsequences (or patterns) occurring in permutations (see [32,41,47]), e.g., the longest increasing subsequences [22]; see the survey [38] for a detailed discussion regarding combinatorial pattern matching problems related to subsequences. Algorithmic problems related to subsequences have also received attention in the context of fine-grained complexity [1,15], where lower bounds for their complexity were shown.

Absent subsequences were first introduced in [36] and appear in the following frameworks: in Simon's congruence (as a distinguishing subsequence, see, e.g., [26,44,50]), k-universality (as an obstacle to obtain a higher universality index) and m-nearly k-universality (where the aim is to find m absent subsequences of length k, see [3,21]), patterns avoided in permutations [32]. In the more applied setting of complex event processing, subsequences (resp., absent subsequences) are used to model sequences of events contained in (resp., avoided by) a string or stream (see, e.g., [24,33,34] and the references therein) formalising the trace of a computation [45,49]. Recall also the study of *missing factors* (substrings) of strings concerned with the set of strings which do not appear as a (contiguous) factor, which are of both theoretical (see, e.g., [7,11,18,19,25]) and practical interest (see [16] and the references therein).

Two classes of absent subsequences were investigated in the seminal paper [36] and subsequent works [3,37,54]. Firstly, the class of *minimal absent subsequences* (MAS) of string w consists of all absent subsequences whose every proper subsequence is, in fact, a subsequence of w. The set of minimal absent subsequences defines the border between absent and present subsequences: every absent subsequence contains a MAS as subsequence. Thus, the set of MASs is fundamental to understanding the set of absent subsequences. Secondly, the class of *shortest absent subsequences* (SAS) contains the absent subsequences of minimal length and thus consists of the extremal elements (w.r.t. length) of the set of absent subsequences; if w is k-universal, i.e., every string of length at most k is a subsequence of w and there is a string of length $k+1$ that is absent from w [29], then any SAS is a MAS of length $k+1$ (and there is no MAS of shorter length). Both combinatorial and algorithmic results regarding these two notions were obtained. In [3,36,37], combinatorial characterisations of the sets of MASs and, respectively, SASs in a string, tight bounds on the possible number of MASs and SASs, as well as compact representations of these sets were shown. In [36,37,54], efficient algorithms testing if a string is a SAS or MAS in a string were given, alongside efficient algorithms for counting and enumerating them.

§ *Our Contribution.* The focus of this paper is the efficient implementation of automata recognising absent subsequences, oriented towards two applications: optimal indexing and enumeration of minimal absent subsequences. Firstly, we revisit and extend the results of [12] regarding (absent) subsequence automata as follows. Given a string w of length n over an alphabet of size $\sigma = O(\log^k n)$, with $k \in O(1)$, we can build in $O(n)$ time data structures which encode the deterministic finite automaton recognising the (absent) subsequences of w, such that the transitions of this automaton can be simulated in $O(1)$ time. Besides the natural application of this result in indexing (absent) subsequences, it can also be used for indexing MASs of w: after preprocessing w in linear time, we can test whether a query-string u is a MAS of w in $O(|u|)$ time. We complement this with an algorithm obtaining a longest MAS of w in $O(n \log \sigma)$ time. Finally, as our main result, we propose an optimal algorithm for the enumeration of all MASs of an input string w of length n over an alphabet of size σ. This algorithm has an $O(n)$-time preprocessing and constant delay, and it relies on an efficient implementation of the MAS-automaton defined in [36]. We emphasise that the enumeration of the set of MASs of a string with optimal preprocessing and $O(1)$ delay is particularly interesting because, as we have already argued, seeing the elements of this set might provide a better understanding of the present/absent subsequences, and, since this set can be exponentially large and we might only be allowed some limited time to run our enumeration algorithm, it still guarantees that we are able to see a number of MASs proportional to the allowed runtime.

§ *Comparison to the State-of-the-Art.* Efficient implementations of the subsequence automaton, and its usage to recognise absent subsequences are discussed in [12,17,54]. More precisely, for a string of length n over an alphabet of size σ, the most efficient algorithms [12] needed either $O(n\sigma^{\epsilon})$-time preprocessing to perform every transition of the automaton in $O(1)$ time, or $O(n)$-time preprocessing to perform every transition of the automaton in $O(\log \log \sigma)$ time. These algorithms are optimal for $\sigma \in O(1)$; we show how an optimal algorithm can be obtained also for alphabets of polylogarithmic size – our result provides the first optimal algorithm for non-constant alphabets. As far as enumeration is concerned, the fastest known algorithm solving this problem required $O(n\sigma)$-time preprocessing with output-linear delay (see [54], where the results of [37] were already improved). Here, we solve this problem optimally. In [54], an algorithm running in $O(n\sigma)$ is given for the computation of the longest minimal absent subsequence; we also improve the respective algorithm significantly here.

Complete proofs for the results of this paper are given in the preprint [42], which includes the extended version of this paper. The preprint [42] also contains an optimal algorithm for the enumeration of the set of SASs.

2 Preliminaries

Let $\mathbb{N}$ be the set of natural numbers. For $m, n \in \mathbb{N}$, we let $[m : n] = \{m, m+1, \dots, n\}$ and $[n] = [1 : n]$. An alphabet Σ is a non-empty finite set of symbols called *letters*. A *string* is a finite sequence of letters from Σ; for the rest of the paper, we assume that all strings we consider are over an alphabet $\Sigma = \{1, 2, \dots, \sigma\}$. The set of all strings over Σ is denoted by Σ^*, and let $\Sigma^+ = \Sigma^* \setminus \{\varepsilon\}$, where ε is the empty string. The *length* of a string $w \in \Sigma^*$ is denoted by $|w|$; let $\Sigma^n = \{w \in \Sigma^* \mid |w| = n\}$. The ith letter of $w \in \Sigma^*$ is denoted by $w[i]$, for $i \in [|w|]$. For $m, n \in \mathbb{N}$, we let $w[m : n] = w[m]w[m+1] \cdots w[n]$ and $|w|_a = |\{i \in [|w|] \mid w[i] = a\}|$. A string $u = w[m : n]$ is a *factor* of w, and we have $w = xuy$ for some $x, y \in \Sigma^*$. If $x = \varepsilon$ (resp., $y = \varepsilon$), u is called a *prefix* (resp., *suffix*) of w. Let $\mathrm{alph}(w) = \{x \in \Sigma \mid |w|_x > 0\}$ be the smallest subset $S \subseteq \Sigma$ such that $w \in S^*$. For $i = 0$, we define $w^0 = \varepsilon$ and, for $i > 0$, we define $w^i = ww^{i-1}$. We can now recall the classical notion of subsequence.

Definition 1. *A string v is a* subsequence *of length $|v| = k$ of string w, where $|w| = n$, if there exist positions $1 \leq i_1 < \ldots < i_k \leq n$, such that $v = w[i_1] \cdots w[i_k]$. For $j \in [k]$, we say that $v[j]$ is* embedded *on position i_j of w.*

If v is a subsequence of w, with $|v| = k$ and $|w| = n$, the *canonical embedding* of v in w is defined as the sequence of positions $1 \leq i_1 < i_2 < \ldots < i_k \leq n$ such that i_1 is the leftmost occurrence of $v[1]$ in w, and, for $j \in [2 : k]$, i_j is the leftmost occurrence of $v[j]$ in $w[i_{j-1} + 1 : n]$.

The main concepts discussed in this paper were introduced in [37]:

Definition 2 ([37]). *A string v is an* absent subsequence *of w if v is not a subsequence of w. An absent subsequence v of w is a* minimal absent subsequence *(for short,* MAS*) of w if every proper subsequence of v is a subsequence of w. We denote the set of all* MASs *of w by* MAS(w).

We recall the sameNext and samePrev data structures defined in [37]. The array sameNext[·] is defined by $\text{sameNext}[i] = \min(\{j > i \mid w[j] = w[i]\} \cup \{n + w[i]\})$, i.e., sameNext[$i$] points to the leftmost position, greater than i, where $w[i]$ occurs (or to $n + w[i]$ if $w[i]$ does not occur after i; this is well-defined, as we work with integer alphabets). Similarly, samePrev[·] is given by $\text{samePrev}[i] = \max(\{j < i \mid w[j] = w[i]\} \cup \{0\})$, i.e., it acts like sameNext on the mirror image of w. Both arrays are computed in time $O(n)$, see [37, Lemma 4.20]. We recall the following theorem from [37].

Theorem 1. *Let $v, w \in \Sigma^*$, $|v| = m + 1$, $|w| = n$, then v is a* MAS *of w if and only if there exist positions $0 = i_0 < i_1 < \ldots < i_m < i_{m+1} = n + 1$ such that $v = w[i_1] \cdots w[i_m]v[m + 1]$ and the following conditions hold:*

1. *$v[r] \notin \mathrm{alph}(w[i_{r-1} + 1 : i_r - 1])$ for all $r \in [m + 1]$,*
2. *$v[r] \in \mathrm{alph}(w[i_{r-2} + 1 : i_{r-1}])$ for all $r \in [2 : m + 1]$.*

For a string w, a subsequence v of w is called MAS-*prefix* of length m of w, if $|v| = m$ and there exist positions $0 = i_0 < i_1 < \ldots < i_m \leq n$ such that $v = w[i_1] \cdots w[i_m]$, $v[r] \notin \mathrm{alph}(w[i_{r-1}+1 : i_r - 1])$ for all $r \in [m]$ (i.e., the sequence $i_1 < \ldots < i_m$ is the canonical embedding of v in w), and $v[r] \in \mathrm{alph}(w[i_{r-2}+1 : i_{r-1}])$ for all $r \in [2 : m]$. The following lemma is immediate.

Lemma 1. *Let $v, w \in \Sigma^*$, $|v| = m$, $|w| = n$, such that v is a* MAS-*prefix of w. Then there exists a* MAS *u which has v as prefix. Moreover, if $w[i] = a$, the string $u = a^{|w[1:i]|_a}$ is a* MAS-*prefix of w whose canonical embedding ends on position i.*

§ *Finite Automata.* A deterministic finite automaton (DFA, for short) is a 5-Tuple $A = (Q, \Sigma, \delta, q_0, F)$, where Q is a finite set of states, Σ is an input alphabet ($|\Sigma| = \sigma$), $\delta : Q \times \Sigma \to Q$ is a transition function, q_0 is the initial state, and $F \subseteq Q$ is the set of final states. See [30] for basic results on finite automata (including, e.g., the definition of the language accepted by a DFA).

§ *Computational Model.* Our algorithms work on the standard unit-cost word RAM with logarithmic word-size $\omega \geq \log n$ (where n is the size of the input), and assume an *integer input alphabet.* More precisely, and very importantly, in all the problems considered here, the input is a string w, of length n, over an alphabet $\Sigma = \{1, \ldots, \sigma\}$, where $\sigma \leq n$. See the extended version of this paper [42].

§ *Directed Acyclic Graphs.* Let us now give a series of definitions related to directed acyclic graphs, which are important in our algorithms. A directed graph is a structure $G = (V, E)$, where V is the set of nodes (or vertices) and $E \subseteq V \times V$ is the set of edges. The direction of the edge $(x, y) \in E$ is said to be from x to y, and x is referred to as the origin of (x, y), while y is the target; we also say that y is a child of x. A (non-trivial) walk of length ℓ in the directed graph G is a sequence of $\ell - 1$ edges $(x_1, x_2), (x_2, x_3), \ldots, (x_{\ell-1}, x_\ell)$, where $\ell \geq 2$; $x_1, \ldots, x_\ell$ are the nodes on that walk, and the walk is called a path if the respective walk contains only pairwise distinct nodes. An $x - y$-path in G is a path connecting nodes x and y. A directed acyclic graph (DAG, for short) G is a directed graph that has no cycles (i.e., if there is an $x - y$-path, then there is no $y - x$-path in G); note that, for DAGs, the notions of path and walk coincide and there is only a finite set of paths between two given nodes.

§ *Enumeration Framework.* While enumeration problems and enumeration algorithms can be defined in a more general setting (see, e.g., [55]), here we focus on enumerating paths between two given nodes in a given DAG, for simplicity.

Let $G = (V, E)$ be a DAG with two nodes $x, y \in V$. We aim to design an algorithm which takes G, x, and y as input, and reports all $x - y$-paths in G, without repetitions. As typical in enumeration (see [35]), our algorithm maintains a data structure $\mathcal{S}$, storing a (succinct) representation of the currently enumerated path (as concatenation of default subpaths, to be explained later), allowing us to directly retrieve the actual path in time proportional to its length. Then, during the execution of the algorithm a procedure `report` is called, each time $\mathcal{S}$ contains the representation of a path which has not been enumerated yet;

we say that, with this call of `report`, the current path is reported. Intuitively, after each path is computed, it is given as input (via `report`) to another process; this can either output it, or process it in some other way. The main motivation of this setting is that we do not want to mix the computation of the enumerated paths with the processing done to them after their reporting. Indeed, in a further processing step we might want to completely print these paths (which would inherently take time linear in the output), we might only want to output the changes applied on the previously enumerated path to get this one, instead of the full path descriptions (which would take as much time as is needed to write these changes), or we might assign each path a score and only output explicitly the ones with the highest score. Using the `report` function we can then pipeline the enumeration and further processing; the runtime of the later processing steps is irrespective of the time needed to compute and enumerate (that is, `report`) all $x - y$-paths.

The *enumeration problem* considered in our setting is the following: Given a DAG $G = (V, E)$, and two nodes $x, y \in V$ of G, enumerate all $x - y$-paths of G. An *enumeration algorithm* solving this problem is an algorithm $\mathcal{A}$ that `reports` every $x - y$-path of G exactly once, and stops once the last path was `reported`.

To analyse the complexity of such an algorithm $\mathcal{A}$, we distinguish between two phases. The *preprocessing phase* consists of the computation done by the algorithm $\mathcal{A}$ before the first call of `report`; the *preprocessing time* is the time needed, in the worst case, by the respective computation, as a function in $|V|+|E|$ (i.e., the size of the DAG $G = (V, E)$). The rest of the computation is the actual *enumeration phase*; the *delay* is the time elapsed, in the worst case, expressed as a function in $|V| + |E|$, between two consecutive calls of `report` (as a side note, our algorithms spend at most $O(1)$ time between the last call of `report` and the end of the algorithm). The algorithm $\mathcal{A}$ is said to have *output-linear delay* if the delay between the calls to `report` corresponding to path s_{i-1} and path s_i is $O(|s_i|)$ (i.e., linear in the length of path s_i; note that this does not include the time needed to output that path, just the time needed to construct its representation and store it in $\mathcal{S}$). In our setting, the best time complexity we can hope for is linear preprocessing time and constant delay. Examples of enumeration algorithms developed in similar settings can be found in [3–5, 55].

§ *Paths in DAGs.* We can now formalise the representation of paths, used to maintain them in our algorithm. Let $G = (V, E)$ be a DAG. We assume that we have a function $de : V \to E \cup \{\bot\}$ which associates to every node $x \in V$ an edge $de(x) = (x, y)$, called the *default edge* of x, or sets $de(x) = \bot$, meaning that there is no default edge leaving node x. Thus, we have at most $|V|$ default edges in a DAG. The path from x to y consisting of only default edges is called *default path*, denoted by $\mathrm{defPath}(x, y)$; clearly, there exists at most one default path from node x to node y. With respect to the function de, defining the default edges of G, an arbitrary $x - y$-path α in G can be *encoded*, recursively, as follows:

- if α is the path of length 0, its encoding is $\Phi(\alpha) = \varepsilon$,
- if α is the (unique) default $x - y$-path, its encoding is $\Phi(\alpha) = \mathrm{defPath}(x, y)$,

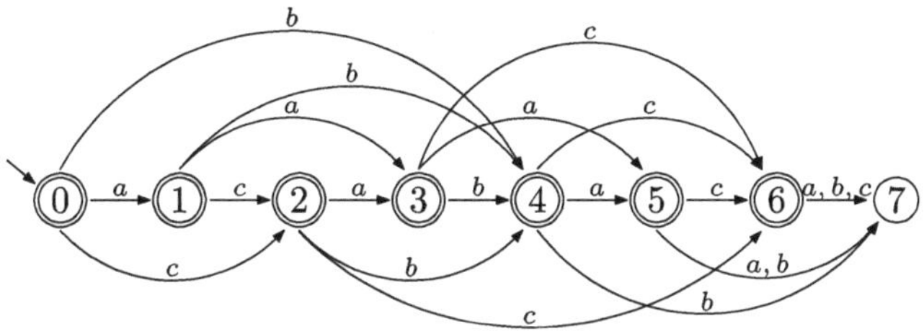

Fig. 1. Subsequence automaton for the string $w = acabac$. States 0 to 6 are accepting, whereas state 7 is not accepting. The absent subsequence automaton corresponding to w has the same states and transitions, but the role of the states is swapped: the only accepting state is state 7.

- if α is the sequence $\beta_1, (z, z'), \beta_2$, where (z, z') is not the default edge $de(z)$, β_1 is an $x - z$-path, and β_2 is a $z' - y$-path, then its encoding is $\Phi(\alpha) = \Phi(\beta_1)(z, z')\Phi(\beta_2)$.

Let us now explain how the solution data structure $\mathcal{S}$ is maintained during the enumeration of the $x - y$-paths in the DAG G. We define $\mathcal{S}$ as a stack, and for $\mathcal{S} = (s_1, \ldots, s_r)$ (with $top(\mathcal{S}) = s_r$), every element s_i, for $i \in [r]$, is the encoding $\Phi(p_i)$ of some path p_i connecting two nodes x_i and x_{i+1} of V, with $x_1 = x$ and $x_{r+1} = y$. The stack $\mathcal{S}$ then stores an encoded representation of the path $p = p_1 p_2 \cdots p_r$ obtained by concatenating the paths $p_1, \ldots, p_r$. Clearly, given stack $\mathcal{S}$ and DAG G, we can directly retrieve the path p in $O(|p|)$ time (i.e., linear in the length of p): follow the edges of $p_1, \ldots, p_r$ in order. Note that, if p only consists of non-default edges, then $|p| \leq |V|$ yields a naive upper bound for the height of $\mathcal{S}$.

3 Finding Single Minimal Absent Subsequences

While the MAS-matching problem, which requires checking, for two given strings u and w, whether u is a MAS of w, can be solved in $O(|u| + |w|)$ time (based on Theorem 1, see [37]), a more interesting problem is the MAS-indexing problem: efficiently preprocess w so that we can answer in time close to $O(m)$ queries asking whether a string of length m is a MAS of w. In the following, we discuss a general, automata-based, framework for indexing minimal absent subsequences (as well as subsequences or absent subsequences). Recall that the input of our problem is a string w, of length n, over the alphabet $\Sigma = \{1, \ldots, \sigma\}$, with $\sigma \leq n$.

§ *Indexing (Minimal Absent) Subsequences Using Deterministic Finite Automata.* For the string w one can define a DFA, called subsequence automaton (or directed acyclic subsequence graph), whose accepted language is the subsequences of w (see [8,17]). Let $A = (Q, \Sigma, \delta, 0, F)$ where $Q = \{0, 1, \ldots, n+1\}$ and $F = Q \setminus \{n+1\}$; the initial state of A is 0 and $n+1$ is a sink-state (i.e., a non-final state, from which there is no outgoing transition to another state). The transition function is defined as $\delta(i, a) = \min(\{j > i \mid w[j] = a\} \cup \{n+1\})$, for

$i \in [0 : n+1]$ and $a \in \Sigma$. In accepting a subsequence u, with $|u| = m$, of w, the DFA A goes through the sequence of states $0, i_1, \ldots, i_m$ where $i_1 < \ldots < i_m$ is the canonical embedding of u in w; if u is not a subsequence of w, A goes through the states $0, i_1, \ldots, i_t, n+1$ where $i_1 < \ldots < i_t$ is the canonical embedding of the longest prefix $u[1 : t]$ of u which is a subsequence of w. Clearly, by switching the final and non-final states of A we obtain a DFA which accepts the set of absent subsequences of w. See Fig. 1 for an example of a (absent) subsequence automaton.

Now, for the indexing of (absent) subsequences of w, in a preprocessing phase, we build the DFA A for w, with $F = Q \setminus \{n+1\}$ if we want to index subsequences, and $F = \{n+1\}$ if we want to index absent subsequences. Then, for a query string s, we read s with the respective DFA and check whether it is accepted or not.

To use the DFA A to index the MAS of w, in the preprocessing phase, we also build, alongside A, (and also in $O(n)$ time) the samePrev$[\cdot]$ array from Sect. 2. For a query string s, for which we want to decide whether $s \in \mathrm{MAS}(w)$, we proceed as follows. We read s letter by letter and store the last two states visited by A. Let i_j denote the state reached after reading $s[1 : j]$ for $j \in [|s|]$ (where $i_0 = 0$). Now, in order to check if s is a MAS of w, after reading $s[j]$, for $j \geq 2$, we have to check whether $i_{j-2} < \mathrm{samePrev}[i_j] \leq i_{j-1}$ holds, to ensure that Theorem 1 holds. If the above inequality holds, we can continue, otherwise $s \notin \mathrm{MAS}(w)$. Furthermore, if the automaton reaches the sink state $n+1$ before the entire string s has been read, then, again, $s \notin \mathrm{MAS}(w)$ and we can terminate. Finally, after reading s completely without an early termination, we decide that $s \in \mathrm{MAS}(w)$ if and only if A is in state $n+1$. The correctness of this approach is immediate from the fact that, when reading s, A goes through the states $0, i_1, \ldots, i_t, n+1$ where $i_1 < \ldots < i_t$ is the canonical embedding of the longest prefix $s[1 : t]$ of s which is a subsequence of w, combined with Theorem 1.

So, we can now focus on the efficient implementation of A. Fundamentally, constructing and using the automaton A efficiently relies on defining data structures allowing us to efficiently answer, for $i \in [n]$ and $a \in \Sigma$, queries next(i, a), returning $\min(\{n+1\} \cup \{j \mid j > i, w[j] = a\})$; clearly, the state $\delta(i, a)$ is the answer to the query next(i, a). In particular, the following theorem holds.

Theorem 2. *After a preprocessing of w running in $T(n, \sigma)$ time, in which data structures allowing us to answer* next(i, a) *queries in $Q(n, \sigma)$ time are computed, we can check if a string u of length m is a minimal absent subsequence (respectively subsequence or absent subsequence) of w in $O(mQ(n, \sigma))$ time.*

§ *Data Structures for Answering* next *Queries.* A series of efficient algorithms for answering next queries were presented in [12]. In the following, we revisit the respective approaches and non-trivially extend them.

We firstly choose a positive integer $\gamma \leq n$ as a parameter of our algorithm. By potentially padding w with at most γ occurrences of a new symbol $\# \notin \Sigma$, we can assume that n is divisible by γ. Also, by traversing the string w left to right, we can build, in $O(n)$ overall time, for each letter $a \in \Sigma$, a sorted list D_a of positions where a occurs in w. Clearly, the lists D_a form a partition of $[1 : n]$.

Now, we partition w into substrings $w[p\gamma+1:(p+1)\gamma]$ of size γ, for $p < n/\gamma$, and split the lists D_a into buckets accordingly (i.e., sublists corresponding to the substrings of w), whilst shrinking the universe of the positions stored in each bucket. More precisely, for each letter a and $p \in [0:(n/\gamma)-1]$, we will have ordered lists, called buckets, $B_{a,p}$ containing the elements of D_a from $[p\gamma+1:(p+1)\gamma]$; moreover, we replace each $j \in B_{a,p}$ by $(j-1) \bmod \gamma$ to shrink the universe of the elements stored in each bucket to $[0:\gamma-1]$. Note that, for any p, the lists $B_{a,p}$, over all $a \in \Sigma$, now form a partition of $[0:\gamma-1]$. Additionally, we store the minimum ($\min_{a,p}$) and maximum ($\max_{a,p}$) value in each bucket $B_{a,p}$; note that bucket $B_{a,p}$ may also be empty. All these steps can be completed in $O(n)$ time (as they only require traversing each list D_a, for every $a \in \Sigma$, once).

Furthermore, for each $a \in \Sigma$ and $p \in [0:(n/\gamma)-1]$, we compute $M[a,p]$ storing the smallest $p' > p$ such that $B_{a,p'}$ is not empty; computing the array M can be done in $O(n\sigma/\gamma)$ time, by a simple dynamic programming approach.

Finally, we build a successor search data structure for each bucket $B_{a,p}$. This data structure allows us to answer, for each bucket $B_{a,p}$, queries $\mathrm{succ}_{a,p}(i)$, retrieving the smallest element $j \in B_{a,p}$ such that $j > i$ (or $\perp$, if there is no such element). The implementation of this step determines the total complexity of our approach, and will be addressed later. Hence, the preprocessing described above, except building structures for successor search, takes $O(n + n\sigma/\gamma)$ time.

After all previous steps are completed, we can answer $\mathrm{next}(i,a)$ queries as follows. Let $p = \lfloor (i-1)/\gamma \rfloor$ and $i' = (i-1) \bmod \gamma$. If $i' < \max_{a,p}$, then let $j = \mathrm{succ}_{a,p}(i')$ and return $\mathrm{next}(i,a) = p\gamma + j + 1$; otherwise, let $p' = M[a,p]$, and return $p'\gamma + \min_{a,p'} + 1$.

Let us address, thus, the construction of structures for successor search. Following [12,53], we take $\gamma = \sigma$ (and note that the preprocessing described above takes $O(n)$ time) and, for a parameter ℓ, we can construct in $O(|B_{a,p}|\sigma^{1/2^\ell})$ time for each bucket $B_{a,p}$, with $a \in \Sigma$ and $p < n/\sigma$, a data structure allowing us to perform successor search in $B_{a,p}$ in $O(\ell+1)$ query time. Since the total size of the buckets is $O(n)$, the time used to construct these structures is $O(n\sigma^{1/2^\ell})$. For $\ell = \log\log\sigma$, we obtain that the preprocessing time is $O(n)$, while the query time (and, as such, the answer time for next queries) is $O(\log\log\sigma)$. Moreover, for every constant $\epsilon > 0$, we can take $\ell \in O(1)$ such that $1/2^\ell < \epsilon$, and obtain $O(n\sigma^\epsilon)$ preprocessing time and $O(1)$ query time. Thus, we can apply Theorem 2, either for $T(n,\sigma) = n$ and $Q(n,\sigma) = \log\log\sigma$, or for $T(n,\sigma) = n\sigma^\epsilon$ and $Q(n,\sigma) = 1$.

We non-trivially extend these ideas to address the case $\sigma \in O(\log n)$; for simplicity, after potentially padding w with occurrences of some fresh letter, we can assume that $n = 2^{2\ell}$, for some $\ell > 0$ (the length of w increases in this step at most fourfold). Let $\gamma = (\log n)/2$; the whole preprocessing, except building the successor search structures, still takes $O(n)$ time. The characteristic bit-array of each $B_{a,p}$ (having 1 on position j iff $j \in B_{a,p}$) has, in this case, $(\log n)/2$ bits, so $B_{a,p}$ can be succinctly represented as the number $n_{a,p} = \sum_{j \in B_{a,p}} 2^j$ (i.e., the number whose binary representation is the characteristic array of $B_{a,p}$). Computing all numbers $n_{a,p}$ takes $O(n)$ time, and we have that $n_{a,p} \leq \sqrt{n} - 1$.

Moreover, in a preprocessing phase we compute for each $x \in [0 : \sqrt{n} - 1]$ a bit array $\mathrm{Bit}_x[0 : (\log n)/2 - 1]$, where $\mathrm{Bit}_x[j]$ is the jth bit in the binary representation of x, as well as an array $\mathrm{Succ}_x[0 : (\log n)/2 - 1]$ where $\mathrm{Succ}_x[i]$ is the smallest $j > i$ such that $\mathrm{Bit}_x[j] = 1$ (or $\perp$, if such a j does not exist). Computing these arrays takes time $O(\sqrt{n} \log^2 n) \subseteq O(n)$, if done naïvely. Now, successor queries for $B_{a,p}$ (and, as such, next queries) can be answered immediately: $\mathrm{succ}_{a,p}(j)$ is $\mathrm{Succ}_{n_{a,p}}[j]$. So, if $\sigma \in O(\log n)$, we can apply Theorem 2, for $T(n, \sigma) = n$ and $Q(n, \sigma) = 1$. The same complexity bounds can be achieved for alphabets of size $\sigma \in O(\log^k n)$, for any constant $k > 1$, by a recursive approach, which non-trivially generalizes and extends the case $k = 1$. We obtain Theorem 3; see [42] for its proof.

Theorem 3. *For an input word of length n, over an alphabet of size $O(\log^k n)$, where $k \in O(1)$, we can compute in $O(n)$ time data structures allowing us to answer* $\mathrm{next}(i, a)$ *queries in $O(1)$ time. In this case, we can check if a string u of length m is a minimal absent subsequence (respectively subsequence or absent subsequence) of w in $O(m)$ time, after a linear time preprocessing.*

§ *Finding a Longest MAS.* Let us briefly consider the extremal elements, w.r.t. length, of $\mathrm{MAS}(w)$. While the shortest elements of $\mathrm{MAS}(w)$ coincide with the shortest absent subsequences (SAS) of w and one such SAS can be obtained trivially (see [37]), computing a longest element of $\mathrm{MAS}(w)$ seems non-trivial. In particular, an algorithm computing such a longest MAS, for string $w \in \Sigma^n$, with $|\Sigma| = \sigma$, in $O(n\sigma)$ time was reported in [54]. We can improve this bound significantly and compute a longest MAS in $O(n \log \sigma)$ time (see [42] for the proof). Further improving this complexity to $O(n)$ remains, however, an open problem.

Theorem 4. *We can compute a longest* MAS *of w in $O(n \log \sigma)$ time.*

4 Finding All Minimal Absent Subsequences

As our main result we now present the first optimal enumeration algorithm, with $O(n)$ time preprocessing and $O(1)$ delay, for the set of $\mathrm{MAS}(w)$ of a given string $w \in \Sigma^*$ of length n, over the alphabet $\{1, \ldots, \sigma\}$ of size $\sigma \leq n$. Considering the results regarding (absent) subsequence indexing for general alphabets in the previous section, where either constructing the automaton or simulating its transitions were too costly, it seems reasonable to assume that enumerating all $\mathrm{MAS}(w)$ with both $O(n)$ time preprocessing and $O(1)$ delay would not be feasible. However, motivated by the recent results of [3] showing that all subsequences of length k can be enumerated optimally with $O(n)$ time preprocessing and $O(1)$ delay, we will show that the same also holds for the enumeration of $\mathrm{MAS}(w)$ by adapting the MAS-automaton defined in [36]. In particular, based on Theorem 1, the authors of [36] define a labelled directed acyclic graph $\mathcal{D}_w$ with the nodes $\{(i, j) \mid 0 \leq j < i \leq n\} \cup \{s, f\}$ (the notation $(0, 0)$ is used instead of s in [36]). The edges (represented as arrows $A \rightarrow B$ between nodes A and B) are defined as follows:

- We have an edge $s \rightarrow (i, 0)$, if there exists $a \in \Sigma$ such that i is the position where a occurs the first time in w. This edge is labelled with $w[i] = a$.
- For $1 \leq j < i < k \leq n$ we have an edge $(i, j) \rightarrow (k, i)$, if there exists $a \in \Sigma$ such that k is the position of the first occurrence of a strictly to the right of position i and $a \in \text{alph}(w[j + 1 : i])$. This edge is labelled with $w[k] = a$.
- We have an edge $(i, j) \rightarrow f$ for each $a \in \text{alph}(w[j + 1 : i])$ such that $a \notin \text{alph}(w[i+1 : n])$. This edge is labelled with a. Note that there can be multiple edges from (i, j) to f, but they have different labels. As we want to refer to $\mathcal{D}_w$ as a DAG, we add to $\mathcal{D}_w$ the nodes $(n + a, i)$ and use them as stand-ins for f, if f was reached by an a-labelled edge $(i, j) \rightarrow f$ for some j; instead of the respective edge, we now have the edge $(i, j) \rightarrow (n + a, i)$, labelled with a. For simplicity of exposure, we refer to $s - (n + a, j)$-paths as $s - f$-paths.

In [36], the DAG $\mathcal{D}_w$ is treated as a DFA, with initial state s and final state f (in our case, the nodes $(n + a, i)$ defined above), and transitions defined by the edges of the graph; see Fig. 2 for an example of a path in $\mathcal{D}_w$. According to [36], there is a bijection between MAS(w) and the set of $s - f$-paths in the DAG $\mathcal{D}_w$ (i.e., the words accepted by the corresponding DFA): the MAS corresponding to an $s - f$-path is obtained by reading the label of that path, and, for each MAS, there is exactly one $s - f$-path whose label is the respective MAS. So, enumerating all elements of MAS(w) is equivalent to enumerating all $s - f$-paths of $\mathcal{D}_w$; this can be done in a canonical way, with output-linear delay, by backtracking (i.e., using a depth-first search starting in node s), but it is quite slow as it requires constructing $\mathcal{D}_w$. Indeed, in [36] it is claimed that this construction may take $O(n^2\sigma)$ time, since $\mathcal{D}_w$ may have $O(n^2)$ nodes and each node may have σ outgoing edges. However, based on Theorem 1, we have that for a fixed i the set $\{(r, i) \mid r \in [n], \exists j : (i, j) \rightarrow (r, i)\}$ has at most σ elements. Therefore, there are $O(n\sigma)$ nodes which are accessible from s, thus providing an improved upper bound of $\mathcal{O}(n\sigma^2)$ on the total size of the graph.

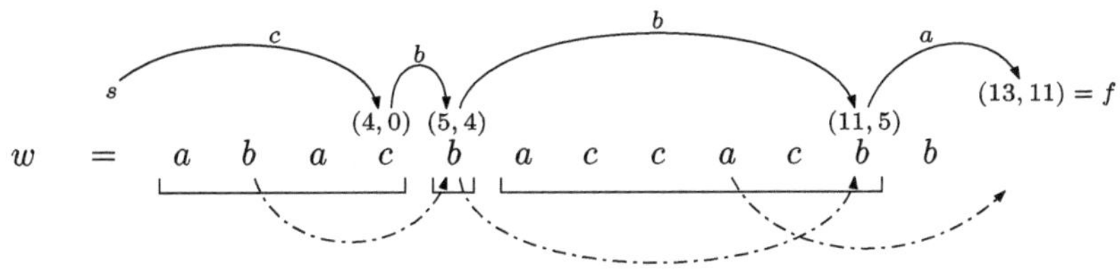

Fig. 2. Path of the MAS *cbba* in the DAG $\mathcal{D}_w$ for given string $w = abacbaccacbb$. The dashed arrows below the string show the letters $x \in \text{alph}(w[j + 1 : i])$ witnessing the transition $(i, j) \xrightarrow{x} (k, i)$ with $w[k] = x$.

We aim here for an optimal enumeration algorithm with $O(1)$ delay and $O(n)$ time preprocessing, so we need to avoid constructing $\mathcal{D}_w$ explicitly, as this would require too much time. Hence, we will first show that we can achieve output-linear delay after only $O(n)$ preprocessing, by not storing $\mathcal{D}_w$ explicitly, but rather constructing its paths one by one, on the fly. This improves the results of [54] (where the results of [36] were already improved), w.r.t. preprocessing time. The full proof of the following theorem can be seen in [42].

Theorem 5. *We can enumerate* MAS(w) *with output-linear delay, after an* $O(n)$ *time preprocessing.*

Proof (Sketch). In this algorithm, we use the range maximum query data structure [9]. We preprocess in $O(n)$ time an array A with n elements, to answer in $O(1)$ time range maximum queries $RMaxQ(i,j)$: given $i,j \in [n]$, return a position $x \in [i:j]$ such that $A[x] \geq A[\ell]$, for all $\ell \in [i:j]$ (if there are multiple answers, we return the smallest one; if $i > j$, $RMaxQ(i,j) = -\infty$). We build $RMaxQ$-data structures for the array sameNext; recall that $\text{sameNext}[i] = \min(\{j > i \mid w[j] = w[i]\} \cup \{n + w[i]\})$, for $i \in [n]$.

Assume that we want to enumerate the elements of MAS(w) starting with letter a, and let i_1 be the position of the first occurrence of a in w. In our enumeration algorithm, we will simulate a backtracking (depth-first) exploration of the graph $\mathcal{D}_w$, identifying the paths starting with $s \to (i_1, 0)$ and ending with a node $(n+c, j)$, for some $j \in [n]$ and $c \in \Sigma$, without building this graph explicitly. In such an exploration, these paths are constructed one by one, as follows: a path $s \to (i_1, 0) \to \ldots \to (i_r, i_{r-1})$ is stored on a stack at some point (with the node (i_r, i_{r-1}) found on the top of the stack), and, for each child ((i_{r+1}, i_r) or $(n+c, i_r)$) of (i_r, i_{r-1}), we put this child on the stack and explore, one by one, in a recursive way, all $s-f$-paths starting with $s \to (i_1, 0) \to \ldots \to (i_r, i_{r-1}) \to (i_{r+1}, i_r)$ (or `report` if the path ends with $(n+c, i_r)$).

To implement this strategy efficiently, we take advantage of the fact that the children of some node (i,j), accessible from s, are determined by the positions i and j only. More precisely, let $\text{alph}(w[j+1:i]) = \{a_1, \ldots, a_q\}$, for some $q \leq \sigma$. One can compute the list of the q children of (i,j) as follows: for $t \in [q]$, (j_t, i) is a child of (i,j), where j_t is the position of the first occurrence of a_t strictly to the right of i, if $a_t \in \text{alph}(w[i+1:n])$; f can also be a child of (i,j) if $\text{alph}(w[j+1:i]) \setminus \text{alph}(w[i+1:n]) \neq \emptyset$, but, for the sake of an uniform presentation, we say, in this case, that $(n + a_t, i)$ (which stands in for f) is a child of (i,j). The main challenges, that we need to address in our algorithm, are how to compute these letters $a_t \in \text{alph}(w[j+1:i])$ to efficiently gain in this way on-the-fly access to the children of (i,j), and how to integrate this in our simulation of the exploration of $\mathcal{D}_w$. We proceed as follows.

The enumeration idea implemented in our algorithm is now relatively simple to explain. Assume that we have constructed the prefix $s \to (i_1, 0) \to \ldots \to (i_r, i_{r-1})$ of some $s-f$-path (and stored its nodes on a stack), which corresponds to the canonical embedding $i_1 < \ldots < i_r$ of a MAS-prefix. We can then identify the possible positions g allowing us to extend this path (respectively, MAS-prefix) with a node (g, i_r) by selecting first the position $g = \text{sameNext}[i_r]$, and then, one by one, the positions $g = \text{sameNext}[g']$ for all $g' \in [i_{r-1}+1 : i_r - 1]$ such that $\text{sameNext}[g'] > i_r$. Indeed, this approach simply selects, one by one, each letter $w[g'] \in \text{alph}(w[i_{r-1}+1 : i_r])$, occurring on some position g', whose next occurrence $g = \text{sameNext}[g']$ is to the right of i_r (i.e., $g > i_r$); the nodes (g, i_r) computed based on these selections are exactly the children of (i_r, i_{r-1}), according to the observation above. Now, the first such letter $w[g']$ that we select is $w[i_r]$, as $\text{sameNext}[i_r] > i_r$ clearly holds; this corresponds to extending the path with the node $(i_{r+1}, i_r) = (\text{sameNext}[i_r], i_r)$. This node is added to the stack, and the process is repeated. Once all paths starting with $s \to (i_1, 0) \to \ldots \to (i_r, i_{r-1}) \to (i_{r+1}, i_r)$ are enumerated (that is, the corresponding MASs are `reported`), we need to find a new letter $w[g']$, as described above. To this end, the main observation is that such letters $w[g'] \neq w[i_r]$, and their corresponding positions g', can be found efficiently by using successive range maximum queries on the array sameNext, on subintervals of $[i_{r-1}+1 : i_r - 1]$, that cover this range completely.

This *RMaxQ*-based search is implemented as follows: we consider $g'_0 = RMaxQ(i_{r-1}+1, i_r-1)$ as the first candidate for g' and, if $\text{sameNext}[g'_0] > i_r$, we recursively enumerate the paths starting with $s \to (i_1, 0) \to \ldots \to (i_r, i_{r-1}) \to (\text{sameNext}[g'_0], i_r)$; then we recursively look for future candidates for g' in $[i_{r-1}+1 : g'_0-1]$ and $[g'_0+1 : i_r-1]$. For the sake of efficiency, we maintain for each node (i_r, i_{r-1}) of our stack a queue of subintervals in which we will look for future candidates for position g', and we insert such an interval $[x : y] \subseteq [i_{r-1}+1 : i_r-1]$ in the respective queue if and only if $\text{sameNext}[RMaxQ(x, y)] > i_r$ (otherwise, that interval contains no valid candidate for the position g'). The first interval considered to be inserted in that queue is $[i_{r-1} + 1 : i_r - 1]$, and when we want to compute a new child of (i_r, i_{r-1}) we simply extract the first interval stored in the queue, do a range maximum query on it, and then split it as explained above (with the two resulting subintervals being further inserted in the queue, in the case the maximum value stored in the corresponding ranges of $\text{sameNext}[\cdot]$ is greater than i_r).

The correctness of this approach follows from the definition of $\mathcal{D}_w$ and Theorem 1, the main point being that once the stack contains the path $s \to (i_1, 0) \to \ldots \to (i_r, i_{r-1})$, all possible edges extending this path are correctly identified, one by one, and the corresponding paths, which extend the current path-prefix by the respective edge, are recursively enumerated.

From a complexity perspective, after a linear time preprocessing, the enumeration algorithm implements a backtracking exploration of the $s-f$-paths in $\mathcal{D}_w$. When this exploration reaches node (i, j), we get in $O(1)$ time its first child, and then explore the $s-f$-paths going through that child. Once an $s-f$-path is completed, to ensure output-linear delay, we first locate (by traversing the stack top to bottom) the node where we have to backtrack (uppermost node with an unexplored child), then `report` the current path, and then directly jump back in the stack to the respective return-node (possible in a static implementation of the stack). When returning to a node (i, j), an unexplored child of (i, j) is retrieved in $O(1)$ time (using *RMaxQ*s). So, while the children of a node are not explicitly stored, we can still retrieve them in order, in $O(1)$ time per child. The statement clearly holds. □

We extend this approach to an optimal enumeration algorithm. First, we define the default edges for the nodes of the DAG $\mathcal{D}_w$, following the approach used in the previous proof: if (i, j) is a node considered in our enumeration algorithm, then its first explored child is $(\text{sameNext}[i], i)$; note that $(n + w[i], i)$ stands in for f, if $\text{sameNext}[i] > n$. So, we define the default edge for (i, j) to be the edge, labelled with $w[i]$, connecting it to $(\text{sameNext}[i], i)$. In this setting, the longest default path starting in a node (i, j) ends with $f = (n+w[i], \ell_{w[i]})$, where ℓ_a is the position of the last occurrence of letter $a \in \Sigma$ in w; for a node (i', j') on this default path, we denote by $\text{defPath}((i, j), (i', j'))$ the default path connecting nodes (i, j) and (i', j'); all edges of this path are labelled with $w[i]$. Moreover, a node (i, j) is a *branching node* if and only if it has more than one outgoing edge. Clearly, (i, j) is branching if and only if $j + 1 < i$, as $w[i] \notin \text{alph}(w[j + 1 : i - 1])$, so there exists a non-default edge from (i, j) to $(\text{sameNext}[i - 1], i) \neq (\text{sameNext}[i], i)$. In this framework, we can state our result; see [42] for the full proof.

Theorem 6. *We can enumerate* MAS(w) *with constant delay, after an* $O(n)$ *time preprocessing.*

Proof (Sketch). We again enumerate the $s - f$-paths in $\mathcal{D}_w$, by efficiently implementing the approach from the previous theorem: we do not construct $\mathcal{D}_w$, but rather construct and explore, by backtracking, its $s - f$-paths on the fly. However, now we work with succinct representations of these paths (based on default edges and default paths), allowing us to move from one path to the next one, in the enumeration, in $O(1)$ time. Other than that, the paths are `reported` in exactly the same order as in the previous algorithm, so the correctness of this approach is immediate. We again construct the array sameNext and *RMaxQ*-data structures for it: these allow us to move from one child of a node (i, j) to the next in $O(1)$ time, while searching for new paths.

For a default path α in $\mathcal{D}_w$ containing, in order, the nodes $(i_1, i_0), \ldots, (i_r, i_{r-1})$, we can identify in $O(1)$ time the rightmost branching node on α, just by checking its first and last node (if $i_r = n + w[i_1]$, we check the second-to-last node, as $(n + w[i_1], i_{r-1}) = f$ cannot be branching). Indeed, we want to find the largest value $\ell \in [r]$ such that $i_\ell - i_{\ell-1} > 1$. This is done as follows: for every position $j \in [n]$, we precompute $\text{lastGap}[j] = [x : y]$, with $x, y \in [n]$ such that $\text{alph}(w[y : j]) = \{w[j]\}$, $y = \text{sameNext}[x]$ and $x + 1 < y$; this can be done in $O(n)$ time. Now, if $\text{lastGap}[i_r] = [x : y]$ with $i_1 < y$, then (y, x) is the last branching node on α; if $i_1 = y$ and $i_0 + 1 < i_1$, then (i_1, i_0) is the last branching node; otherwise, the default path α contains no branching node. Similarly, for $\ell > 1$, the predecessor of $(i_\ell, i_{\ell-1})$ on α is either $(i_{\ell-1}, i_{\ell-1} - 1)$ if $w[i_{\ell-1} - 1] = w[i_{\ell-1}]$ or $(i_{\ell-1}, x)$ if $\text{lastGap}[i_{\ell-1}] = [x : i_{\ell-1}]$, so can also be retrieved in $O(1)$ time. Further, let ℓ_a be the position of the last occurrence of letter $a \in \Sigma$ in w; the values ℓ_a, for all $a \in \Sigma$, can be precomputed in $O(n)$ time.

Now, we move on to the enumeration. In order to implement the approach of Theorem 5 efficiently, we store the currently enumerated path as a sequence of default paths and branching nodes (seen as default paths of length 0) on a stack $\mathcal{S}$. We say that $\mathcal{S}$ stores *path-objects*: every path-object corresponds to a default path defPath$((i,j),(i',j'))$ from (i,j) to (i',j') (if $(i,j) = (i',j')$, this corresponds to the single node (i,j)); if $i' = n+a$, we use $(n+a,j')$ as stand-in for f, where $j' = \ell_{w[i]}$ is the last occurrence of $w[i]$ in w (if $(i,j) \neq (n+a,j')$, then $a = w[i]$). We do not explicitly store non-default edges; instead, if $s_t =$ defPath$((i_1,j_1),(i_2,j_2))$ is directly below $s_{t+1} =$ defPath$((i_3,j_3),(i_4,j_4))$ in $\mathcal{S}$, then the path implicitly contains the edge $(i_2,j_2) \rightarrow (i_3,j_3)$, labelled with $w[i_3]$ (or a, if $i_3 = n+a$); note that $i_2 = j_3$. If the path-object s_t corresponds to a default path containing branching nodes, which have not been explored yet, we *mark* s_t. We can retrieve the topmost marked path-object in $O(1)$ time, using a second stack containing pointers to the marked elements of $\mathcal{S}$.

To enumerate the elements of MAS(w) starting with letter a, let i_1 be the position of the first occurrence of a in w. The first enumerated path corresponds to s followed by defPath$((i_1,0),(n+a,\ell_a))$, marked if it contains a branching node. This path is `reported`. In a general step, we retrieve from the stack $\mathcal{S}$ (in $O(1)$ time) the topmost marked element $s_t =$ defPath$((i,j),(i',j'))$. We then retrieve the rightmost branching node (i'',j'') on s_t. We remove s_t from $\mathcal{S}$ (and, implicitly, everything above it, by setting the top of $\mathcal{S}$ to point to s_{t-1}; implementing $\mathcal{S}$ statically, this takes $O(1)$ time) and push a new path-object $s'_t =$ defPath$((i,j),(z,v))$, where (z,v) is the predecessor of (i'',j'') on defPath$((i,j),(i',j'))$ (or we do not push anything if $(i'',j'') = (i,j)$); this is marked if it contains a branching node. Then we push defPath$((i'',j''),(i'',j''))$; for this node, we maintain a list of intervals as in the proof of the previous theorem, allowing us to explore the non-default edges leaving it, using range maximum queries on sameNext; initially this list only contains the interval $[j''+1 : i''-1]$, which contains the positions leading to all children of (i'',j''). We then retrieve a non-default edge $((i'',j''),(g',i''))$, and update the list of intervals of (i'',j'') as described in the proof of Theorem 5: we use $RMaxQ$ on the first interval of the list, to retrieve the non-default child (g',i''), then we split this interval around the position returned by the query, and add the two halves to the list if they still contain positions leading to other non-default edges; as long as this list of intervals is non-empty, the node defPath$((i'',j''),(i'',j''))$ is marked. Finally, we push defPath$((g',i''),(n+w[g'],\ell_{w[g']}))$ on the stack (marked if it contains a branching node), and `report`. The process is then repeated.

Intuitively speaking, we efficiently implement the backtracking exploration from Theorem 5 as follows: for an $s-f$-path stored on $\mathcal{S}$, we find in $O(1)$ time the last branching node with unexplored edges, and jump back on the path to this node. Using $RMaxQ$, we find a new child of this branching node and then follow the default path from this child to f, so the delay between two `reports` is constant. In every step we maintain information regarding the branching nodes (and list of their unexplored children, encoded as intervals) that still have to be considered. The correctness of the algorithm clearly follows from Theorem 5. □

Theorem 6 gives an optimal enumeration algorithm for MAS(w) for a string w; this completely solves the problem of efficiently enumerating MAS(w). Moreover, the enumeration is done in such a way that, w.r.t. the currently enumerated

MAS s, the next MAS we output is the remaining one that shares the longest common prefix with s, or one of these if there are more with this property. Enumerating MAS(w) in lexicographic order, as efficiently, remains an open problem.

Acknowledgements. The authors are supported by the German Research Foundation (DFG) through the projects 466789228 (Heisenberg) and 562495345.

References

1. Abboud, A., Backurs, A., Williams, V.V.: Tight hardness results for LCS and other sequence similarity measures. In: IEEE 56th Annual Symposium on Foundations of Computer Science, FOCS 2015, Berkeley, CA, USA, 17-20 October, 2015, pp. 59–78 (2015). https://doi.org/10.1109/FOCS.2015.14
2. Adamson, D., Fleischmann, P., Huch, A., Koß, T., Manea, F., Nowotka, D.: k-universality of regular languages. In: Iwata, S., Kakimura, N. (eds.) 34th International Symposium on Algorithms and Computation (ISAAC 2023). LIPIcs, vol. 283, pp. 4:1–4:21 (2023). https://doi.org/10.4230/LIPIcs.ISAAC.2023.4, full version: https://arxiv.org/abs/2311.10658
3. Adamson, D., Fleischmann, P., Huch, A., Manea, F., Sarnighausen-Cahn, P., Wiedenhöft, M.: Tight bounds for the number of absent subsequences. In: Jez, A., Otop, J. (eds.) Fundamentals of Computation Theory - 25th International Symposium, FCT 2025, Wrocław, Poland, September 15-17, 2025, Proceedings. LNCS, vol. 16106, pp. 15–29. Springer (2025). https://doi.org/10.1007/978-3-032-04700-7_2
4. Adamson, D., Gawrychowski, P., Manea, F.: Enumerating m-length walks in directed graphs with constant delay. In: Soto, J.A., Wiese, A. (eds.) LATIN 2024: Theoretical Informatics - 16th Latin American Symposium, Puerto Varas, Chile, March 18-22, 2024, Proceedings, Part I. Lecture Notes in Computer Science, vol. 14578, pp. 35–50. Springer (2024). https://doi.org/10.1007/978-3-031-55598-5_3
5. Amarilli, A., Monet, M.: Enumerating regular languages with bounded delay. In: Berenbrink, P., Bouyer, P., Dawar, A., Kanté, M.M. (eds.) 40th International Symposium on Theoretical Aspects of Computer Science, STACS 2023, March 7-9, 2023, Hamburg, Germany. LIPIcs, vol. 254, pp. 8:1–8:18. Schloss Dagstuhl - Leibniz-Zentrum für Informatik (2023). https://doi.org/10.4230/LIPICS.STACS.2023.8
6. Artikis, A., Margara, A., Ugarte, M., Vansummeren, S., Weidlich, M.: Complex event recognition languages: tutorial. In: Proceedings of the 11th ACM International Conference on Distributed and Event-based Systems, DEBS 2017, Barcelona, Spain, June 19-23, 2017, pp. 7–10 (2017). https://doi.org/10.1145/3093742.3095106
7. Badkobeh, G., Charalampopoulos, P., Pissis, S.P.: Internal shortest absent word queries. In: 32nd Annual Symposium on Combinatorial Pattern Matching, CPM 2021, July 5-7, 2021, Wrocław, Poland. LIPIcs, vol. 191, pp. 6:1–6:18. Schloss Dagstuhl - Leibniz-Zentrum für Informatik (2021). https://doi.org/10.4230/LIPICS.CPM.2021.6
8. Baeza-Yates, R.A.: Searching subsequences. Theor. Comput. Sci. **78**(2), 363–376 (1991). https://doi.org/10.1016/0304-3975(91)90358-9

9. Bender, M.A., Farach-Colton, M.: The LCA problem revisited. In: Gonnet, G.H., Panario, D., Viola, A. (eds.) LATIN 2000: Theoretical Informatics, 4th Latin American Symposium, Punta del Este, Uruguay, April 10-14, 2000, Proceedings. LNCS, vol. 1776, pp. 88–94. Springer (2000). https://doi.org/10.1007/10719839_9
10. Bergroth, L., Hakonen, H., Raita, T.: A survey of longest common subsequence algorithms. In: de la Fuente, P. (ed.) Seventh International Symposium on String Processing and Information Retrieval, SPIRE 2000, A Coruña, Spain, September 27-29, 2000, pp. 39–48 (2000). https://doi.org/10.1109/SPIRE.2000.878178
11. Bernardini, G., Marchetti-Spaccamela, A., Pissis, S.P., Stougie, L., Sweering, M.: Constructing strings avoiding forbidden substrings. In: 32nd Annual Symposium on Combinatorial Pattern Matching, CPM 2021, July 5-7, 2021, Wrocław, Poland. LIPIcs, vol. 191, pp. 9:1–9:18. Schloss Dagstuhl - Leibniz-Zentrum für Informatik (2021). https://doi.org/10.4230/LIPICS.CPM.2021.9
12. Bille, P., Farach-Colton, M.: Fast and compact regular expression matching. Theor. Comput. Sci. **409**(3), 486–496 (2008). https://doi.org/10.1016/j.tcs.2008.08.042
13. Bille, P., Gørtz, I.L., Skjoldjensen, F.R.: Subsequence automata with default transitions. J. Discrete Algorithms **44**, 48–55 (2017). https://doi.org/10.1016/J.JDA.2017.05.001
14. Bille, P., Gørtz, I.L., Vildhøj, H.W., Wind, D.K.: String matching with variable length gaps. Theor. Comput. Sci. **443**, 25–34 (2012). https://doi.org/10.1016/j.tcs.2012.03.029
15. Bringmann, K., Künnemann, M.: Multivariate fine-grained complexity of longest common subsequence. In: Proc. SODA 2018, pp. 1216–1235 (2018). https://doi.org/10.1137/1.9781611975031.79
16. Charalampopoulos, P., Crochemore, M., Fici, G., Mercaş, R., Pissis, S.P.: Alignment-free sequence comparison using absent words. Inf. Comput. **262**, 57–68 (2018). https://doi.org/10.1016/j.ic.2018.06.002
17. Crochemore, M., Melichar, B., Troníček, Z.: Directed acyclic subsequence graph – overview. J. Discrete Algorithms **1**(3–4), 255–280 (2003). https://doi.org/10.1016/S1570-8667(03)00029-7
18. Crochemore, M., Mignosi, F., Restivo, A.: Automata and forbidden words. Inf. Process. Lett. **67**(3), 111–117 (1998). https://doi.org/10.1016/S0020-0190(98)00104-5
19. Crochemore, M., Mignosi, F., Restivo, A., Salemi, S.: Data compression using antidictionaries. Proc. IEEE **88**(11), 1756–1768 (2000). https://doi.org/10.1109/5.892711
20. Fazekas, S.Z., Koß, T., Manea, F., Mercas, R., Specht, T.: Subsequence matching and analysis problems for formal languages. In: 35th International Symposium on Algorithms and Computation, ISAAC 2024, December 8-11, 2024, Sydney, Australia. LIPIcs, vol. 322, pp. 28:1–28:23. Schloss Dagstuhl - Leibniz-Zentrum für Informatik (2024). https://doi.org/10.4230/LIPICS.ISAAC.2024.28
21. Fleischmann, P., Haschke, L., Höfer, J., Huch, A., Mayrock, A., Nowotka, D.: Nearly k-universal words - investigating a part of Simon's congruence. Theor. Comput. Sci. **974**, 114113 (2023). https://doi.org/10.1016/J.TCS.2023.114113
22. Fredman, M.L.: On computing the length of longest increasing subsequences. Discrete Math. **11**(1), 29–35 (1975). https://doi.org/10.1016/0012-365X(75)90103-X
23. Freydenberger, D.D., Gawrychowski, P., Karhumäki, J., Manea, F., Rytter, W.: Testing k-binomial equivalence. https://arxiv.org/abs/1509.00622, pp. 239–248 (2015). https://doi.org/10.48550/arXiv.1509.00622

24. Frochaux, A., Kleest-Meißner, S.: Puzzling over subsequence-query extensions: Disjunction and generalised gaps. In: Kimelfeld, B., Martinez, M.V., Angles, R. (eds.) Proceedings of the 15th Alberto Mendelzon International Workshop on Foundations of Data Management (AMW 2023), Santiago de Chile, Chile, May 22-26, 2023. CEUR Workshop Proceedings, vol. 3409. CEUR-WS.org (2023). https://ceur-ws.org/Vol-3409/paper3.pdf
25. Fujishige, Y., Tsujimaru, Y., Inenaga, S., Bannai, H., Takeda, M.: Computing DAWGs and minimal absent words in linear time for integer alphabets. In: 41st International Symposium on Mathematical Foundations of Computer Science (MFCS 2016). Schloss Dagstuhl-Leibniz-Zentrum fuer Informatik (2016). https://doi.org/10.4230/LIPIcs.MFCS.2016.38
26. Gawrychowski, P., Kosche, M., Koß, T., Manea, F., Siemer, S.: Efficiently testing Simon's congruence. In: 38th International Symposium on Theoretical Aspects of Computer Science, STACS 2021, March 16-19, 2021, Saarbrücken, Germany (Virtual Conference), pp. 34:1–34:18 (2021). https://doi.org/10.4230/LIPIcs.STACS.2021.34
27. Giatrakos, N., Alevizos, E., Artikis, A., Deligiannakis, A., Garofalakis, M.: Complex event recognition in the Big Data era: a survey. VLDB J. **29**(1), 313–352 (2020). https://doi.org/10.1007/s00778-019-00557-w
28. Halfon, S., Schnoebelen, P., Zetzsche, G.: Decidability, complexity, and expressiveness of first-order logic over the subword ordering. In: Proceedings of the 32nd Annual ACM/IEEE Symposium on Logic in Computer Science, LICS '17, pp. 1–12 (2017). https://doi.org/10.1109/LICS.2017.8005141
29. Hébrard, J.J.: An algorithm for distinguishing efficiently bit-strings by their subsequences. Theor. Comput. Sci. **82**(1), 35–49 (1991). https://doi.org/10.1016/0304-3975(91)90170-7
30. Hopcroft, J.E., Ullman, J.D.: Introduction to Automata Theory, Languages and Computation. Addison-Wesley (1979)
31. Karandikar, P., Schnoebelen, P.: The height of piecewise-testable languages and the complexity of the logic of subwords. Log. Methods Comput. Sci. **15**(2) (2019). https://doi.org/10.23638/LMCS-15(2:6)2019
32. Kitaev, S.: Patterns in permutations and words. Springer (2011). https://doi.org/10.1007/978-3-642-17333-2
33. Kleest-Meißner, S., Sattler, R., Schmid, M.L., Schweikardt, N., Weidlich, M.: Discovering event queries from traces: Laying foundations for subsequence-queries with wildcards and gap-size constraints. In: Olteanu, D., Vortmeier, N. (eds.) 25th International Conference on Database Theory, ICDT 2022, March 29 to April 1, 2022, Edinburgh, UK (Virtual Conference). LIPIcs, vol. 220, pp. 18:1–18:21. Schloss Dagstuhl - Leibniz-Zentrum für Informatik (2022). https://doi.org/10.4230/LIPICS.ICDT.2022.18
34. Kleest-Meißner, S., Sattler, R., Schmid, M.L., Schweikardt, N., Weidlich, M.: Discovering multi-dimensional subsequence queries from traces - from theory to practice. In: König-Ries, B., Scherzinger, S., Lehner, W., Vossen, G. (eds.) Datenbanksysteme für Business, Technologie und Web (BTW 2023), 20. Fachtagung des GI-Fachbereichs „Datenbanken und Informationssysteme" (DBIS), 06.-10, März 2023, Dresden, Germany, Proceedings. LNI, vol. P-331, pp. 511–533. Gesellschaft für Informatik e.V. (2023). https://doi.org/10.18420/BTW2023-24
35. Knuth, D.E.: The Art of Computer Programming, Volume 4A: Combinatorial Algorithms, Part 1. Addison-Wesley (2014)

36. Kosche, M., Koß, T., Manea, F., Siemer, S.: Absent subsequences in words. In: Bell, P.C., Totzke, P., Potapov, I. (eds.) Reachability Problems - 15th International Conference, RP 2021, Liverpool, UK, October 25-27, 2021, Proceedings. LNCS, vol. 13035, pp. 115–131 (2021). https://doi.org/10.1007/978-3-030-89716-1_8
37. Kosche, M., Koß, T., Manea, F., Siemer, S.: Absent subsequences in words. Fundam. Informaticae **189**(3–4), 199–240 (2022). https://doi.org/10.3233/FI-222159
38. Kosche, M., Koß, T., Manea, F., Siemer, S.: Combinatorial algorithms for subsequence matching: a survey. In: Bordihn, H., Horváth, G., Vaszil, G. (eds.) Proceedings 12th International Workshop on Non-Classical Models of Automata and Applications, NCMA 2022, Debrecen, Hungary, August 26-27, 2022. EPTCS, vol. 367, pp. 11–27. (2022). https://doi.org/10.4204/EPTCS.367.2
39. Li, C., Wang, J.: Efficiently Mining Closed Subsequences with Gap Constraints, pp. 313–322. SIAM (2008). https://doi.org/10.1137/1.9781611972788.28
40. Li, C., Yang, Q., Wang, J., Li, M.: Efficient mining of gap-constrained subsequences and its various applications. ACM Trans. Knowl. Discov.Data **6**(1), 2:1–2:39 (2012). https://doi.org/10.1145/2133360.2133362
41. Linton, S., Ruškuc, N., Vatter, V. (eds.): Permutation Patterns. London Mathematical Society Lecture Note Series, vol. 376, Cambridge (2010). https://doi.org/10.1017/CBO9780511902499
42. Manea, F., Ringleb, T., Siemer, S., Winkler, M.: Efficiently finding all minimal and shortest absent subsequences in a string. CoRR (2025). https://doi.org/10.48550/ARXIV.2504.21471
43. Mateescu, A., Salomaa, A., Yu, S.: Subword histories and Parikh matrices. J. Comput. Syst. Sci. **68**(1), 1–21 (2004). https://doi.org/10.1016/j.jcss.2003.04.001
44. Pin, J.É.: The influence of Imre Simon's work in the theory of automata, languages and semigroups. Semigroup Forum **98**(1), 1–8 (2019). https://doi.org/10.1007/s00233-019-09999-8
45. Riddle, W.E.: An approach to software system modelling and analysis. Comput. Lang. **4**(1), 49–66 (1979). https://doi.org/10.1016/0096-0551(79)90009-2
46. Rigo, M., Salimov, P.: Another generalization of abelian equivalence: binomial complexity of infinite words. Theor. Comput. Sci. **601**, 47–57 (2015). https://doi.org/10.1016/j.tcs.2015.07.025
47. Romik, D.: The surprising mathematics of longest increasing subsequences. Cambridge (2015). https://doi.org/10.1017/CBO9781139872003
48. Salomaa, A.: Connections between subwords and certain matrix mappings. Theor. Comput. Sci. **340**(2), 188–203 (2005). https://doi.org/10.1016/j.tcs.2005.03.024
49. Shaw, A.C.: Software descriptions with flow expressions. IEEE Trans. Softw. Eng. **4**(3), 242–254 (1978). https://doi.org/10.1109/TSE.1978.231501
50. Simon, I.: Hierarchies of events with dot-depth one - Ph.D. thesis, University of Waterloo (1972)
51. Simon, I.: Piecewise testable events. In: Brakhage, H. (ed.) Automata Theory and Formal Languages, pp. 214–222. LNCS, Springer, Heidelberg (1975). https://doi.org/10.1007/3-540-07407-4_23
52. Simon, I.: Words distinguished by their subwords (extended abstract). In: Proc. WORDS 2003. TUCS General Publication, vol. 27, pp. 6–13 (2003)
53. Thorup, M.: Space efficient dynamic stabbing with fast queries. In: Larmore, L.L., Goemans, M.X. (eds.) Proceedings of the 35th Annual ACM Symposium on Theory of Computing, June 9-11, 2003, San Diego, CA, USA, pp. 649–658. ACM (2003). https://doi.org/10.1145/780542.780636

54. Troníček, Z.: On problems related to absent subsequences. In: Wu, W., Guo, J. (eds.) Combinatorial Optimization and Applications - 17th International Conference, COCOA 2023, Hawaii, HI, USA, December 15-17, 2023, Proceedings, Part II. LNCS, vol. 14462, pp. 351–363. Springer (2023). https://doi.org/10.1007/978-3-031-49614-1_26
55. Wasa, K.: Enumeration of enumeration algorithms. CoRR (2016). https://doi.org/10.48550/arXiv.1605.05102
56. Zetzsche, G.: The complexity of downward closure comparisons. In: Chatzigiannakis, I., Mitzenmacher, M., Rabani, Y., Sangiorgi, D. (eds.) 43rd International Colloquium on Automata, Languages, and Programming (ICALP 2016). LIPIcs, vol. 55, pp. 123:1–123:14 (2016). https://doi.org/10.4230/LIPIcs.ICALP.2016.123
57. Zetzsche, G.: Separability by piecewise testable languages and downward closures beyond subwords. In: Dawar, A., Grädel, E. (eds.) Proceedings of the 33rd Annual ACM/IEEE Symposium on Logic in Computer Science, LICS 2018, Oxford, UK, July 09-12, 2018, pp. 929–938. ACM (2018). https://doi.org/10.1145/3209108.3209201
58. Zhang, H., Diao, Y., Immerman, N.: On complexity and optimization of expensive queries in complex event processing. In: International Conference on Management of Data, SIGMOD 2014, Snowbird, UT, USA, June 22-27, 2014, pp. 217–228 (2014). https://doi.org/10.1145/2588555.2593671

Measure Many Quantum Finite Automata on Infinite Words

Abhisek Midya[1(✉)] and A. Baskar[2]

[1] Department of Computer Science and Engineering, Dayananda Sagar University, Bengaluru, India
abhisekmidyacse@gmail.com

[2] Department of CS and IS, BITS Pilani, K. K. Birla Goa Campus, Sancoale, Goa, India
abaskar@goa.bits-pilani.ac.in

Abstract. We define a quantum computational model, called *Measure-Many Quantum Büchi Automata* (MMQBA), which extends *Measure-Many Quantum Finite Automata* (MMQFA) to the infinite-word setting with a Büchi acceptance condition. In MMQBA, the quantum state evolves through unitary transformations followed by repeated projective measurements. An infinite word is accepted with respect to a cut-point $p \in (0, 1]$ if (i) the run visits accepting states infinitely often, (ii) the limiting cumulative acceptance probability is at least p, and (iii) the limiting cumulative rejection probability is strictly less than p. We formalize the semantics of MMQBA and establish a language-theoretic characterization showing that MMQBA languages are precisely of the form $\lim(L(M, p))$ for an MMQFA M. We further develop a structural analysis of the non-halting subspace and prove that certain ω-languages cannot be recognized by MMQBA. Finally, we show that the class of MMQBA-recognizable languages is not closed under union, intersection and complement.

Keywords: Quantum finite automata · Infinite words · Büchi acceptance condition

1 Introduction

Quantum finite automata (QFA) combine principles of quantum mechanics with classical finite automata and provide a theoretical model for finite-memory quantum computation. In one-way QFA models (1QFA), the input head moves only to the right and measurements may be performed either only at the end of the computation (measure-once) or after reading each input symbol (measure-many). Although several variants of QFA achieve exponential state savings compared with classical automata, many bounded-error models recognize only a proper subset of regular languages.

Büchi automata [15] are classical acceptors of infinite words (ω-words) and play a central role in the analysis of systems with non-terminating behavior.

T. J. Smith (Ed.): CIAA 2026, LNCS 16695, pp. 171–184, 2026.
https://doi.org/10.1007/978-3-032-31176-4_12

They are widely used in formal verification and model checking. Several related acceptance conditions have been studied, including Muller, Rabin, and Streett automata [16], as well as probabilistic variants of ω-automata [3,12]. For a broad overview of quantum finite automata and related models in quantum computing, we refer the reader to the survey by Ambainis and Yakaryılmaz [2].

Quantum automata over finite words have been investigated for more than two decades; see, for example, [1,5,13,14]. However, quantum automata on infinite words have received relatively limited attention. Preliminary definitions of quantum Büchi, Streett, and Rabin automata were introduced in [8]. In particular, [8] studied measure-once quantum Büchi automata (MOQBA) and proposed the notion of measure-many quantum Büchi automata (MMQBA). Their acceptance criterion requires the cumulative acceptance probability to converge to 1. Although this implicitly ensures infinitely many accepting visits, it does not restrict the rejection behavior of the computation.

In this paper we introduce a refined definition of measure-many quantum Büchi automata (MMQBA). In our model, an infinite word is accepted if (i) accepting states are visited infinitely often, (ii) the cumulative acceptance probability converges to at least a cutpoint p, and (iii) the cumulative rejection probability remains strictly below p. This formulation provides a more robust semantics by preventing cases where Büchi recurrence holds while significant rejection probability accumulates. We first establish a limit characterization connecting MMQBA-recognizable ω-languages with languages recognized by measure-many quantum finite automata (MMQFA). This characterization provides a structural connection between finite-word and infinite-word quantum automata, analogous to classical limit constructions for ω-languages, while preserving the distinct infinite-computation behavior captured by MMQBA. We then prove a non-recognizability result showing that certain languages cannot be recognized by MMQBA with cutpoint greater than $\frac{1}{2}$, and further investigate structural properties of the non-halting subspace as well as closure properties of the model. Our approach differs from the quantum Büchi automaton model of [17], which is based on disturbing or non-disturbing measurements in a measure-once setting rather than cumulative cutpoint semantics. In contrast, our model incorporates repeated measurements and cumulative acceptance/rejection probabilities, making it closer in spirit to measure-many quantum finite automata and better suited for studying limit behavior of infinite computations. Similar to classical Büchi automata, acceptance depends on infinite recurrence behavior along infinite runs; however, in the quantum setting this behavior is further governed by cumulative acceptance and rejection probabilities arising from repeated measurements.

The results developed herein contribute to the theoretical foundations of quantum automata operating on infinite words. Such models provide a mathematical framework for studying quantum computational processes that evolve indefinitely under repeated measurements. Understanding these foundations may be relevant to areas such as the formal analysis and verification of quantum systems, the semantics and verification of quantum programs as well as quantum cryptographic protocols [6,7,10], and the formal analysis of quantum games [9,11].

In Sect. 2, we recall the framework of MMQFA, which forms the basis for our later development. Section 3 presents the formal definition of MMQBA together with examples. Sections 4 and 5 discuss the structural properties and language characterization of MMQBA respectively. Section 6 investigates closure properties. Finally, Sect. 7 concludes with a summary and future research directions.

A full version of this paper, containing all the complete proofs and additional examples, is available at: https://github.com/abhisekmidyacse/ciaa2026-paper/blob/main/CIAA_2026_complete_version.pdf

2 Preliminaries

Definition 1. *MMQFA [1] MMQFA is a tuple* $M = (Q, \Sigma, \delta, q_0, Q_{acc}, Q_{rej})$ *where*

- Q *is a finite set of states,*
- Σ *is an input alphabet,*
- $\delta : Q \times \Gamma \times Q \rightarrow \mathbb{C}_{[0,1]}$ *is the transition function*
- $q_0 \in Q$ *is a starting state and*
- $Q_{acc} \subseteq Q$ *is a set of accepting states and*
- $Q_{rej} \subseteq Q$ *is a set of rejecting states.*

The states in Q_{acc} and Q_{rej} are referred to as halting states and the states in $Q_{non} = Q - (Q_{acc} \cup Q_{rej})$ are referred to as non-halting states. Let $\#, \$ \notin \Sigma$ and $\Gamma = \Sigma \cup \{\#, \$\}$ where $\#, \$$ are left and right end markers respectively.

Recall notation $l_2(Q)$ (the space mappings from Q to $\mathbb{C}$ with l_2 norm), and emphasize that superpositions of M have norm 1. $|q\rangle$ represents the unit vector for $q \in Q$, which has a value of 1 at q and 0 elsewhere. It is possible to express every element of $l_2(Q)$ as a linear combination of vectors $|q\rangle$. We will use ψ to denote elements of $l_2(Q)$ and S_U to denote the span of basis states in $U \subseteq Q$. In other words $S_U = span(\{|q\rangle : q \in U\})$.

We define $\delta : Q \times \Gamma \times Q \longrightarrow \mathbb{C}_{[0,1]}$. The amplitude of $|q_2\rangle$ in the superposition of states to which M proceeds from $|q_1\rangle$ after reading σ is represented by the value $\delta(q_1, \sigma, q_2)$. For $\sigma \in \Gamma$, V_σ is a linear transformation on $l_2(Q)$ defined by

$$V_\sigma(|q_1\rangle) = \sum_{q_2 \in Q} \delta(q_1, \sigma, q_2)|q_2\rangle$$

We require all V_σ to be unitary[1]. The computation of M starts in the superposition $|q_0\rangle$. Then transformations corresponding to the left end marker $\#$, the letters(σ) of the input word x and the right end marker $\$$ are applied. The transformation corresponding to $\sigma \in \Gamma$ consists of the following steps.

[1] A matrix U is unitary if $U^\dagger U = I$, where $U^\dagger$ is the conjugate transpose of U. Unitary matrices preserve the norm of quantum states.

- First, V_σ is applied. The new superposition ψ' is $V_\sigma(\psi)$, where ψ is the superposition before this step.
- Then, ψ' is measured with respect to the observable $S_{acc} \oplus S_{rej} \oplus S_{non}$, where $\oplus$ denotes the orthogonal direct sum of subspaces. Here, $S_{acc} = \text{span}(\{|q\rangle : q \in Q_{acc}\})$, $S_{rej} = \text{span}(\{|q\rangle : q \in Q_{rej}\})$, and $S_{non} = \text{span}(\{|q\rangle : q \in Q_{non}\})$.
  The input is accepted with probability $\|P_{acc}\psi'\|^2$, rejected with probability $\|P_{rej}\psi'\|^2$, and the computation continues from $P_{non}\psi'$ with probability $\|P_{non}\psi'\|^2$. $P_p, p \in \{acc, rej, non\}$ the projection operator into the subspace S_p, and and let $P_{\text{halt}} := P_{acc} + P_{rej}$ denote the projection onto the halting subspace.
  Note that $P_{non}\psi'$ is generally not normalized; its squared norm represents the remaining non-halting probability.

In order to define formally the overall probability with which an input is accepted (rejected) by a $MMQFA$ M, we define the set $V_M = l_2(Q) \times C \times C$ of so-called "total states" of M, that will be used only with the following interpretation. M is at any time during the computation in the state (ψ, p_a, p_r) if so far in its computation M accepted the input with probability p_a, rejected with probability p_r and neither with probability $1 - p_a - p_r = ||\psi||^2$, and $|\psi\rangle$ is its current, unnormalized state. For each $\sigma \in \Gamma$ the evolution of M, with respect to the total state, on an input σ is given by the operator T_σ defined in the following way:

$$T_\sigma(\psi, p_a, p_r) \longrightarrow (P_{non}V_\sigma\psi, p_a + ||P_{acc}V_\sigma\psi||^2, p_r + ||P_{rej}V_\sigma\psi||^2).$$

For $x = \sigma_1\sigma_2 \ldots \sigma_n \in \Sigma^*$ let $T_{\#x\$} = T_\$ T_{\sigma_n} T_{\sigma_{n-1}} \ldots T_{\sigma_1} T_\#$. If $T_{\#x\$}(|q_0\rangle, 0, 0) = (\psi, p_a, p_r)$, then we say that M accepts x with probability p_a and rejects with probability p_r. We say that $L(M, p)$ is the language accepted by M with cutpoint p, that is, $L(M, p) := \{x \in \Sigma^* \mid M$ accepts x with probability at least $p\}$.

3 Measure Many Quantum Büchi Automata (MMQBA)

Definition 2. *MMQBA is a tuple* $M = (Q, \Sigma, \delta, q_0, Q_{acc}, Q_{rej})$ *where*

- Q *is a finite set of states,*
- Σ *is an input alphabet,*
- $q_0 \in Q$ *is a starting state,*
- $Q_{acc} \subseteq Q$ *is a set of accepting states,*
- $Q_{rej} \subseteq Q$ *is a set of rejecting states,*
- $\delta : Q \times \Gamma \times Q \rightarrow \mathbb{C}_{[0,1]}$ *is the transition function*

The state space of M *is the Hilbert space* $\mathbb{C}^Q$ *with basis* $\{|q\rangle : q \in Q\}$.

Let the left end marker $\#$ does not belong to Σ and $\Gamma = \Sigma \cup \{\#\}$. Let $Q_{halt} = Q_{acc} \cup Q_{rej}$ and $|Q_{halt}| = n$.

A computation of M on the input word $X = \#\sigma_1\sigma_2\cdots \in \Gamma^\omega$ proceeds as follows: We require all V_{σ_i} to be unitary. The computation of M starts in the superposition $|q_0\rangle$. A unitary transition corresponding to the current input symbol is performed. After every transition, M measures its state with respect to the observable as follows:

$$\bigoplus_{i=1}^{n} S_i \oplus S_{non} \text{ where } S_i = span\{|q_i\rangle\} : q_i \in Q_{halt} \text{ and } S_{non} = span\{|q\rangle : q \notin Q_{halt}\}$$

After each unitary transition, a measurement is performed. If the resulting state lies in some halting subspace S_i, corresponding to $q_i \in Q_{\text{halt}}$, then the computation halts. Otherwise, the state lies in S_{non}, and the computation proceeds to the next input symbol from the resulting (generally unnormalized) state in S_{non}.

The computation of M on input $x = \sigma_1\sigma_2\sigma_3\cdots \in \Sigma^\omega$ starting from an arbitrary initial state $\psi_0 \in S_{\text{non}}$ is the sequence $\psi_0,\ \psi_1,\ \psi_1',\ \psi_2,\ \psi_2', \ldots$ defined as follows.

First define the post-measurement state at step 0 by $\psi_0' := P_{\text{non}}(\psi_0)$. Since $\psi_0 \in S_{\text{non}}$, we have $\psi_0' = \psi_0$. For every $j \geq 1$, define: $\psi_j := V_{\sigma_j}(\psi_{j-1}')$ (pre-measurement state), $\psi_j' := P_{\text{non}}(\psi_j)$ (post-measurement non-halting state). We now define how an infinite word is recognized by an MMQBA M for a fixed cutpoint $p \in (0, 1)$.

At any time during the computation M is in the total state (ψ, acc, rej), where ψ is a superposition of non-halting states, acc is the cumulative acceptance probability, and rej is the cumulative rejection probability. For each $\sigma \in \Gamma$, the evolution on input σ is given by the operator T_σ defined as

$$T_\sigma(\psi, acc, rej) \longrightarrow \big(P_{\text{non}}V_\sigma\psi,\ acc + \|P_{acc}V_\sigma\psi\|^2,\ rej + \|P_{rej}V_\sigma\psi\|^2\big),$$

where P_{acc}, P_{rej}, and P_{non} are the projections onto the accepting, rejecting, and non-halting subspaces, respectively, and $P_{\text{halt}} := P_{acc} + P_{rej}$ denote the projection onto the halting subspace.

Let $x = \#\sigma_1\sigma_2\cdots \in \Gamma^\omega$ be an input word with the left end marker $\#$, where $\Gamma = \Sigma \cup \{\#\}$, and let x_j denote its prefix of length j over Σ. The total state after reading x_j is $T_{\#x_j}(|q_0\rangle, 0, 0) = (|\psi_j'\rangle, acc_j, rej_j)$ where acc_j and rej_j denote the cumulative acceptance and rejection probabilities up to step j, respectively, and $|\psi_j'\rangle$ is the post-measurement non-halting state after applying P_{non}. The cumulative acceptance and rejection probabilities are defined by $acc_j = \sum_{i=1}^{j} \|P_{acc}(\psi_i)\|^2$ and $rej_j = \sum_{i=1}^{j} \|P_{rej}(\psi_i)\|^2$.

The MMQBA M accepts x with probability p if the run of M on x satisfies all of the following:

Büchi condition: $\|P_{acc}(\psi_j)\|^2 > 0$ for infinitely many j. That is, M visits accepting halting states infinitely often.

Cutpoint acceptance: $\lim_{j\to\infty} acc_j \geq p$. The cumulative acceptance probability in the run converges to the cutpoint p.

Cutpoint rejection: $\lim_{j\to\infty} rej_j < p$. The cumulative rejection probability in the run remains strictly below the cutpoint p.

The ω-language recognized by M with cutpoint p, denoted by $L(M,p)$, is the set of all infinite words accepted by M with probability at least p. Formally, $L(M,p) := \{x \in \Sigma^\omega \mid M \text{ accepts } x \text{ with probability at least } p\}$.

Remark 1. Let M be an MMQBA and $x = x_1x_2x_3\cdots \in \Sigma^\omega$ an input word. Let ψ_i denote the pre-measurement state after reading the i-th symbol. Then the cumulative acceptance probabilities satisfy $acc_{i+1} > acc_i$ iff $P_{acc}(\psi_{i+1}) \neq 0$.

We illustrate the model by the following example.

Example 1. Let $L = a(a+b)^\omega$, $\Sigma = \{a,b\}$.

We define the MMQBA $M = (Q, \Sigma, \delta, q_0, Q_{acc}, Q_{rej})$ where $Q = \{q_0, q_1, q_2\}, Q_{acc} = \{q_1\}, Q_{rej} = \{q_2\}$, cumulative acceptance probability is 0.8 and q_0 is the initial state.

We use the basis order (q_0, q_1, q_2) and define the unitary matrices by giving their columns (the j-th column is $V_\sigma|q_{j-1}\rangle$ for the corresponding symbol). The matrices are:

$$V_a = \begin{pmatrix} \frac{1}{\sqrt{3}} & \frac{\sqrt{2}}{\sqrt{3}} & 0 \\ -\frac{\sqrt{2}}{\sqrt{3}} & \frac{1}{\sqrt{3}} & 0 \\ 0 & 0 & 1 \end{pmatrix}, \qquad V_b = \begin{pmatrix} \frac{1}{3} & \frac{2}{3} & \frac{2}{3} \\ \frac{2}{3} & \frac{1}{3} & -\frac{2}{3} \\ \frac{2}{3} & -\frac{2}{3} & \frac{1}{3} \end{pmatrix}.$$

We briefly explain the behavior of M. For any input $x \in L = a(a+b)^\omega$, the first symbol a ensures that a positive amount of probability is transferred to the accepting subspace S_{acc}. Subsequent transitions maintain a bias towards acceptance, and the cumulative acceptance probability converges to a value approximately 0.8, which is strictly greater than $\frac{1}{2}$.

On the other hand, for inputs $x \notin L$, this acceptance bias is not established, and the cumulative acceptance probability does not exceed the cutpoint, while rejection dominates. Hence, M recognizes L according to the definition of MMQBA.

We now present a second MMQBA construction recognizing the same ω-language $a(a+b)^\omega$ using four states instead of three. This demonstrates that MMQBA representations for the same language are not necessarily unique and motivates further investigation of state complexity and minimization problems for MMQBA.

Example 2. Let $L = a(a+b)^\omega$, $\Sigma = \{a,b\}$.

We define the MMQBA $M = (Q, \Sigma, \delta, q_0, Q_{acc}, Q_{rej})$ where $Q = \{q_0, q_1, q_2, q_3\}, Q_{acc} = \{q_2\}, Q_{rej} = \{q_3\}$, cumulative acceptance probability is

0.88 and q_0 is the initial state. The matrices are:

$$V_a = \begin{pmatrix} 0 & 0 & 0 & 1 \\ \frac{2}{3} & \frac{1}{\sqrt{2}} & \frac{1}{3\cdot\sqrt{2}} & 0 \\ \frac{2}{3} & -\frac{1}{\sqrt{2}} & \frac{1}{3\cdot\sqrt{2}} & 0 \\ \frac{1}{3} & 0 & -\frac{4}{3\cdot\sqrt{2}} & 0 \end{pmatrix}, \qquad V_b = \begin{pmatrix} 0 & 0 & 0 & 1 \\ 0 & \frac{1}{\sqrt{2}} & -\frac{1}{\sqrt{2}} & 0 \\ 0 & \frac{1}{\sqrt{2}} & \frac{1}{\sqrt{2}} & 0 \\ 1 & 0 & 0 & 0 \end{pmatrix}.$$

4 Structural Properties of MMQBA

This section establishes structural properties of measure-many quantum Büchi automata (MMQBA) that are used in subsequent language-theoretic results, focusing on the behavior of the non-halting subspace during infinite computations. We show that this subspace admits a decomposition into components with distinct long-term behaviors, enabling a characterization of how halting probability is generated and how acceptance probability accumulates along a run. Intuitively, this parallels the classical setting where non-accepting states can be divided into those from which acceptance remains possible and those forming trap states. The quantum analogue, formalized in the following lemma, decomposes the non-halting subspace into one component where the computation evolves without ever producing halting probability, and another from which probability mass gradually leaks into the halting subspace, thereby enabling acceptance or rejection.

Lemma 1 (Non-halting subspace decomposition). *Let M be an MMQBA with state space $\mathbb{C}^Q$, and let S_{non} denote the non-halting subspace. Then there exist subspaces $S^1, S^2 \subseteq S_{non}$ such that $S_{non} = S^1 \oplus S^2$ and the following properties hold. For any $\psi \in S_{non}$ and any infinite word $x = \sigma_1\sigma_2\cdots \in \Sigma^\omega$, let $\psi_0, \psi_1, \psi_1', \psi_2, \psi_2', \ldots$ denote the computation of M on input x starting from ψ.*

1. *For every $\psi \in S^1$, the computation never leaves S^1, and no halting probability is ever produced. Formally, $\forall k \geq 1$, $\psi_k' \in S^1$ and $\sum_{k=1}^{\infty} \|P_{halt}(\psi_k)\|^2 = 0$.*
2. *For every $\psi \in S^2$, the non-halting component converges to zero. Formally, $\lim_{k\to\infty} \|\psi_k'\| = 0$.*

Note that if the initial state were contained entirely in S^1, then by Lemma 1 the computation would remain inside S^1 and no halting probability would ever be generated. Consequently the automaton would accept no word, i.e., $L(M,p) = \emptyset$.

Corollary 1. *Let $\psi = \phi_1 + \phi_2$, where $\phi_1 \in S^1$ and $\phi_2 \in S^2$. Let the run of M on $x = \sigma_1\sigma_2\sigma_3\cdots$ be defined as in an MMQBA, and let ψ_k denote the pre-measurement state at step k. Then, for all $k \geq 1$, $\|P_{halt}(\psi_k)\|^2 = \|P_{halt}(V_{\sigma_k} P_{non} V_{\sigma_{k-1}} \cdots P_{non} V_{\sigma_1}(\phi_2))\|^2$.*

Theorem 1. *Let $x = \sigma_1\sigma_2\cdots \in \Sigma^\omega$ and M be an MMQBA. If $\|P_{acc}(\psi_j)\|^2 > 0$ for infinitely many j and $\sum_{j=1}^{\infty} \|P_{rej}(\psi_j)\|^2 = 0$, then the cumulative acceptance probability converges to 1, that is $\lim_{j\to\infty} acc_j = 1$.*

Proof sketch. We show that the cumulative probability of accepting prefixes converges to 1. Let $\psi_0 = |q_0\rangle$ be the initial quantum state. We recall the pre-measurement state as ψ_j and the post-measurement non-halting state as ψ'_j. The halting probabilities at step j are defined by

$$\alpha_j := \|P_{acc}(\psi_j)\|^2,\ \rho_j := \|P_{rej}(\psi_j)\|^2,\ acc_j := \sum_{i=1}^{j} \alpha_i,\ rej_j := \sum_{i=1}^{j} \rho_i$$

For the first step, since $\psi_0 \in S_{\text{non}}$, $\|\psi'_1\|^2 = \|\psi_0\|^2 - \alpha_1 - \rho_1$. For every $j \geq 2$, the evolution satisfies the norm conservation identity $\|\psi'_j\|^2 = \|\psi'_{j-1}\|^2 - \alpha_j - \rho_j$. By induction on j, it follows that $\|\psi'_j\|^2 = 1 - acc_j - rej_j$.

We are given: (i) $\alpha_j > 0$ for infinitely many $j \in \mathbb{N}$ (i.e., accepting states are visited infinitely often), and (ii) $\sum_{j=1}^{\infty} \rho_j = 0$ (i.e., no amplitude is ever lost to rejecting states). From (ii) it follows that $\rho_j = 0$ for all j, so $\text{rej}_j = 0$, and hence $\|\psi'_j\|^2 = 1 - \text{acc}_j$.

The total acceptance probability is distributed across infinitely many steps. As the computation never halts (it resumes from the non-halting subspace at each step), this mass gradually accumulates in accepting states over time rather than collapsing at a finite point. Thus, the cumulative acceptance probability $acc_j = \sum_{i=1}^{j} \alpha_i$ tends to 1 as j tends to ∞, i.e., $\lim_{j\to\infty} acc_j = 1$.

□

Remark 2. The converse of Theorem 1 does not hold. That is, a run may accumulate total acceptance probability 1 without visiting accepting states infinitely often. Indeed, consider the following MMQBA. Let $\Sigma = \{a\}$ and define an automaton $M = (Q, \Sigma, \delta, q_0, Q_{acc}, Q_{rej})$ with $Q = \{q_0, q_1\}$, $Q_{acc} = \{q_1\}$, and $Q_{rej} = \emptyset$. Define the unitary transition $V_a = \begin{bmatrix} 0 & 1 \\ 1 & 0 \end{bmatrix}$. On the input $x = a^\omega$, the automaton transitions as $V_a|q_0\rangle = |q_1\rangle$. Thus the automaton halts at the first step with acceptance probability 1, and the computation terminates immediately. Consequently, $\lim_{j\to\infty} acc_j = 1$. However, the accepting state is visited only once. Hence the Büchi condition $\|P_{acc}(\psi_j)\| > 0$ for infinitely many j is violated. This shows that convergence of cumulative acceptance probability to 1 does not imply infinite visits to accepting states.

5 Limit Characterization and Non-recognizability of MMQBA

In this section we investigate the relationship between MMQFA and MMQBA recognizability via the limit operator. We first establish a structural lemma concerning the behavior of non-halting subspaces under a fixed transition, which serves as the main technical tool for the subsequent arguments. We then prove a limit-characterization theorem. Finally, we introduce a class of languages whose limit ω-languages are not recognizable by MMQBA.

Definition 3. *Let M be an MMQBA with non-halting subspace S_{non}, and let $\sigma \in \Sigma$. A subspace $S \subseteq S_{non}$ is called a* σ-cycle subspace *of M if $V_\sigma(S) \subseteq S$, that is, $\forall \psi \in S, V_\sigma(\psi) \in S$.*

Lemma 2 (No-entry into a σ-cycle subspace). *Let M be an MMQBA with basis state set Q and $S \subseteq S_{non}$ be a σ-cycle subspace. For any basis vector $|r\rangle \in Q$ with $|r\rangle \notin S$, then $P_S\big(V_\sigma|r\rangle\big) = 0$.*

The previous lemma formalizes a crucial structural phenomenon: once a non-halting subspace becomes invariant under a transition V_σ, no computation starting outside that subspace can ever enter it. This restriction on the movement of amplitude between subspaces will play a decisive role in proving non-recognizability results later in this section.

Definition 4 (Limit of a language). *[4] For $L \subseteq \Sigma^*$, define $\lim(L) = \{x \in \Sigma^\omega \mid x$ has infinitely many prefixes in $L\}$.*

Theorem 2 (limit-characterization). *Let $L' \subseteq \Sigma^\omega$. There exist an MMQBA M' and a cutpoint $p' > \frac{1}{2}$ such that $L' = L(M', p')$ iff there exist an MMQFA M and a cutpoint $p > \frac{1}{2}$ such that $L' = \lim\big(L(M,p)\big)$*

Proof sketch. We prove both directions.

($\Rightarrow$) Let $M = (Q, \Sigma, \delta, q_0, Q_{acc}, Q_{rej})$ be an MMQFA with cutpoint $p > \frac{1}{2}$, and let $L = L(M,p)$. We construct an MMQBA $M' = (Q', \Sigma, \delta', q_0', Q'_{acc}, Q'_{rej})$ with cutpoint $p' > \frac{1}{2}$ such that $L(M', p') = \lim(L)$.

The construction keeps the same state space and accepting/rejecting subspaces, i.e., $Q' = Q$, $Q'_{acc} = Q_{acc}$, $Q'_{rej} = Q_{rej}$, and $q_0' = q_0$. The difference is that an MMQBA has no right end-marker.

The construction distributes the effect of the end-marker $V_\$$ over infinitely many steps. Fix $\varepsilon \in (0,1)$. For each prefix u, if $u \in L$, then a fraction ε of the current non-halting probability is transferred to the halting subspaces, while a fraction $(1-\varepsilon)$ remains in S_{non}. This redistribution preserves total probability and the structure of superpositions, and hence can be realized by unitary transitions.

If $x \in \lim(L)$, then x has infinitely many prefixes $u_j \in L$. Let r_j denote the non-halting probability after processing u_j, then

$$r_j \leq (1-\varepsilon)^j \longrightarrow 0.$$

Thus the total halting probability tends to 1. Moreover, each such prefix contributes acceptance with bias towards S_{acc}, and hence acceptance dominates in the limit. Therefore, $x \in L(M', p')$ for some $p' > \frac{1}{2}$.

($\Leftarrow$) Let $M' = (Q, \Sigma, \delta', q_0, Q_{acc}, Q_{rej})$ be an MMQBA recognizing $\lim(L)$ with cutpoint $p' > \frac{1}{2}$. We construct an MMQFA $M = (Q, \Sigma, \delta, q_0, Q_{acc}, Q_{rej})$, again preserving the state space and partitions. Fix $\varepsilon \in (0,1)$ such that $(1-\varepsilon) > \frac{1}{2}$.

The transitions of M are defined so that, after reading any $u \in \Sigma^*$,

$$\|P_{non}(\psi_u)\|^2 \geq (1-\varepsilon) > \tfrac{1}{2}.$$

Such transitions preserve total probability and the structure of superpositions, and hence can be implemented by unitary operators. Thus most of the probability remains in the non-halting subspace.

Upon reading the end-marker \$, the remaining non-halting probability is entirely transferred to the halting subspaces, so that the final acceptance-rejection distribution reflects whether $u \in L$ or $u \notin L$. Since more than half of the total probability resides in the non-halting component before applying $V_\$$, this majority determines the final decision, yielding a cutpoint $p > \frac{1}{2}$.

This establishes the equivalence. □

Definition 5 ($\sigma -$ *block* forcing language). *Let Σ be an alphabet and $\sigma \in \Sigma$. A language $L \subseteq \Sigma^*$ is called a $\sigma -$ block forcing language if every infinite word $x \in \lim(L)$ has the following form, such that, $x = \sigma^{k_1} u\sigma\sigma^{k_2} u\sigma\sigma^{k_3} u\sigma \ldots$ where $k_i \geq 1$ for all $i \geq 1$, $u \in \Sigma^+$, $u \notin \sigma^+$. (u contains at least one symbol different from σ)*

Example 3. We present an example of σ-block forcing languages. and fix $\sigma = a$. Consider the language $L = (a^+bca)^*$. Then $\lim(L) = (a^+bca)^\omega$. Every infinite word in $\lim(L)$ has the form $a^{k_1}bca\, a^{k_2}bca\, a^{k_3}bca \cdots$ where $k_i \geq 1$ for all $i \geq 1$. Here $u = bc \in \Sigma^+$ and $u \notin a^+$. Hence L is an a-block forcing language.

The key property of a σ-block forcing language is that every $x \in \lim(L)$ contains infinitely many long σ-blocks separated by fixed non-σ segments. In an MMQBA, repeated application of V_σ over each block forces the finite-dimensional computation to eventually stabilize within a V_σ-invariant non-halting subspace. However, the block-forcing structure requires acceptance leakage at the end of every block, compelling the computation to exit any previously occupied invariant subspace. By the no-entry Lemma 2, re-entry is impossible, thus requiring infinitely many pairwise disjoint invariant subspaces—contradicting finite dimensionality.

Theorem 3 (Non recognizability of σ-block forcing language). *If $L \subseteq \Sigma^*$ is a σ-block forcing language then there exists no MMQBA M' with cutpoint $p' > \frac{1}{2}$ such that $L(M', p') = \lim(L)$.*

Proof sketch. Assume for contradiction that an MMQBA $(Q, \Sigma, \delta, q_0, Q_{acc}, Q_{rej})$ with cutpoint $p' > \frac{1}{2}$ recognizes $\lim(L)$. Since L is a σ-block forcing language, every word x in $\lim(L)$ has the form $\sigma^{k_1} u\sigma\sigma^{k_2} u\sigma\sigma^{k_3} u\sigma \cdots$ where $k_i \geq 1$, $u \in \Sigma^+$, and $u \notin \sigma^+$. Hence the transition V_σ is applied infinitely many times during the run of M' on x.

Let S_{non} denote the non-halting subspace and $Q_{\text{non}} = Q \setminus Q_{\text{halt}}$. For each σ-block in the run, let $\psi_{i,m}$ denote the non-halting state obtained after the m-th application of V_σ in the i-th block. Since infinitely many non-halting states appear during the run and Q_{non} is finite, some basis states must occur with nonzero amplitude infinitely often. We define

$B_\infty := \{q \in Q_{\text{non}} : q$ has nonzero amplitude in infinitely many states $\psi_{i,m}\}$ and

$$S^{(1)} = \text{span}\{ |q\rangle : q \in B_\infty \} \subseteq S_{\text{non}}.$$

Then $S^{(1)} \neq \{0\}$. Moreover, if $q \in B_\infty$, then q appears in infinitely many states $\psi_{i,m}$ and $\psi_{i,m+1} = P_{\text{non}}(V_\sigma \psi_{i,m})$. If $V_\sigma|q\rangle$ had a component on some basis state $r \notin B_\infty$, then r would also appear in infinitely many subsequent non-halting states, contradicting the definition of B_∞. Hence $V_\sigma(S^{(1)}) \subseteq S^{(1)}$,so $S^{(1)}$ is a nonzero V_σ-cycle subspace.

Next we show that vectors in $S^{(1)}$ produce no acceptance under V_σ. If some $\psi \in S^{(1)}$ satisfied $\|P_{acc}(V_\sigma\psi)\|^2 > 0$, then along the infinite input σ^ω acceptance leakage would occur infinitely many times, forcing the cumulative acceptance probability to exceed the cutpoint. This would imply $\sigma^\omega \in \lim(L)$, contradicting the definition of a σ-block forcing language. Hence $\|P_{acc}(V_\sigma\psi)\|^2 = 0$ for all $\psi \in S^{(1)}$.

Now consider the configuration after reading the prefix $\sigma^{k_1}u$ of x. If the computation halted at this point, the remaining infinite input could not be processed, contradicting the semantics of MMQBA. Hence the computation continues from a non-halting state.

Consider the next symbol σ in the prefix $\sigma^{k_1}u\sigma$. Since this prefix occurs in some word of L, the acceptance-rejection bias satisfies $\|P_{acc}(V_\sigma\psi_1)\|^2 > \|P_{rej}(V_\sigma\psi_1)\|^2$ and in particular a positive accepting component is produced. After measurement the non-halting residue $\psi_1' = P_{\text{non}}(V_\sigma\psi_1)$ is nonzero. Since vectors of $S^{(1)}$ produce no accepting component under V_σ, we must have $\psi_1' \notin S^{(1)}$.

Because $S^{(1)}$ is V_σ-invariant, once the computation leaves $S^{(1)}$ it can never re-enter it. During the next σ-block σ^{k_2} the same argument produces another nonzero V_σ-cycle subspace $S^{(2)} \subseteq S_{\text{non}}$ with $S^{(2)} \cap S^{(1)} = \{0\}$. Repeating this reasoning for each block σ^{k_m} yields infinitely many pairwise disjoint nonzero subspaces $S^{(1)}, S^{(2)}, S^{(3)}, \ldots$ satisfying $V_\sigma(S^{(m)}) \subseteq S^{(m)}$. This contradicts the finite dimension of S_{non}. Hence no MMQBA with cutpoint greater than $\frac{1}{2}$ can recognize $\lim(L)$. □

Corollary 2. *If $L \subseteq \Sigma^*$ is a σ-block forcing language, then there exists no MMQFA M with cutpoint $p > \frac{1}{2}$ such that $L(M,p) = L$.*

Corollary 3. *If $L \subseteq \Sigma^*$ is a finite language, then there exist an MMQFA M and an MMQBA M' such that $L(M,1) = L$ and $L(M',1) = \lim(L)$.*

6 Closure Properties of MMQBA

We present closure properties of MMQBA in this section. The class of languages recognized by MMQBA is not closed under union, intersection and complement. For completeness.

Theorem 4. *The class of languages recognized by MMQBA is not closed under union.*

Proof sketch. Let $L_1 = (ab)^\omega$ and $L_2 = (aab)^\omega$, both recognizable by MMQBA, these constructions are straight forward. Assume for contradiction that there exists an MMQBA M with cutpoint $p > \frac{1}{2}$ recognizing $L = L_1 \cup L_2$.

By Lemma 1, the non-halting subspace decomposes as $S_{\text{non}} = S^1 \oplus S^2$, and the initial state must have a nonzero component in S^2. Consider the state ψ_{aa} reached after reading the prefix aa from the initial state.

For the run on $x_1 = (ab)^\omega$, the prefix aa does not occur. If ψ_{aa} had a nonzero component in S^2, then continuing the computation from ψ_{aa} along $(ab)^\omega$ would again produce infinitely many accepting projections, causing words with prefix aa to be accepted. Hence $\psi_{aa} \in S^1$.

However, in the word $x_2 = (aab)^\omega$ the prefix aa occurs infinitely often, and acceptance requires that the computation after each occurrence continues to generate accepting amplitude. Thus the state after reading aa must lie in S^2, i.e., $\psi_{aa} \in S^2$.

This contradicts $S^1 \cap S^2 = \{0\}$. Therefore no MMQBA can recognize $L_1 \cup L_2$, and the class of MMQBA-recognizable ω-languages is not closed under union. □

Theorem 5. *The class of languages recognized by MMQBA is not closed under intersection.*

Proof sketch. Let $\Sigma = \{a, b\}$ and define $L_a = \lim\{\, w \mid w$ has an odd number of a's$\}$ and $L_b = \lim\{\, w \mid w$ has an odd number of b's$\}$. Both languages are recognizable by MMQBA, these are trivial constructions.

Let $L = L_a \cap L_b$. Suppose, for contradiction, that there exists an MMQBA M with cutpoint $p > \frac{1}{2}$ such that $L(M,p) = L$. Then for all $k, \ell \geq 1$ we have $(a^k b^\ell)^\omega \in L(M,p)$, while $a^{2t} b^\omega \notin L(M,p)$ for every $t \geq 1$.

Analyzing the run of M on $(a^k b^\ell)^\omega$, the non-halting space decomposes as $S_{\text{non}} = S^1 \oplus S^2$. States reached after the a-segments lie in S^1, while accepting amplitude is generated from S^2 during the b-segments. The run therefore repeatedly alternates $(S^1 \to S^2 \to S^1 \dots)$ along the periodic input.

Using this structure one can construct $t \geq 1$ such that the run of M on $a^{2t} b^\omega$ still accumulates acceptance probability above the cutpoint and satisfies the Büchi condition. Hence $a^{2t} b^\omega \in L(M,p)$, contradicting the definition of L. Therefore the class of languages recognized by MMQBA is not closed under intersection. □

Theorem 6. *The class of languages recognized by MMQBAs with cutpoint p is not closed under complement.*

Proof sketch. Assume for contradiction that the class of languages recognized by MMQBA is closed under complement. Let $L_a = \lim\{\, w \in \Sigma^* \mid w$ contains an odd number of a's $\}$, and suppose there exists an MMQBA M recognizing $\overline{L_a}$. Using the acceptance condition of M, we construct an infinite word $x = aab^{k_1} aab^{k_2} aab^{k_3} \cdots$ such that after reading each block aab^{k_i} the run of M reaches an accepting configuration. Consequently the run of M on x contains infinitely many accepting visits and satisfies the MMQBA acceptance conditions, so M accepts x. However, the prefixes ending at the first a of each block contain $1, 3, 5, \ldots$ occurrences of a. Hence x has infinitely many prefixes with an odd number of a's, implying $x \in L_a$. Thus M accepts a word belonging to L_a, contradicting the assumption that M recognizes $\overline{L_a}$. Therefore the class of languages recognized by MMQBA is not closed under complement. □

7 Conclusion

In this work, we introduced Measure-Many Quantum Büchi Automata (MMQBA) and developed a formal semantic framework for their operation on infinite words. The model combines Büchi recurrence with quantitative cutpoint conditions, capturing the cumulative effect of repeated projective measurements during an infinite computation. We established a language-theoretic characterization showing that MMQBA-recognizable ω-languages are precisely of the form $\lim(L(M, p))$ for an MMQFA M. As part of the analysis, we developed structural results describing the behavior of the non-halting subspace under repeated applications of unitary transitions. Using these ideas, we proved a non-recognizability theorem showing that certain ω-languages cannot be recognized by MMQBA. In addition, we showed that the class of MMQBA-recognizable languages is not closed under union, intersection, complement.

We also presented two constructions of MMQBA recognizing the same ω-language $a(a + b)^\omega$ using different Hilbert-space dimensions. This observation suggests natural questions regarding the minimization of MMQBA representations and the further development of MMQBA theory in analogy with classical Büchi automata. Understanding how the dimension of the underlying Hilbert space affects recognizability, developing systematic techniques for reducing the dimension of equivalent automata, investigating algorithmic decision problems such as emptiness, universality, and equivalence for MMQBA-recognizable ω-languages, and exploring possible applications of MMQBA in modeling recurring behaviors in infinite quantum computations and quantum games remain interesting directions for future research.

References

1. Ambainis, A., Freivalds, R.: 1-way quantum finite automata: strengths, weaknesses and generalizations. In: Proceedings 39th Annual Symposium on Foundations of Computer Science (Cat. No. 98CB36280), pp. 332–341. IEEE (1998)
2. Ambainis, A., Yakaryılmaz, A.: Automata and quantum computing. In: Handbook of Automata Theory, pp. 1457–1493. European Mathematical Society-EMS-Publishing House GmbH (2021)
3. Baier, C., Grosser, M.: Recognizing/spl omega/-regular languages with probabilistic automata. In: 20th Annual IEEE Symposium on Logic in Computer Science (LICS'05), pp. 137–146. IEEE (2005)
4. Baier, C., Katoen, J.P.: Principles of model checking. MIT press (2008)
5. Brodsky, A., Pippenger, N.: Characterizations of 1-way quantum finite automata. SIAM J. Comput. **31**(5), 1456–1478 (2002)

6. Brunet, O., Jorrand, P.: Dynamic quantum logic for quantum programs. Int. J. Q. Inf. **2**(01), 45–54 (2004)
7. Chadha, R., Mateus, P., Sernadas, A.: Reasoning about imperative quantum programs. Electr. Notes Theor. Comput. Sci. **158**, 19–39 (2006)
8. Dzelme-Berzina, I.: Quantum finite automata and logic. Ph.D. thesis, PhD thesis, Ph. D. thesis, University of Latvia, Riga (2010)
9. Eisert, J., Wilkens, M., Lewenstein, M.: Quantum games and quantum strategies. Phys. Rev. Lett. **83**(15), 3077 (1999)
10. Feng, Y., Duan, R., Ji, Z., Ying, M.: Proof rules for the correctness of quantum programs. Theoret. Comput. Sci. **386**(1–2), 151–166 (2007)
11. Gutoski, G., Watrous, J.: Toward a general theory of quantum games. In: Proceedings of the thirty-ninth annual ACM symposium on Theory of computing, pp. 565–574 (2007)
12. Katoen, J.P.: The probabilistic model checking landscape. In: Proceedings of the 31st Annual ACM/IEEE Symposium on Logic in Computer Science, pp. 31–45 (2016)
13. Kondacs, A., Watrous, J.: On the power of quantum finite state automata. In: Proceedings 38th Annual Symposium on Foundations of Computer Science, pp. 66–75. IEEE (1997)
14. Moore, C., Crutchfield, J.P.: Quantum automata and quantum grammars. Theoret. Comput. Sci. **237**(1–2), 275–306 (2000)
15. Richard, B.: On a decision method in restricted second order arithmetic. In: Proceedings of the International Congress on Logic, Stanford University Press (1962)
16. Thomas, W.: Automata on infinite objects. In: Formal Models and Semantics, pp. 133–191. Elsevier (1990)
17. Wang, Q., Ying, M.: Quantum büchi automata. Theoret. Comput. Sci. **1012**, 114740 (2024)

Complexity of Linear Subsequences of k-Automatic Sequences

Delaram Moradi[1], Narad Rampersad[2], and Jeffrey Shallit[1](✉)

[1] School of Computer Science, University of Waterloo, 200 University Ave. W., Waterloo, ON N2L 3G1, Canada
{delaram.moradi,shallit}@uwaterloo.ca

[2] Department of Math/Stats, University of Winnipeg, 515 Portage Ave., Winnipeg, MB R3B 2E9, Canada
n.rampersad@uwinnipeg.ca

Abstract. We construct automata with input(s) in base k recognizing some basic relations and study their number of states. We also consider some basic operations on k-automatic sequences $(h(i))_{i\geq 0}$ and discuss their state complexity. We find a relationship between subword complexity of the interior sequence $(h'(i))_{i\geq 0}$ and state complexity of the linear subsequence $(h(ni+c))_{i\geq 0}$. We resolve a recent question of Zantema and Bosma about linear subsequences of k-automatic sequences with input in most-significant-digit-first format. We also discuss the state complexity and runtime complexity of using a reasonable interpretation of Büchi arithmetic to actually construct some of the studied automata recognizing relations or carrying out operations on automatic sequences.

1 Introduction

Determining the number of states in a minimal automaton for a formal language or automatic sequence is an extensively studied area of formal languages and automata theory called *state complexity*. The reader can refer to works by Yu et al. [23], Gao et al. [11], and Zantema and Bosma [24] for some examples. This paper includes discussions and results on the state complexity of specific languages and automatic sequences with input in base k.

The k-automatic sequences for integers $k \geq 2$ is an interesting class of sequences introduced by Cobham [9,10] and has been studied for more than 50 years; see [2] for a monograph on automatic sequences. A sequence $(h(i))_{i\geq 0}$ is said to be k-*automatic* if there is a deterministic finite automaton with output (DFAO) computing the i-th term, given the base-k representation of i as input. A DFAO is a 6-tuple $M = (Q, \Sigma, \delta, q_0, \Delta, \tau)$ where

– Q is a finite nonempty set of states;

D. Moradi—Research supported by NSERC grant RGPIN-2024-03725.
N. Rampersad—Research supported by NSERC grants RGPIN-2019-04111 and RGPIN-2025-04076.

T. J. Smith (Ed.): CIAA 2026, LNCS 16695, pp. 185–199, 2026.
https://doi.org/10.1007/978-3-032-31176-4_13

- $\Sigma = \Sigma_k = \{0, 1, \ldots, k-1\}$ is the input alphabet;
- $\delta : Q \times \Sigma \to Q$ is the transition function, with domain extended to $Q \times \Sigma^*$ in the usual way;
- q_0 is the initial state;
- Δ is the finite nonempty output alphabet; and
- $\tau : Q \to \Delta$ is the output function.

Recall that if $x = e_{i-1} \cdots e_0$ is a finite word, then $[x]_k$ denotes the integer value of x when considered as a base-k expansion; that is, $[x]_k = \sum_{0 \le j < i} e_j k^j$. Thus, for example, $[101011]_2 = 43$ for the most-significant-digit-first (msd-first) format.

We say a DFAO M *generates* the sequence $\mathbf{h} = (h(i))_{i \ge 0}$ if $\tau(\delta(q_0, x)) = h(i)$ for all $i \ge 0$ and all words $x \in \Sigma_k^*$ such that $[x]_k = i$. In particular, M is expected to give the correct result *no matter how many leading zeros are appended to the input.* A DFAO with base-k input is called a k-DFAO.

For an automatic sequence $\mathbf{h} = (h(i))_{i \ge 0}$ generated by the minimal DFAO $(Q, \Sigma_k, \delta, q_0, \Delta, \tau)$, there is a unique associated *interior sequence* $\mathbf{h}' = (h'(i))_{i \ge 0}$ with output values from the states of the minimal DFAO generating $(h(i))_{i \ge 0}$, defined by taking the minimal DFAO and replacing τ by the identity map ι on Q. Here "unique" means up to renaming of the letters. Therefore, $(h(i))_{i \ge 0}$ is the image of $(h'(i))_{i \ge 0}$ under the coding $q \to \tau(q)$.

A classical result of Cobham [10, p. 174] states that if $(h(i))_{i \ge 0}$ is k-automatic, then so is the linear subsequence $(h(ni + c))_{i \ge 0}$ for integers $n \ge 1$ and $c \ge 0$. In this paper, we are interested in the state complexity of linear subsequences of automatic sequences and related topics. This topic was previously studied by Zantema and Bosma [24]. One of our main results is a novel relation between the state complexity of linear subsequences and the subword complexity of an automatic sequence. We use this result to expand on the work of Zantema and Bosma on the state complexity of linear subsequences of automatic sequences and we answer one of their open questions. We then apply our results to the specific case of linear subsequences of the famous 2-automatic Thue–Morse sequence.

One can also consider automata where the least significant digit is read first. For least-significant-digit first (lsd-first) input, Boudet and Comon [6] and for msd-first input, Wolper and Boigelot [22] provided constructions for base-2 automata recognizing arithmetical relations. Their constructions relate to some of our results in Sect. 3; however, our results are for all bases k and for specific relations.

Boudet and Comon [6] and Wolper and Boigelot [22] also discussed the size of the automaton for an arbitrary formula in Presburger arithmetic. Furthermore, Amat et al. [3] have developed a tool named `Patapsco` that computes a lower bound for the number of states in the automaton for a given Presburger arithmetic formula. This tool computes a lower bound on automaton size without actually constructing the automaton.

The last major topic of this paper is to study automaton size for formulas of Büchi arithmetic, which is an extension of Presburger arithmetic and has been implemented (for example) in the `Walnut` software system [17] for solving problems in combinatorics on words. Users of this system want to know how long some of the basic expressions can take to evaluate, and how big the resulting automata are. In this paper, we consider the automata for specific relations and automatic sequences. Here we have to examine different considerations than simply state complexity (which measures the size of the *smallest possible* finite automaton for a language or sequence), because we need to know how large the intermediate automata can be in the constructions implied by Büchi arithmetic. In particular, we need to consider (a) how an expression in Büchi arithmetic is translated to an automaton, (b) what are the sizes of the intermediate automata created in this translation, and (c) what the total running time of the procedure is.

Remark 1. In this paper we sometimes provide state complexity and runtime complexity bounds using big-O notation, and sometimes these bounds involve several variables. For these bounds we use the convention that the bounds hold for all sufficiently large values of all the variables.

The full version of this paper, with all missing proofs, is available at https://arxiv.org/abs/2512.10017. In two companion papers, we address similar topics with inputs in other numeration systems [15,16].

2 Background

For the remainder of this paper, we assume that the inputs to automata are represented in base k and the automatic sequences are k-automatic for some constant integer $k \geq 2$. Furthermore, we use Σ_k for the set $\{0, \ldots, k-1\}$.

If $x = a_1 \cdots a_i$ and $y = b_1 \cdots b_i$ are words of the same length over the alphabet Σ_k, then by $x \times y$ we mean the word $[a_1, b_1] \cdots [a_i, b_i]$ over the alphabet $\Sigma_k \times \Sigma_k$, and similarly for more than two inputs. If the automata we discuss have multiple inputs, we assume they are encoded in parallel in this fashion.

We can specify the order the input is read: *lsd-first* stands for least significant digit first and *msd-first* stands for most significant digit first. We design our msd-first (resp., lsd-first) automata such that they have the correct behaviour or output no matter how many leading zeros (resp., trailing zeros) there are in the input.

In this paper, we occasionally refer to $O(\log i)$ where i can be 0 or 1. In these cases, we adopt the usual convention that $O(\log i) = O(1)$ for $i \in \{0, 1\}$.

The *canonical base-k representation* of an integer i is the word x, without leading zeros, such that $[x]_k = i$. Furthermore, we use the notation $(i)_k$ for the canonical representation of i in base k. The canonical representation of 0 is λ, the empty word. Sometimes, if the context is clear, we just use $[x]$ instead of $[x]_k$ and (i) instead of $(i)_k$. We say a word z is a *right extension* of a word x if x is a prefix of z.

The definition of the state complexity of a formal language is the number of states in the minimal DFA recognizing it and we define the state complexity of an automatic sequence as the number of states in the minimal DFAO generating it.

Let $\mathbf{x}$ be any infinite word. The number of distinct length-n subwords present in $\mathbf{x}$ is denoted by the subword complexity function $\rho_{\mathbf{x}}(n)$. Here by a *subword* we mean a contiguous block of symbols within another word; this concept is also known as a *factor*.

We now recall a known result about the subword complexity of automatic sequences.

Theorem 1. *Let $\mathbf{x}$ be a k-automatic sequence generated by an m-state DFAO $M = (Q, \Sigma, \delta, q_0, \Delta, \tau)$ with msd-first input and let $\hat{\mathbf{M}}$ be the sequence generated by the intrinsic DFAO $\hat{M} = (Q, \Sigma, \delta, q_0, Q, \iota)$, where ι is the identity map. Then $\rho_{\mathbf{x}}(n) \leq \rho_{\hat{\mathbf{M}}}(n) \leq kn\rho_{\hat{\mathbf{M}}}(2) \leq knm^2$ for all $n \geq 1$.*

Proof. See [2, Theorem 10.3.1]. We point out that the statement of the theorem in [2] only gives the inequality $\rho_{\mathbf{x}}(n) \leq knm^2$, but the proof actually presented there provides the other inequality. □

3 Recognizing Relations

We say a DFA recognizes a relation (such as the addition relation $i + j = l$) if it takes the values of the variables in parallel, represented in some base k, and accepts if and only if the relation holds. In this section we review some of the basic arithmetic operations, what is known about their state complexity, and some new results.

3.1 Addition

The following is well-known:

Theorem 2. *For both lsd-first and msd-first input, and for all integers $k \geq 2$, there is a 2-state automaton that recognizes the relation $[x]_k + [y]_k = [z]_k$.*

Now we consider what happens when we specify one summand to be a fixed constant.

Theorem 3. *Let $c \geq 1$ be a fixed integer constant. For both lsd- and msd-first input, there exists an automaton with $O(\log c)$ states accepting $x \times y$ where $[x]_k + c = [y]_k$.*

Proof. We first design the automaton M_c accepting $[x]_k + c = [y]_k$ for msd-first input. The automaton for the special case of base 2 can be created using the construction provided by Wolper and Boigelot [22], which is similar to what we do here.

On input $x \times y$, we need to keep track of the difference $D_k(x, y) = [y]_k - [x]_k$ and accept the input if and only if the difference is c. Each state of the automaton is a possible difference d, but we need to find a suitable range I for d so that the automaton is actually finite. We have

$$[yb]_k - [xa]_k = k([y]_k - [x]_k) + b - a. \tag{1}$$

Therefore, if $D(x, y)$ is outside the range $[0, c]$ it will not return to it for any right extension of $x \times y$, and we can take I to be the interval $[0, c]$. The transition function is based on Eq. (1). After reading each input letter pair, the difference either stays the same or increases (unless it is no longer in the range I). Furthermore, if we consider a state $d \neq 0$, there are only two other states that have a direct transition to d, namely $\lfloor d/k \rfloor$ and $\lceil d/k \rceil$.

So we construct the automaton $M_c = (Q, \Sigma_k^2, \delta, q_0, A)$ where $\delta(d, [a, b]) = kd + b - a$, $q_0 = 0$, and $A = \{c\}$. To create the set Q, we start with $\{c\}$. Then we add the states $c' = \lfloor c/k \rfloor$ and $c'' = \lceil c/k \rceil$ to Q. We know $|c'' - c'| \leq 1$; therefore, the set $\{\lfloor c'/k \rfloor, \lfloor c''/k \rfloor, \lceil c'/k \rceil, \lceil c''/k \rceil\}$ has size at most 2. We repeat the same step (adding the floor and ceiling of division by k to Q) until reaching the state 0. This step is repeated $O(\log_k c)$ times and there are always at most two new states created at each step. Therefore, the number of states in M_c is $O(\log_k c)$.

The automaton M_c for lsd-first input just implements the simple manual addition method and we omit the details (see Boudet and Comon [6] for the special case of base 2). □

Recall the definition of the k-adic valuation: $\nu_k(n) = e$ if $k^e \mid n$ but $k^{e+1} \nmid n$. Note that $\nu_k(c)$ is the number of consecutive 0s at the end of $(c)_k$.

We also use the following notation: if some condition is described inside brackets, the brackets are Iverson brackets. The Iverson bracket $[P]$ equals 1 if the condition P is satisfied, and 0 otherwise. A more careful analysis of the number of states in the previous theorem gives the next result, which appears to be new.

Theorem 4. *Let $c \geq 1$ be a fixed constant. For msd-first input, the number of states in the minimal DFA recognizing $x \times y$ where $[x]_k + c = [y]_k$ is exactly*

$$2|(c)_k| + 1 - \nu_k(c) - [(c)_k \text{ starts with } 1] - [k = 2, (c)_2 \text{ starts with } 10, \\ \text{and } c \text{ is not a power of } 2].$$

3.2 Multiplication

Previous work related to the next result can be found in [6,8,21,22].

Theorem 5. *Let $n \geq 1$, $0 \leq c < n$ be fixed constants, and let $k \geq 2$ be an integer. For both lsd- and msd-first input, there is an n-state automaton accepting $x \times y$ where $n[x]_k + c = [y]_k$.*

4 Linear Subsequences of Automatic Sequences

We now turn to examining the state complexity of transformations on automatic sequences.

Zantema and Bosma [24, Theorem 16] showed that for lsd-first input, if $(h(i))_{i\geq 0}$ is a k-automatic sequence generated by a DFAO of m states, then $(h(i+1))_{i\geq 0}$ is generated by a DFAO of at most $2m$ states. Furthermore, in the same theorem they showed that for msd-first input, the sequence $(h(i+1))_{i\geq 0}$ is generated by a DFAO of at most m^2 states and there is an example where this bound is tight. Additionally, they showed [24, Theorem 17] that the same bounds hold for $(h(i-1))_{i\geq 0}$ where $h(0)$ is any chosen letter from the sequence alphabet.

However, they did not show that their upper bound of $2m$ states for $(h(i+1))_{i\geq 0}$ in the lsd-first case is tight. We suspect it is not. In the following result, we show there are examples attaining the very slightly weaker bound of $2m-1$ states.

Theorem 6. *For lsd-first input, for each $m \geq 2$, there exists a m-state DFAO M generating the sequence $(h(i))_{i\geq 0}$ such that at least $2m-1$ states are required in the DFAO M' generating $(h(i+1))_{i\geq 0}$.*

Proof. We use an adaptation of the Myhill-Nerode theorem [13, Sect. 3.4] for DFAOs. The adaptation for DFAOs is also used by Zantema and Bosma [24] and the state equivalence for DFAOs is defined by van Spaendonck [19].

Let $k = 2$, $m \geq 3$, and define the lsd-first DFAO M generating $(h(i))_{i\geq 0}$ as follows.

$$Q = \{0, \ldots, m-1\},$$

$$\delta(q, a) = \begin{cases} (q+1) \bmod m, & \text{if } a = 1; \\ q+1, & \text{if } a = 0 \text{ and } q \in \{0, \ldots, m-3\}; \\ q, & \text{if } a = 0 \text{ and } q \in \{m-2, m-1\}; \end{cases}$$

$$q_0 = 0,$$

$$\Delta = \Sigma_2,$$

$$\tau(q) = \begin{cases} 1, & \text{if } q = m-1; \\ 0, & \text{otherwise.} \end{cases}$$

Let M' be the lsd-first DFAO generating the sequence $(h(i+1))_{i\geq 0}$. We show that there are $2m-1$ words w_i such that for all w_i, w_j where $i \neq j$, there exists w_{ij} such that the output of M' on input $w_i w_{ij}$ is different from the output of M' on input $w_j w_{ij}$. In this case, we say w_i and w_j are *distinguishable*.

The $2m-1$ pairwise distinguishable words are the following. 0^i where $1 \leq i \leq m-2$; 1^i where $0 \leq i \leq m-2$; $0^{m-2}1$; and $0^{m-2}11$. We need to show each pair of words is distinguishable. There are eight cases to consider.

Case 1: 0^i and 0^j where $1 \leq i < j \leq m-2$. Concatenate $0^{m-j-2}1$. The new words are $0^{i+m-j-2}1$ and $0^{m-2}1$. The output of the automaton M' on these words is then the same as the output of the automaton M on inputs $10^{i-j+m-3}1$ and $10^{m-3}1$, which are 0 and 1. Therefore, the two words are distinguishable.

Proving the rest of the cases follows the same logic.

Case 2: 1^i and 1^j where $0 \le i < j \le m-2$. Concatenate $1^{m-j-2}0$.
Case 3: $0^{m-2}1$ and $0^{m-2}11$. Concatenate λ.
Case 4: 0^i and 1^j where $1 \le i \le m-2$ and $0 \le j \le m-2$. Concatenate $1^{2m-i}0$.
Case 5: 0^i and $0^{m-2}1$ where $1 \le i \le m-2$. Concatenate λ.
Case 6: 0^i and $0^{m-2}11$ where $1 \le i \le m-2$. Concatenate 1^{m-i-1}.
Case 7: 1^i and $0^{m-2}1$ where $0 \le i \le m-2$. Concatenate 1^{m-i-1}.
Case 8: 1^i and $0^{m-2}11$ where $0 \le i \le m-2$. Concatenate 1^{m-i-2}.

Therefore, the DFAO M' for $(h(i+1))_{i \ge 0}$ requires at least $2m-1$ states. □

We now turn to linear subsequences of automatic sequences. More specifically, we consider the following problem. Let $(h(i))_{i \ge 0}$ be a k-automatic sequence generated by a DFAO of m states. Let $n \ge 1$ and $c \ge 0$ be integer constants. How many states are needed to generate $(h(ni+c))_{i \ge 0}$?

Zantema and Bosma [24] studied the specific case where $n = k$ and $0 \le c < k$. They showed that for this case and for both msd-first and lsd-first input, the subsequence $(h(ki+c))_{i \ge 0}$ is generated by a DFAO of at most m states.

In [2, Proof of Theorem 6.8.1], it is proved that if $(h(i))_{i \ge 0}$ is a k-automatic sequence generated by an m-state DFAO with lsd-first input, then for $n \ge 1, c \ge 0$ the linear subsequence $(h(ni+c))_{i \ge 0}$ is generated by an lsd-first DFAO with $m(n+c)$ states. (The theorem is phrased in terms of the k-kernel, but the number of elements of the k-kernel of a sequence is the same as the number of states in a minimal lsd-first DFAO generating the sequence.)

In the special case that $0 \le c < n$, this lsd-first bound was improved slightly to at most mn states by Zantema and Bosma [24]. Furthermore, they provided examples where the same bound does not hold for msd-first input. They stated obtaining a bound in the msd-first case as an open question. We address this in the following theorem, which provides a close and novel connection between state complexity and subword complexity.

Theorem 7. *Let $n \ge 1$, $c \ge 0$ be fixed integer constants. Let $\mathbf{h} = (h(i))_{i \ge 0}$ be an automatic sequence generated by a DFAO with msd-first input, and let $\mathbf{h}' = (h'(i))_{i \ge 0}$ be the interior sequence of $\mathbf{h}$.*

(a) If $c < n$, then there is a DFAO of at most $\rho_{\mathbf{h}'}(n)$ states generating $(h(ni+c))_{i \ge 0}$ with msd-first input.
(b) If $c \ge n$, then there is a DFAO of at most $\rho_{\mathbf{h}'}(c+1)$ states generating $(h(ni+c))_{i \ge 0}$ with msd-first input.

Proof. Let $(h(i))_{i \ge 0}$ be generated by a minimal DFAO $M = (Q, \Sigma_k, \delta, q_0, \Delta, \tau)$.

(a) Suppose $c < n$. We want to create a DFAO generating $(h(ni+c))_{i \ge 0}$. The idea is that we construct M' such that on input x it reaches a state given by the following subword of the interior sequence $(h'(i))_{i \ge 0}$:

$$[h'(n[x]_k), h'(n[x]_k + 1), \ldots, h'(n[x]_k + n - 1)].$$

Then, since $c \le n-1$, we only need to obtain $h'(n[x]_k + c)$ by mapping the subword $[r_0, \ldots, r_{n-1}]$ to r_c.

Let $S_n(h')$ be the set of length-n subwords of the sequence $(h'(i))_{i \ge 0}$. We can define $M' = (Q', \Sigma_k, \delta', q_0', \Delta, \tau')$ as follows:

$$\begin{aligned} Q' &= \{[r_0, \ldots, r_{n-1}] \in S_n(h')\}, \\ \delta'([r_0, \ldots, r_{n-1}], a) &= [s_0, \ldots, s_{n-1}] \\ &\quad \text{where } s_d = \delta(r_j, b),\ j = \lfloor (na+d)/k \rfloor,\ b = (na+d) \bmod k, \\ q_0' &= [h'(0), h'(1), \ldots, h'(n-1)], \\ \tau'([r_0, \ldots, r_{n-1}]) &= \tau(r_c). \end{aligned}$$

The states of M' are the length-n subwords of the interior sequence $(h'(i))_{i \ge 0}$. Note that not all states are necessarily reachable from the initial state.

First, let us see that

$$s_d = \delta(r_j, b), \quad j = \lfloor (na+d)/k \rfloor, \quad b = (na+d) \bmod k$$

(i.e., $kj + b = na + d$) in the transition function is well-defined. We have $0 \le a \le k-1$, and the definition of b implies that $0 \le b < k$. We need to verify that $0 \le j \le n-1$. We do this as follows: $j = \lfloor (na+d)/k \rfloor$ implies that $j \le (n(k-1)+d)/k$. Since $d < n$, we see this implies that $kj < nk$. Hence $j < n$, as desired.

Now let us prove by induction on input length $|x|$ that

$$\delta'(q_0', x) = [h'(n[x]_k), \ldots, h'(n[x]_k + n - 1)]. \tag{2}$$

The base case is $x = \lambda$. Here the result follows by the definition of q_0'.

Otherwise, assume Eq. (2) is true for $|x|$; we prove it for $|x|+1$. Write $y = xa$. So $[y]_k = k[x]_k + a$. Then $\delta'(q_0', y) = \delta'(\delta'(q_0', x), a)$. By induction $\delta'(q_0', x) = [r_0, r_1, \ldots r_{n-1}]$ where $r_j = h'(n[x]_k + j)$. By the definition of δ' we have $\delta'(q_0', y) = [s_0, s_1 \ldots, s_{n-1}]$ where

$$\begin{aligned} s_d = \delta(r_j, b) &= \delta(h'(n[x]_k + j), b) = h'(k(n[x]_k + j) + b) \\ &= h'(kn[x]_k + na + d) = h'(n(k[x]_k + a) + d) \\ &= h'(n[y]_k + d), \end{aligned}$$

as desired.

(b) Suppose $c \ge n$. The proof proceeds exactly as in (a), except now the states are the length-$(c+1)$ (instead of length-n) subwords of $(h'(i))_{i \ge 0}$. □

Corollary 1. *Let $n \ge 1$, $c \ge 0$ be fixed integer constants. Let $(h(i))_{i \ge 0}$ be an automatic sequence generated by a DFAO of m states with msd-first input.*

(a) If $c < n$, then there is a DFAO of at most knm^2 states generating the automatic sequence $(h(ni + c))_{i \ge 0}$ with msd-first input.

(b) If $c \ge n$, then there is a DFAO of at most $k(c+1)m^2$ states generating the automatic sequence $(h(ni + c))_{i \ge 0}$ with msd-first input.

Proof. Apply Theorem 1. □

We do not know if the bounds in Corollary 1 are tight. This seems worthy of further study, so we state it as an open problem.

Open Problem 1. *Find a class of automatic sequences $(h_m(i))_{i\geq 0}$, generated by a DFAO of m states with msd-first input, such that there is a constant C so the linear subsequences $(h_m(ni))_{i\geq 0}$ have state complexity at least Cnm^2 for infinitely many n.*

4.1 Connection with Subword Complexity

We now consider an interesting application of this connection between state complexity and subword complexity. Given an automatic sequence $\mathbf{h} = (h(i))_{i\geq 0}$ we would like to obtain an upper bound on the subword complexity of its linear subsequence $(h(ni))_{i\geq 0}$.

Theorem 8. *Let $\mathbf{h} = (h(i))_{i\geq 0}$ be a k-automatic sequence and $\mathbf{h}' = (h'(i))_{i\geq 0}$ its interior sequence. Let $\mathbf{h}_n$ denote the linear subsequence $(h(ni))_{i\geq 0}$ and similarly for $\mathbf{h}'_n$. Then $\rho_{\mathbf{h}_n}(\ell) \leq k\ell\rho_{\mathbf{h}'}(2n)$ for all $\ell, n \geq 1$.*

Proof. Clearly $\rho_{\mathbf{h}_n}(\ell) \leq \rho_{\mathbf{h}'_n}(\ell)$. Let $M = (Q, \Sigma_k, \delta, q_0, \Delta, \tau)$ be an automaton generating $\mathbf{h}'_n$ built by the construction in Theorem 7, and let $\hat{\mathbf{M}}$ be the sequence generated by the intrinsic DFAO $\hat{M} = (Q, \Sigma_k, \delta, q_0, Q, \iota)$ where $\iota : Q \to Q$ is the identity map. By Theorem 1, we have $\rho_{\mathbf{h}'_\mathbf{n}}(\ell) \leq k\ell\rho_{\hat{\mathbf{M}}}(2)$. However, each length-2 subword of $\hat{\mathbf{M}}$ corresponds, from the construction of Theorem 7, to a distinct length-$2n$ subword of $\mathbf{h}'$. Thus $\rho_{\hat{\mathbf{M}}}(2) \leq \rho_{\mathbf{h}'}(2n)$. Putting this all together gives

$$\rho_{\mathbf{h}_n}(\ell) \leq \rho_{\mathbf{h}'_n}(\ell) \leq k\ell\rho_{\hat{\mathbf{M}}}(2) \leq k\ell\rho_{\mathbf{h}'}(2n),$$

as desired. □

4.2 Thue-Morse Sequence

We now apply some of the previous theorems to the famous Thue-Morse sequence $\mathbf{t} = (t(i))_{i\geq 0}$. The Thue-Morse sequence is defined as the fixed point of the morphism $0 \to 01$, $1 \to 10$; alternatively, $t(i)$ is equal to the number of 1's modulo 2 in the binary representation of i [1].

First we apply Theorem 8 to the Thue-Morse sequence.

Corollary 2. *Let $\mathbf{t}_n$ denote the linear subsequence $(t(ni))_{i\geq 0}$. We have $\rho_{\mathbf{t}_n}(\ell) \leq \frac{40}{3}\ell n$ for $\ell, n \geq 1$.*

Proof. Results from Avgustinovich [4] imply that $\rho_{\mathbf{t}}(n) \leq \frac{10}{3}n$. So by Theorem 8, we have $\rho_{\mathbf{t}_n}(\ell) \leq 2 \cdot \ell \cdot \frac{10}{3} \cdot 2n = \frac{40}{3}\ell n$. □

Now we apply Theorem 7 to the Thue-Morse sequence $(t(i))_{i\geq 0}$ and $c = 0$.

Theorem 9. *For all integers $n \geq 1$, the number of states in the minimal DFAO generating $(t(ni))_{i\geq 0}$ with msd-first input is $\rho_{\mathbf{t}}(n/2^{\nu_2(n)})$.*

Theorem 9 follows easily from the next result.

Theorem 10. *For all odd integers $n \geq 1$, the construction provided in Theorem 7 is optimal for $(t(ni))_{i\geq 0}$.*

Proof. Consider the automaton created by the construction for $(t(ni))_{i\geq 0}$. On input x the automaton reaches the state $[t(n[x]), \ldots, t(n[x]+n-1)]$ where the output is $t(n[x])$. Furthermore, by [5, Lemma 12] we know all length-n subwords of the Thue-Morse sequence are reachable states in the automaton.

Now suppose we have two distinct states in this automaton reachable by x and y, say $[t(n[x]), \ldots, t(n[x]+n-1)] \neq [t(n[y]), \ldots, t(n[y]+n-1)]$. We need to show the states are distinguishable in the Myhill-Nerode sense. Since these states are distinct, there exists j, $0 \leq j < n$, such that $t(n[x]+j) \neq t(n[y]+j)$.

Choose $m = 1 + \lfloor \log_2 n \rfloor$, so that $2^m > n$. Define $c = \lceil 2^m j/n \rceil$. Now $0 \leq 2^m j/n \leq 2^m(n-1)/n = 2^m - 2^m/n \leq 2^m - 1$, so $0 \leq c < 2^m$, and hence there exists a length-m binary word z such that $[z] = c$.

Write $nc = 2^m j + r$. Then $0 \leq r < n < 2^m$. Hence

$$\begin{aligned} n[xz] &= n(2^m[x] + [z]) = 2^m n[x] + n[z] = 2^m na + nc \\ &= 2^m na + 2^m j + r = 2^m(na+j) + r. \end{aligned}$$

Since $r < 2^m$, it follows that $t(n[xz]) \equiv t(na+j) + t(r) \pmod 2$. In exactly the same fashion, we get $t(n[yz]) \equiv t(nb+j) + t(r) \pmod 2$. Since by assumption $t(na+j) \neq t(nb+j)$, we get $t(n[xz]) \neq t(n[yz])$, as desired. □

Let $\mathrm{sc}(h(ni))$ denote the number of states in a minimal automaton generating the sequence $(h(ni))_{i\geq 0}$.

Corollary 3. *For all integers $r, p \geq 1$, we have*

$$\begin{cases} \mathrm{sc}(t(2ri)) = \mathrm{sc}(t(ri)), \\ \mathrm{sc}(t((2^p + r)i)) = 3 \cdot 2^p + 4(r-1), & \text{if } 1 < r < 2^{p-1} \text{ and } r \bmod 2 = 1; \\ \mathrm{sc}(t((2^p + 2^{p-1} + r)i)) = 5 \cdot 2^p + 2(r-1), & \text{if } 1 < r < 2^{p-1} \text{ and } r \bmod 2 = 1. \end{cases}$$

Next we apply Theorem 7 to the Thue-Morse sequence $(t(i))_{i\geq 0}$ and $n = 1$.

Theorem 11. *For all integers $c \geq 1$, the number of states in the minimal automaton generating $(t(i+c))_{i\geq 0}$ with msd-first input is at most $\frac{10}{3}c$.*

Unlike the case of Theorem 10, where the construction in Theorem 7 gives the minimal automaton, in the case of Theorem 11, the construction in Theorem 7 does not necessarily give the minimal automaton, so the upper bound in Theorem 11 is not tight. The next theorem gives a weak lower bound on the state complexity of $(t(i+c))_{i\geq 0}$.

Theorem 12. *The minimal DFAO computing* $(t(i+c))_{i \geq 0}$ *with msd-first input has more than* $c^{0.694}$ *states for infinitely many* c.

The exact number of states in the minimal automaton generating $(t(i+c))_{i \geq 0}$ for small values of c is given as sequence A382296 of the OEIS [18]. We state the following as an open problem.

Open Problem 2. *What is a good formula for the exact number of states in the minimal automaton generating* $(t(i+c))_{i \geq 0}$ *with msd-first input, as a function of* c*?*

5 Construction by Büchi Arithmetic

We now re-examine the previous relations and operations in light of their implementation in an interpretation of Büchi arithmetic (as, for example, in the system `Walnut` [17]) to find the time required for constructing their automata.

Büchi arithmetic is the logical theory of the natural numbers together with addition and the function $V_k(n)$, the largest power of k dividing n, for some fixed $k \geq 2$. This theory is powerful enough to express finite automata [7]; in what follows we treat DFAO's computing automatic sequences as more or less a primitive object, without requiring an explicit expression in terms of $\langle \mathbb{N}, +, V_k \rangle$.

In its current version, `Walnut` does not use the algorithms for creating the automata we discussed in previous sections. Instead, more general techniques, suitable for any addable numeration system, are used. This means that the size of intermediate automata may differ from the minimal automata we constructed, and the running time can similarly be larger. Here the total number of transitions of an automaton M is used as a measure for the computational complexity of constructing M.

In our analyses in this section, we focus on msd-first input. We need to make sure the final automata always have the correct output regardless of the number of leading zeros read; this is accomplished as follows. After creating an automaton M with initial state q_0 and transition function δ, we change the initial state to the state q_0' where $\delta(q_0, [0^j \times \cdots \times 0^j]) = q_0'$ for some j and $\delta(q_0', [0 \times \cdots \times 0]) = q_0'$. The rationale is that our automata should always have the same correct output regardless of the number of leading zeros read and by creating the automaton in this manner we can always assume we have considered enough leading zeros for the automaton computations to be correct. A similar idea for base-2 is mentioned by Wolper and Boigelot [22].

Let the msd-first 2-state automaton for recognizing addition from Theorem 2 be M_{add}. We use $M_{\text{add}}(x, y, z)$ to show the output of M_{add} on input $x \times y \times z$, which is 1 if and only if $[x]_k + [y]_k = [z]_k$; we use a similar notation to refer to the output of other automata on a specific input. Recall that M_{add} states correspond to the difference $[z]_k - ([x]_k + [y]_k)$.

In `Walnut` each DFA or DFAO created is minimized. If n is the number of states in an automaton (and the number of transitions is $O(n)$), the minimization steps in the implementation can use an algorithm running in $O(n \log n)$ time by

Hopcroft [12] or Valmari [20]. The algorithm by Hopcroft can be applied to DFAOs with slight modifications as described by van Spaendonck [19].

We start with an analysis of how an automaton for a statement like $[x]_k = c$ is created, where c is a fixed natural number constant. Recall that in Walnut, constants are entered as numbers written in decimal notation (base 10). There are well-known efficient algorithms for converting a number from base 10 to base k for an integer $k \geq 2$; see, for example, [14, §4.4], but these are not used by Walnut. Instead, a more general technique, suitable for any numeration system, is used.

Lemma 1. *The minimal automaton for recognizing the relation $[x]_k = c$ with msd-first input has $\lfloor \log_k c \rfloor + 2$ states.*

Proof. Such an automaton consists of a linear chain of nodes labeled with the base-k representation of c, which has length $\lfloor \log_k c \rfloor + 1$. □

Theorem 13. *Let $c \geq 0$ be a fixed constant. The automaton recognizing the relation $[x]_k = c$ with msd-first input can be computed in $O((\log^2 c)(\log \log c))$ time.*

Proof. We recursively create the automaton $M_{=c}$.

The base case is $c = 0$ and the automaton has 1 state.

If c is even, we recursively compute a minimal DFA $M_{=(c/2)}$ recognizing $[y]_k = c/2$ on input y and then use the first-order expression

$$\exists y \; M_{=(c/2)}(y) \wedge M_{\text{add}}(y, y, x)$$

which is translated into an automaton. We know the automaton for M_{add} has 2 states and by Lemma 1, the automaton $M_{=(c/2)}$ has $\lfloor \log_k (c/2) \rfloor + 2$ states. First we apply the product construction to $M_{=(c/2)}$ and M_{add} such that the $M_{=(c/2)}$ part processes y and the M_{add} part processes $y \times y \times x$; the resulting automaton has $2(\lfloor \log_k (c/2) \rfloor + 2)$ states. Each state in the product automaton is of the form $[q, p]$ where on input $x \times y$, q is a state from the minimal automaton $M_{=(c/2)}$ representing the number of letters from $(c/2)_k$ matched with y by Lemma 1 and $p \in \{0, 1\}$ is a state from M_{add} equal to $[x]_k - 2[y]_k$. Then we apply the $\exists y$ quantifier by removing the component corresponding to y from the transitions and we get an NFA. In the NFA there are no two same transitions from a single state leading to different states. So after subset construction the resulting DFA is the same as the NFA. Then we minimize and the resulting DFA is the automaton from Lemma 1 with $O(\log c)$ states.

If c is odd, we recursively compute a DFA $M_{=(c-1)}$ recognizing $[y]_k = c - 1$ on input y and then use the first-order expression

$$\exists y \; M_{=(c-1)}(y) \wedge M_{\text{add}}(y, 1, x)$$

which is translated into an automaton. Following a construction similar to the one described for the case where c is even, we get the DFA from Lemma 1 for constant c with $O(\log c)$ states.

Taking minimization steps into account, we get the following recursive formula for the time complexity:

$$T(c) = \begin{cases} O((\log c)(\log\log c)) + T(c/2), & \text{if } c \bmod 2 = 0; \\ O((\log c)(\log\log c)) + T(c-1), & \text{if } c \bmod 2 = 1. \end{cases}$$

So the total time is $O((\log^2 c)(\log\log c))$. □

Theorem 14. *Let $c \geq 0$ be a fixed constant. The automaton recognizing the relation $[x]_k + c = [z]_k$ with msd-first input can be computed in $O((\log^2 c)(\log\log c))$ time.*

We continue by constructing the automaton M_n recognizing $n[x]_k = [z]_k$ on input $x \times z$ where n is a fixed constant. We have $M_n(x, z) = 1$ if and only if $n[x]_k = [z]_k$. There are several possible recursions for creating this automaton; for example, we could use the following:

- $\exists y\ M_{n-1}(x, y) \wedge M_{\text{add}}(x, y, z)$.
- $\exists y_1, y_2\ M_{\lfloor n/2 \rfloor}(x, y_1) \wedge M_{\lceil n/2 \rceil}(x, y_2) \wedge M_{\text{add}}(y_1, y_2, z)$.

However, there is a third recursive implementation that is more efficient than the two above. The base case is $n = 1$.

If n is even, we recursively compute a DFA $M_{n/2}$ recognizing $[y]_k = (n/2)[x]_k$ on input $x \times y$ and then use the first-order expression $\exists y\ M_{n/2}(x, y) \wedge M_{\text{add}}(y, y, z)$, which is translated into an automaton by a direct product construction for the $\wedge$, and projection of the second coordinate to collapse transitions on y.

If n is odd, we recursively compute a DFA $M_{n-1}(x, y)$ recognizing $[y]_k = (n-1)[x]_k$ and use the expression $\exists y\ M_{n-1}(x, y) \wedge M_{\text{add}}(x, y, z)$, which is similarly translated into an automaton.

In both cases, the translation can potentially generate a nondeterministic automaton (by the $\exists y$ projection). In this case, the subset construction is used to determinize the resulting automaton. Thus, at least in principle, this particular translation mechanism could take exponential time. However, we show that in this case, the subset construction does not cause an exponential blowup in the number of states.

Theorem 15. *Let $n \geq 1$ be a fixed constant. The DFA M_n with msd-first input for recognizing $[y]_k = n[x]_k$ constructed recursively has at most n states corresponding to differences $[y]_k - n[x]_k$ in the range $[0, n-1]$.*

Theorem 16. *Let $n \geq 1$ be a fixed constant. The automaton for recognizing $[y]_k = n[x]_k$ on msd-first input can be computed in $O(n \log^2 n)$ time.*

Theorem 17. *Let $n \geq 1$, $0 \leq c < n$ be fixed constants. The automaton recognizing the relation $n[x]_k + c = [z]_k$ with msd-first input can be computed in $O(n \log(n) \log(n \log(n)))$ time.*

Theorem 18. *Let $n \geq 1$, $0 \leq c < n$ be fixed constants. For msd-first input, given an m-state DFAO generating $(h(i))_{i \geq 0}$, we can compute the DFAO generating $(h(ni + c))_{i \geq 0}$ in $O(n \log n \log(n \log n) + nm^2 \log(nm^2))$ time.*

References

1. Allouche, J.P., Shallit, J.: The ubiquitous Prouhet-Thue-Morse sequence. In: Ding, C., Helleseth, T., Niederreiter, H. (eds.) Sequences and Their Applications, pp. 1–16. Discrete Mathematics and Theoretical Computer Science, Springer (1999). https://doi.org/10.1007/978-1-4471-0551-0_1
2. Allouche, J.P., Shallit, J.: Automatic sequences: theory, applications, Generalizations (2003). https://doi.org/10.1017/CBO9780511546563
3. Amat, N., Ganty, P., Mansutti, A.: How big is the automaton? Certified lower bounds on the size of Presburger DFAs, preprint. https://patapsco.software.imdea.org/. Accessed 11 December 2025
4. Avgustinovich, S.V.: The number of distinct subwords of fixed length in the Morse-Hedlund sequence. Sibirskii Zhurnal Issledovaniya Operatsii **1**(2), 3–7 (1994). in Russian. English translation in A. D. Korshunov, editor, Discrete Analysis and Operations Research, volume 355 of Mathematics and Its Applications, pp. 1–5. Kluwer Academic Publishers, 1996
5. Blanchet-Sadri, F., D. Currie, J., Rampersad, N., Fox, N.: Abelian complexity of fixed point of morphism $0 \mapsto 012$, $1 \mapsto 02$, $2 \mapsto 1$. Electr. J. Combinatorics **14**, #A11 (2014). https://math.colgate.edu/~integers/vol14.html
6. Boudet, A., Comon, H.: Diophantine equations, Presburger arithmetic and finite automata. In: Kirchner, H. (ed.) Trees in Algebra and Programming—CAAP '1996, pp. 30–43. Lecture Notes in Computer Science, Springer (1996). https://doi.org/10.1007/3-540-61064-2_27
7. Bruy'ere, V., Hansel, G., Michaux, C., Villemaire, R.: Logic and p-recognizable sets of integers. Bull. Belgian Math. Soc. - Simon Stevin **1**(2), 191–238 (1994). https://doi.org/10.36045/bbms/1103408547
8. Charlier, É., Cisternino, C., Massuir, A.: Minimal automaton for multiplying and translating the Thue-Morse set. Electr. J. Combinatorics **28**(3), #P3.12 (2021). https://www.combinatorics.org/ojs/index.php/eljc/article/view/v28i3p12
9. Cobham, A.: On the base-dependence of sets of numbers recognizable by finite automata. Math. Syst. Theory **3**(2), 186–192 (1969). https://doi.org/10.1007/BF01746527
10. Cobham, A.: Uniform tag sequences. Math. Syst. Theory **6**(1-2), 164–192 (1972). https://doi.org/10.1007/BF01706087
11. Gao, Y., Moreira, N., Reis, R., Yu, S.: A survey on operational state complexity. J. Autom. Lang. Comb. **21**(4), 251–310 (2016)
12. Hopcroft, J.: An $n \log n$ algorithm for minimizing states in a finite automaton. In: Kohavi, Z., Paz, A. (eds.) Theory of Machines and Computations, pp. 189–196. Academic Press (1971). https://www.sciencedirect.com/science/article/pii/B9780124177505500221
13. Hopcroft, J.E., Ullman, J.D.: Introduction to Automata Theory, Languages, and Computation. Addison-Wesley Series in Computer Science, Addison-Wesley Publishing Company (1979)
14. Knuth, D.E.: Seminumerical Algorithms, The Art of Computer Programming, vol. 2. Addison-Wesley, 3rd edn. (1997). https://learning.oreilly.com/library/view/the-art-of/9780321635778/
15. Moradi, D., Popoli, P., Shallit, J., Vukusic, I.: State complexity of shifts of the Fibonacci word (2026), arxiv preprint arXiv:2603.18858. Available at https://arxiv.org/abs/2603.18858

16. Moradi, D., Rampersad, N., Shallit, J.: Complexity of linear subsequences of Fibonacci-automatic sequences (2026). arxiv preprint arXiv:2603.21645. Available at https://arxiv.org/abs/2603.21645
17. Mousavi, H.: Automatic theorem proving in Walnut (2021). http://arxiv.org/abs/1603.06017v2, arXiv:1603.06017v2
18. Sloane, N.J.A., et al.: The On-Line Encyclopedia of Integer Sequences. https://oeis.org/. Accessed 8 December 2025
19. Spaendonck, F.v.: Automatic Sequences: The Effect of Local Changes on Complexity. Master's thesis, Radboud University, Netherlands (2020)
20. Valmari, A.: Fast brief practical DFA minimization. Inf. Process. Lett. **112**(6), 213–217 (2012). https://www.sciencedirect.com/science/article/pii/S0020019011003267
21. Waxweiler, L.: Caractère reconnaissable d'ensembles de polynômes à coefficients dans un corps fini. Docteur en sciences dissertation, Université de Liège, Belgium (2009)
22. Wolper, P., Boigelot, B.: On the Construction of Automata from Linear Arithmetic Constraints. In: Graf, S., Schwartzbach, M. (eds.) Tools and Algorithms for the Construction and Analysis of Systems, pp. 1–19. Lecture Notes in Computer Science, Springer (2000). https://doi.org/10.1007/3-540-46419-0_1
23. Yu, S., Zhuang, Q., Salomaa, K.: The state complexities of some basic operations on regular languages. Theoretical Comput. Sci. **125**(2), 315–328 (1994). https://www.sciencedirect.com/science/article/pii/030439759200011F
24. Zantema, H., Bosma, W.: Complexity of automatic sequences. Inf. Comput. **288**, 104710 (2022). https://www.sciencedirect.com/science/article/pii/S0890540121000250

Minimal Deterministic Finite Automata from Alternating and Boolean Automata

Hellis Tamm[1(✉)] and Brink van der Merwe[2,3]

[1] Department of Software Science, Tallinn University of Technology, Tallinn, Estonia
hellis@ioc.ee

[2] Department of Computer Science, Stellenbosch University, Stellenbosch, South Africa
abvdm@cs.sun.ac.za

[3] National Institute for Theoretical and Computational Sciences, Stellenbosch, South Africa

Abstract. We consider the question of what conditions are required such that determinising an alternating or a Boolean automaton yields a minimal deterministic finite automaton. For $\cup$-NFAs, we recall the classical atomic characterisation of Brzozowski and Tamm. For $\cap$-NFAs, we show that the corresponding statement should be formulated using a support-based notion of left languages. We present a two-step determinisation procedure for alternating automata via disjunctive and conjunctive normal forms, and introduce minterm and maxterm automata as canonical constructions linking Boolean automata to $\cup$- and $\cap$-NFAs.

1 Introduction

Determinisation of finite automata remains one of the most studied and technically subtle topics in automata theory. While the classical subset construction provides a complete and well-understood transformation from nondeterministic finite automata (NFAs, hereafter also called $\cup$-NFAs) to deterministic ones (DFAs), its generalisation to alternating finite automata (AFAs) introduces both logical and combinatorial challenges. An AFA combines existential and universal branching, naturally representing its transition structure as a Boolean formula over successor states. Consequently, the determinisation of an AFA can be viewed as a Boolean functional lifting of the subset construction, where states of the resulting DFA correspond to equivalence classes of Boolean formulas over the original state set.

Historically, the notion of AFA was introduced by Chandra, Kozen, and Stockmeyer [5], while Brzozowski and Leiss [3] introduced an equivalent concept of a Boolean finite automaton (BFA). Often, the notions of AFA and BFA are used interchangeably. However, in the current work, we treat AFAs as the subclass of BFAs whose transitions use only positive connectives ($\vee$ and $\wedge$), excluding negation.

H. Tamm—Supported by the Estonian Research Council grant PRG1210.

T. J. Smith (Ed.): CIAA 2026, LNCS 16695, pp. 200–215, 2026.
https://doi.org/10.1007/978-3-032-31176-4_14

Despite the theoretical elegance of Boolean automata and their connections to logic and circuit complexity, very little is known about what conditions guarantee that the DFA obtained by determinising an AFA or a BFA is minimal. For $\cup$-NFAs, minimality after determinisation has been characterised in terms of the atomic structure of the language [4], and a similar characterisation has been obtained for symmetric difference NFAs ($\oplus$-NFAs) [7].

The atomic theory of regular languages, developed by Brzozowski and Tamm [4] and further studied in the lattice-theoretic framework of Bell, Smertnig, and Tamm [1], has by now become a standard tool for understanding the state complexity of NFA-to-DFA conversion. Yet, despite the prominence of alternating and Boolean automata in logic, verification, and circuit complexity (see e.g. [3,5,6]), no analogous atomic characterisation of minimality after determinisation is currently known beyond the $\cup$-NFA and $\oplus$-NFA cases. Filling this gap is the main motivation for the present work: we identify the precise sense in which the atomic theory does—and does not—extend to $\cap$-NFAs, AFAs, and BFAs, and we identify the canonical $\cup$-NFA (the minterm automaton) through which a BFA's determinisation can always be analysed. The minterm and maxterm constructions provide a uniform bridge between Boolean automata and the established atomic theory, and clarify which structural features of an automaton are responsible for non-minimal determinisation.

The present work begins with the line of reasoning for $\cup$-NFAs and re-examines how far it extends. However, for $\cap$-NFAs, the naive use of the singleton-evaluation of left languages (Definition 4) turns out to be misleading. Instead, the correct dual notion is support-based and is derived from the corresponding $\cup$-NFA via the negation diagram in Fig. 1.

For general AFAs, this points to a broader picture, but the exact criterion is yet to be discovered.

We explore the duality between $\cup$- and $\cap$-NFAs via language complementation and negation of Boolean formulas, and the possibility of viewing determinisation as a two-step process: writing formulas in DNF first yields an intermediate $\cup$-NFA on conjunctive clauses, while writing them in CNF first yields an intermediate $\cap$-NFA on disjunctive clauses. This viewpoint extends naturally to the minterm and maxterm automata constructions used later for BFAs.

As related work, Fellah, Jürgensen, and Yu [6] have studied constructions between BFAs, NFAs and DFAs, as well as closure constructions and normal forms for BFAs. Our minterm automaton is similar to one of their NFA constructions.

The remainder of this paper is organised as follows. Section 2 provides definitions and background material. Section 3 formalises the duality between $\cup$- and $\cap$-NFAs, while Sect. 4 introduces a two-step determinisation perspective for AFAs. Section 5 develops minterm and maxterm automata as canonical constructions for BFAs. Section 6 discusses minimality results of DFAs obtained by determinising AFAs and BFAs, and Sect. 7 concludes the paper.

2 Background and Definitions

We collect in this section the notions used throughout the paper. Most are standard, and we adopt the conventions of [1,4,8]. The first three definitions fix the four classes of automata that we work with—BFAs, AFAs, ∪-NFAs and ∩-NFAs—together with the convention that an AFA is a BFA whose transitions use only positive Boolean connectives. Definitions 4 and 5 are less standard outside the atomic-theory literature, so we will accompany them with a brief intuitive description and a comparison with the classical NFA case. The remaining definitions concern determinisation and reversal, both of which we use repeatedly.

Definition 1 (Boolean automaton). *A* Boolean finite automaton *(BFA) is a tuple $\mathcal{A} = (Q, \Sigma, \delta, \varphi_0, F)$, where Q is a finite, non-empty set of states; Σ is a finite input alphabet; $\delta : Q \times \Sigma \to \mathcal{B}(Q)$ is the transition function, mapping each pair (q, a) to a Boolean formula over Q constructed using $\vee$, $\wedge$, and $\neg$; $\varphi_0 \in \mathcal{B}(Q)$ is the* initial Boolean formula *over Q; and $F \subseteq Q$ is the set of accepting states.*

For a Boolean formula $\varphi \in \mathcal{B}(Q)$ and a subset $S \subseteq Q$, define the evaluation as $[\![\varphi]\!](S) = \texttt{true}$ if φ evaluates to true when all $q \in S$ are set to true and all $q \notin S$ are set to false, and $\texttt{false}$ otherwise.

The extension of δ to formulas is defined recursively by substitution: $\delta^*(q, \varepsilon) = q$ and $\delta^*(q, aw) = \delta^*(\delta(q, a), w)$. For any Boolean formula ψ over Q, $\delta^*(\psi, w)$ is obtained by substituting each state q in ψ with $\delta^*(q, w)$. An input word $w \in \Sigma^*$ is accepted by $\mathcal{A}$ iff $[\![\delta^*(\varphi_0, w)]\!](F) = \texttt{true}$.

Definition 2 (Alternating automaton). *An* alternating finite automaton *(AFA) is a Boolean automaton $\mathcal{A} = (Q, \Sigma, \delta, \varphi_0, F)$ whose transition function and initial formula contain only positive connectives: $\delta : Q \times \Sigma \to \mathcal{B}^+(Q)$ and $\varphi_0 \in \mathcal{B}^+(Q)$, where $\mathcal{B}^+(Q)$ denotes the set of positive Boolean formulas built from Q using $\wedge$ and $\vee$.*

Definition 3 (∪-NFA, ∩-NFA, and DFA).

- *A ∪-NFA (union-NFA) is an AFA in which every transition formula and the initial formula are constructed using only disjunction.*
- *An ∩-NFA (intersection-NFA) is an AFA in which all transition formulas and the initial formula are constructed using only conjunction.*
- *A* DFA *is an AFA in which all transition formulas and the initial formula consist of single states.*

Remark 1 (Implicit sink states). In a ∪-NFA, if a transition $\delta(q, a)$ is undefined, the convention is to treat the implicit sink state as *rejecting*. This reflects the semantics of disjunction: an empty disjunction evaluates to `false`. Dually, in an ∩-NFA, an undefined transition corresponds to an empty conjunction, which evaluates to `true`; hence the implicit sink state is *accepting*.

Two notions central to the analysis below are the *left* and *right* languages of a state. Intuitively, the right language of a state q records which suffixes lead to acceptance *from* q, while the left language records which prefixes lead the automaton *to* a configuration in which q matters for acceptance. For an ordinary NFA, "q matters" means simply "q is reachable on input w"; for $\cup$-NFAs this remains the standard reachability condition. For BFAs, however, the situation is subtler, since acceptance depends on a Boolean evaluation of the reached formula, and we have to decide which set of states to evaluate the reached formula on. The next definition adopts a singleton-evaluation convention; we will see later (Sect. 3) that for $\cap$-NFAs this convention is *not* the right one and must be replaced by a support-based variant.

Definition 4 (Left and right languages). *Let $\mathcal{A} = (Q, \Sigma, \delta, \varphi_0, F)$ be a BFA recognising a language L.*

The right language *of a state $q \in Q$ is $RL(\mathcal{A}, q) = \{w \in \Sigma^* \mid [\![\delta^*(q, w)]\!](F) = \mathtt{true}\}$. That is, $RL(\mathcal{A}, q)$ is the set of strings accepted when starting from q as if it were the sole initial state. The* left language *of a state $q \in Q$ is $LL(\mathcal{A}, q) = \{w \in \Sigma^* \mid [\![\delta^*(\varphi_0, w)]\!](\{q\}) = \mathtt{true}\}$.*

For a $\cup$-NFA $\mathcal{N} = (Q, \Sigma, \delta, I, F)$, these definitions simplify to the classic ones: $RL(\mathcal{N}, q) = \{w \mid \delta^*(q, w) \cap F \neq \varnothing\}$ and $LL(\mathcal{N}, q) = \{w \mid q \in \delta^*(I, w)\}$, where $\delta^*(I, w)$ denotes the set of states reachable from the set of initial states I on input w. However, for an $\cap$-NFA $\mathcal{N}_\cap$, the singleton-evaluation notion $LL(\mathcal{N}_\cap, q)$ need not coincide with the more natural support-based notion obtained from negation duality; this will be addressed in Sect. 3.

We now turn to two further notions that are central to the atomic theory of regular languages [4]: *quotients* and *atoms*. Intuitively, a left quotient $x^{-1}L$ collects all suffixes that complete the prefix x to a word of L, and the (finitely many) distinct left quotients are exactly the right languages of the states of the minimal DFA. Similarly, a right quotient Ly^{-1} contains all prefixes that appear in front of suffix y of a word in L. *Atoms* are the finest distinctions one can make using these quotients: a left atom is a non-empty Boolean combination $\bigcap_{i \in S} L_i \cap \bigcap_{i \notin S} \overline{L_i}$, picking, for each left quotient L_i, whether to include it or its complement. Equivalently, two strings lie in the same left atom iff they belong to the same set of left quotients. Right atoms are defined dually using right quotients. The atoms partition Σ^* into the smallest classes that are not refinable by any further quotient, and they will play the role of "elementary regions" in our minimality criteria. As a small example, for $L = \Sigma^* a \Sigma^*$ over $\Sigma = \{a, b\}$ there are two left quotients $L_0 = L$ and $L_1 = \Sigma^*$, and two left atoms: the set $L = \Sigma^* a \Sigma^*$ of words containing an a, and the set b^* of words avoiding a. Incidentally, in this example, the right quotients of L are the same as the left quotients, and the right atoms of L are equal to the left atoms.

Definition 5 (Left/right quotients and atoms). *Let $L \subseteq \Sigma^*$ be a regular language.*

A left quotient *(or simply* quotient*) of L is a language of the form $x^{-1}L = \{w \in \Sigma^* \mid xw \in L\}$ for some $x \in \Sigma^*$. The set of left quotients of L is finite and equals the set of right languages of the states of the minimal DFA for L.*

A right quotient *of L is a language of the form $Ly^{-1} = \{w \in \Sigma^* \mid wy \in L\}$ for some $y \in \Sigma^*$.*

Let $L_0, L_1, \ldots, L_{k-1}$ be the left quotients of L. A left atom *of L is a non-empty language of the form*

$$A_S^L = \bigcap_{i \in S} L_i \cap \bigcap_{i \notin S} \overline{L_i},$$

where $S \subseteq \{0, \ldots, k-1\}$. The left atoms partition Σ^* into the finest classes that cannot be further refined by the left quotients.

Dually, let $R_0, R_1, \ldots, R_{\ell-1}$ be the right quotients of L. A right atom *of L is a non-empty language of the form $A_T^R = \bigcap_{j \in T} R_j \cap \bigcap_{j \notin T} \overline{R_j}$, where $T \subseteq \{0, \ldots, \ell-1\}$.*

It is known [1,4] that the left languages of the minimal DFA of L are exactly the right atoms of L, and dually for left atoms under reversal.

Example 1 (A small running example). Let $\Sigma = \{a, b\}$ and $L = \Sigma^* a \Sigma^*$. Consider the $\cup$-NFA $\mathcal{N} = (\{q_0, q_1\}, \{a, b\}, \delta, q_0, \{q_1\})$ with $\delta(q_0, a) = q_0 \vee q_1$, $\delta(q_0, b) = q_0$, and $\delta(q_1, a) = \delta(q_1, b) = q_1$. Reading w from $\{q_0\}$, the set of reachable states is $\{q_0\}$ if $w \in b^*$, and $\{q_0, q_1\}$ otherwise. The right languages are $RL(\mathcal{N}, q_0) = L$ and $RL(\mathcal{N}, q_1) = \Sigma^*$, and the left languages are $LL(\mathcal{N}, q_0) = \Sigma^*$ and $LL(\mathcal{N}, q_1) = L$. The right quotients of L are L and Σ^*, and the right atoms of L are $L = \Sigma^* a \Sigma^*$ and b^*. One verifies that $LL(\mathcal{N}, q_0) = L \cup b^*$ and $LL(\mathcal{N}, q_1) = L$ are unions of right atoms, so by Proposition 1 below, $\mathcal{N}^D$ is minimal—and indeed it has the expected two states.

By contrast, consider the $\cup$-NFA $\mathcal{N}' = (\{q_0, q_a, q_b\}, \{a, b\}, \delta', q_0, \{q_a, q_b\})$ for $L' = \{a, b\}$, with $\delta'(q_0, a) = q_a$ and $\delta'(q_0, b) = q_b$ (all other transitions undefined). Here $LL(\mathcal{N}', q_a) = \{a\}$ and $LL(\mathcal{N}', q_b) = \{b\}$ are both proper subsets of the right atom $\{a, b\}$ of L', so neither is a union of right atoms; the subset construction leaves q_a and q_b as equivalent DFA states, and the resulting DFA is not minimal.

We will revisit $\mathcal{N}$ in Examples 2 and 4, and $\mathcal{N}'$ in Example 5.

Definition 6 (Determinisation of BFAs). *Let $\mathcal{A} = (Q, \Sigma, \delta, \varphi_0, F)$ be a BFA with $Q = \{q_0, \ldots, q_{n-1}\}$.* Determinisation *of $\mathcal{A}$ produces a DFA $\mathcal{A}^D = (Q_D, \Sigma, \delta_D, s_0, F_D)$ as follows. Each state of $\mathcal{A}^D$ is an equivalence class of Boolean formulas over Q under logical equivalence. We represent each equivalence class by a canonical formula, for instance, in disjunctive normal form (DNF) or conjunctive normal form (CNF).*

- *The initial state is $s_0 = [\varphi_0]$, the equivalence class of the initial formula.*
- *The transition function is $\delta_D([\varphi], a) = [\varphi']$, where φ' is obtained by substituting each q_i in φ with $\delta(q_i, a)$ and simplifying to the canonical form.*
- *A state $[\varphi]$ is accepting iff $[\![\varphi]\!](F) = \mathtt{true}$.*

For a $\cup$*-NFA, this reduces to the classic subset construction, since disjunctions of states correspond to subsets. For an* $\cap$*-NFA, states of* $\mathcal{A}^D$ *correspond to subsets under the dual interpretation: a state* $S \subseteq Q$ *is accepting iff* $S \subseteq F$.

Definition 7 (Reverse of $\cup$**-NFAs).** *Let* $\mathcal{N} = (Q, \Sigma, \delta, I, F)$ *be a* $\cup$*-NFA with* $Q = \{q_0, \ldots, q_{n-1}\}$. *The* reverse *of* $\mathcal{N}$ *is a* $\cup$*-NFA* $\mathcal{N}^R = (Q, \Sigma, \delta_R, F, I)$, *where* $q_j \in \delta_R(q_i, a)$ *if and only if* $q_i \in \delta(q_j, a)$, *for* $q_i, q_j \in Q$ *and* $a \in \Sigma$. *If* $\mathcal{N}$ *accepts the language* L, *then* $\mathcal{N}^R$ *accepts the* reverse language L^R.

Next, we reformulate the result presented in [4]. Intuitively, the proposition below says that the subset construction applied to a $\cup$-NFA $\mathcal{N}$ collapses no two states of $\mathcal{N}^D$ exactly when each state of $\mathcal{N}$ "sees" a piece of the language built out of the elementary regions (right atoms): if every left language of $\mathcal{N}$ is a union of right atoms, then no two reachable subsets can have the same right language, and the DFA $\mathcal{N}^D$ obtained by determinising $\mathcal{N}$ cannot be reduced further.

Proposition 1 (Minimal DFA by determinisation). *For any* $\cup$*-NFA* $\mathcal{N}$, *the DFA* $\mathcal{N}^D$ *is minimal if and only if the left languages of* $\mathcal{N}$ *are unions of right atoms.*

Proof. In [4] it was shown that for any $\cup$-NFA $\mathcal{N}$ of a language L, $\mathcal{N}^D$ is minimal if and only if the right languages of $\mathcal{N}^R$ are unions of left atoms of L^R. Since the right languages of $\mathcal{N}^R$ are the reversed left languages of $\mathcal{N}$, and the left atoms of L^R are exactly the reversed right atoms of L, we conclude that $\mathcal{N}^D$ is minimal if and only if the left languages of $\mathcal{N}$ are unions of right atoms of L. □

Our goal is to obtain a characterisation similar to that of Proposition 1, for AFAs and BFAs.

3 Applying Negation to $\cup$-NFAs to Study $\cap$-NFAs

Our strategy in this section is to lift the classical $\cup$-NFA characterisation of Proposition 1 to $\cap$-NFAs by way of negation duality. We first show that complementing the language and negating all Boolean formulas turns a $\cup$-NFA into an $\cap$-NFA recognising the complement (Proposition 2), so determinisation and complementation commute. From this it follows that minimality is preserved on both sides (Corollary 1). Finally, we transport the classical left-language notion through the diagram of Fig. 1 to obtain the correct dual, support-based, notion of left language for $\cap$-NFAs (Definition 8), and the resulting minimality criterion (Corollary 2). A reader may keep the following picture in mind: "what works for $\cup$-NFAs, works for $\cap$-NFAs, provided we read all left languages from the supporting subset rather than from a single state".

There is a natural duality between $\cup$-NFAs and $\cap$-NFAs obtained by complementing the language and negating all Boolean formulas. Given a $\cup$-NFA $\mathcal{N} = (Q, \Sigma, \delta, \varphi_0, F)$ recognising L, define the *dual* $\cap$-NFA

$$\overline{\mathcal{N}} = (Q,\, \Sigma,\, \overline{\delta},\, \overline{\varphi_0},\, Q \setminus F),$$

where $\overline{\delta}(q,a)$ is obtained by replacing every $\vee$ in $\delta(q,a)$ with $\wedge$ (and vice versa), and $\overline{\varphi_0}$ is obtained from φ_0 in the same way. By De Morgan's laws, for any $S \subseteq Q$,

$$[\![\overline{\delta}(q,a)]\!](S) = \neg\, [\![\delta(q,a)]\!](Q \setminus S).$$

It follows that $\overline{\mathcal{N}}$ is an $\cap$-NFA recognising the complement language $\overline{L}$.

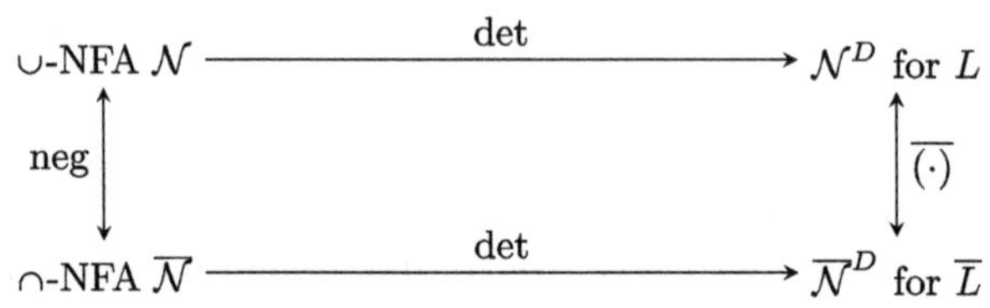

Fig. 1. Commutative diagram relating $\cup$-NFAs, $\cap$-NFAs, and DFAs. Horizontal arrows denote determinisation; vertical arrows denote negation (language complementation). The diagram commutes: determinising and then complementing yields the same DFA as negating and then determinising.

Proposition 2 (Commutativity of negation and determinisation). *Let $\mathcal{N}$ be a $\cup$-NFA recognising L. Then $\overline{\mathcal{N}^D} \cong \overline{\mathcal{N}}^D$, where $\overline{\mathcal{N}^D}$ is the DFA obtained by complementing the accept/reject status of every state of $\mathcal{N}^D$.*

The intuition is that for both a $\cup$-NFA and its negated $\cap$-NFA dual, the subset construction visits exactly the same sequence of subsets of Q on every input—the underlying reachable structure only depends on the substitution rules "replace each state q by the set of states appearing in $\delta(q,a)$", and this set does not change under De Morgan duality. Hence, the only difference between $\mathcal{N}^D$ and $\overline{\mathcal{N}}^D$ is which of these subsets are declared accepting, and that flip is precisely the language complementation.

Proof. Let $\mathcal{N} = (Q, \Sigma, \delta, \varphi_0, F)$ be a $\cup$-NFA recognising L. Both $\overline{\mathcal{N}^D}$ and $\overline{\mathcal{N}}^D$ recognise $\overline{L}$. By the subset construction, $\mathcal{N}^D$ has states $\{S \subseteq Q \mid S \text{ is reachable}\}$, with S accepting iff $S \cap F \neq \varnothing$. Then $\overline{\mathcal{N}^D}$ has the same states, with S accepting iff $S \cap F = \varnothing$, i.e. $S \subseteq Q \setminus F$. Since the set of accepting states of the $\cap$-NFA $\overline{\mathcal{N}}$ is $Q \setminus F$, any state S of the DFA $\overline{\mathcal{N}}^D$ is accepting iff $S \subseteq Q \setminus F$.

The transition function is identical in both cases: reading a from S yields the set $S' = \{q' \mid q' \text{ occurs as a state in } \delta(q,a) \text{ for some } q \in S\}$, and this set is the same whether we track disjunctions ($\cup$-NFA) or conjunctions ($\cap$-NFA) because the underlying state transitions are shared. Hence, the DFAs $\overline{\mathcal{N}^D}$ and $\overline{\mathcal{N}}^D$ are isomorphic. $\square$

Corollary 1. *The result of determinising a $\cup$-NFA $\mathcal{N}$ is a minimal DFA if and only if the result of determinising the $\cap$-NFA $\overline{\mathcal{N}}$ is a minimal DFA.*

Proof. A DFA is minimal if and only if its complement is minimal, since complementation preserves the number of states and the Myhill–Nerode equivalence classes. By Proposition 2, $\overline{\mathcal{N}}^D \cong \overline{\mathcal{N}^D}$, and the result follows. □

The commutative diagram also suggests the correct statewise notion on the $\cap$-side. Recall that, by Proposition 2, the DFAs $\mathcal{N}^D$ and $\overline{\mathcal{N}}^D$ share the same underlying state set: in both, each state is a reachable subset $S \subseteq Q$, and only the accept/reject labelling differs. We refer to this shared subset structure as the *common determinised subset structure* of $\mathcal{N}$ and $\overline{\mathcal{N}}$. If $w \in \Sigma^*$ and $S_w \subseteq Q$ is the subset reached after reading w in this common determinised subset structure (the same subset for both $\mathcal{N}^D$ and $\overline{\mathcal{N}}^D$), then S_w records the *support* of the computation after w. This motivates the following definition.

In one sentence: the support-based left language of a state q in an $\cap$-NFA is the set of words w such that, when we run the subset construction, the state q *occurs in* the subset reached after w (its "support"), as opposed to the singleton-evaluation notion that asks whether the reached formula is true under the assignment in which only q is true. The next definition makes this precise; Example 2 below shows the difference on a concrete two-state example.

Definition 8 (Support-based left languages for $\cap$-NFAs). *Let $\mathcal{N}$ be a $\cup$-NFA and let $\overline{\mathcal{N}}$ be its dual $\cap$-NFA. For a state $q \in Q$, the* support-based left language *of q in $\overline{\mathcal{N}}$ is*

$$LL^{\dagger}(\overline{\mathcal{N}}, q) := LL(\mathcal{N}, q).$$

Equivalently, if S_w is the subset reached after reading w in the common determinised subset structure of $\mathcal{N}$ and $\overline{\mathcal{N}}$ (i.e. the shared subset reached in both $\mathcal{N}^D$ and $\overline{\mathcal{N}}^D$ of Fig. 1), then

$$w \in LL^{\dagger}(\overline{\mathcal{N}}, q) \iff q \in S_w.$$

Corollary 2 (Support-based minimality criterion for $\cap$-NFAs). *For an $\cap$-NFA $\mathcal{N}_\cap$, the DFA $\mathcal{N}_\cap^D$ is minimal if and only if the support-based left languages of $\mathcal{N}_\cap$ are unions of right atoms.*

Proof. Let $\mathcal{N}_\cap$ be an $\cap$-NFA for L, and let $\mathcal{N}$ be the dual $\cup$-NFA for $\overline{L}$ such that $\mathcal{N}_\cap = \overline{\mathcal{N}}$. By Corollary 1, the DFA $\mathcal{N}_\cap^D$ is minimal if and only if $\mathcal{N}^D$ is minimal. By Proposition 1, the latter holds if and only if $LL(\mathcal{N}, q)$ is a union of right atoms for every state q. By Definition 8, this is equivalent to requiring that each $LL^{\dagger}(\mathcal{N}_\cap, q)$ is a union of right atoms. Finally, passing from L to $\overline{L}$ does not change the family of right atoms, since complementing the right quotients only swaps complemented and uncomplemented choices in the definition of the atom. □

Example 2 (Negation duality and support). Let $\Sigma = \{a, b\}$ and consider the $\cup$-NFA $\mathcal{N} = (\{q_0, q_1\}, \{a, b\}, \delta, q_0, \{q_1\})$ from Example 1 with $\delta(q_0, a) = q_0 \vee q_1$, $\delta(q_0, b) = q_0$, and $\delta(q_1, a) = \delta(q_1, b) = q_1$. The language accepted by $\mathcal{N}$ is $L = \Sigma^* a \Sigma^*$.

The dual $\cap$-NFA $\overline{\mathcal{N}} = (\{q_0, q_1\}, \{a, b\}, \overline{\delta}, q_0, \{q_0\})$ has transitions $\overline{\delta}(q_0, a) = q_0 \wedge q_1$, $\overline{\delta}(q_0, b) = q_0$, and $\overline{\delta}(q_1, a) = \overline{\delta}(q_1, b) = q_1$, and accepts the language $\overline{L} = b^*$.

The DFA $\mathcal{N}^D = (\{\{q_0\}, \{q_0, q_1\}\}, \{a, b\}, \delta_D, \{q_0\}, \{\{q_0, q_1\}\})$ with transitions $\delta_D(\{q_0\}, a) = \delta_D(\{q_0, q_1\}, a) = \delta_D(\{q_0, q_1\}, b) = \{q_0, q_1\}$ and $\delta_D(\{q_0\}, b) = \{q_0\}$ has two states, which equals the number of left quotients of L, so it is minimal.

The right quotients of L are $R_0 = L$ and $R_1 = \Sigma^*$, the right atoms are $A^R_{\{0,1\}} = R_0 \cap R_1 = L$ and $A^R_{\{1\}} = \overline{R_0} \cap R_1 = b^*$, and the left languages of $\mathcal{N}$ are $LL(\mathcal{N}, q_0) = \Sigma^* = A^R_{\{0,1\}} \cup A^R_{\{1\}}$ and $LL(\mathcal{N}, q_1) = L = A^R_{\{0,1\}}$. Hence, the left languages of $\mathcal{N}$ are right atoms, and by Proposition 1, the DFA $\mathcal{N}^D$ is minimal.

By Corollary 1, the DFA $\overline{\mathcal{N}}^D$ is also a minimal 2-state DFA, now for b^*. The support-based left languages of $\overline{\mathcal{N}}$ are $LL^\dagger(\overline{\mathcal{N}}, q_0) = \Sigma^*$ and $LL^\dagger(\overline{\mathcal{N}}, q_1) = \Sigma^* a \Sigma^*$, namely, the left languages of the dual $\cup$-NFA transported through the diagram. These are unions of the right atoms b^* and $\Sigma^* a \Sigma^*$ of b^*.

By contrast, the singleton-evaluation left languages from Definition 4 are $LL(\overline{\mathcal{N}}, q_0) = b^*$ and $LL(\overline{\mathcal{N}}, q_1) = \varnothing$, since after reading a word containing an a the reached formula is $q_0 \wedge q_1$, which is false under the singleton valuation $\{q_1\}$. This is why, for $\cap$-NFAs, the support-based notion is the correct dual analogue of the classical left language.

4 Two-Step Determinisation of AFAs

When an AFA uses both $\vee$ and $\wedge$ in its transitions, its determinisation can be decomposed into two phases by exploiting normal forms.

Definition 9 (Two-step determinisation via DNF). *Consider an AFA $\mathcal{A} = (Q, \Sigma, \delta, \varphi_0, F)$.*

Step 1: DNF Expansion to a $\cup$-NFA. *Write each transition formula $\delta(q, a)$ in disjunctive normal form (DNF). Each conjunction (clause) $c = q_{i_1} \wedge \cdots \wedge q_{i_k}$ is interpreted as a state of a new automaton. The initial formula φ_0 in DNF is a disjunction of such clauses, which become the initial states. The result is a $\cup$-NFA $\mathcal{A}_\cup$: its states are the reachable conjunctive clauses, and the transition from a clause c on input a is the disjunction of all clauses obtained by substituting each variable in c with the corresponding transition formula and re-expanding into DNF. A clause-state c is accepting iff $[\![c]\!](F) = \texttt{true}$.*

Step 2: $\cup$-Determinisation. *Apply the ordinary subset construction to $\mathcal{A}_\cup$ to obtain a DFA. A subset S of conjunctive clauses is accepting iff $S \cap F_\cup \neq \varnothing$, where $F_\cup$ is the set of accepting clause-states.*

Dually, one may start from conjunctive normal form (CNF): each disjunctive clause becomes a state of an intermediate $\cap$-NFA, after which one applies $\cap$-determinisation.

Proposition 3. *The DFA obtained by the two-step process is isomorphic to the DFA obtained by direct determinisation of the AFA, provided that logically equivalent formulas are identified at each step.*

Intuitively, both procedures end up labelling DFA states by the same equivalence classes of positive Boolean formulas: the one-shot determinisation manipulates these formulas directly, while the two-step procedure first chooses a normal form (DNF or CNF) and then runs the subset construction over the resulting clause-NFA. As long as logically equivalent formulas are identified, the two routes only differ in book-keeping.

Proof. Both processes compute the same set of reachable equivalence classes of positive Boolean formulas over Q. In direct determinisation, each DFA state is an equivalence class $[\varphi]$. In the two-step process via DNF, the intermediate $\cup$-NFA represents formulas as disjunctions of conjunctive clauses, and the subsequent subset construction collects these clauses into sets whose disjunction is logically equivalent to φ. Since two formulas label the same DFA state iff they are logically equivalent, the final DFA states are identical regardless of the decomposition path. The CNF/$\cap$-NFA presentation is dual. □

We can also state the following theorem:

Theorem 1. *Given an AFA $\mathcal{A}$, the DFA $\mathcal{A}^D$ is minimal if and only if the left languages of the states of the $\cup$-NFA $\mathcal{A}_\cup$ are unions of right atoms.* □

Proof. The theorem follows from Propositions 1 and 3.

Theorem 1 presents a criterion for obtaining a minimal DFA by determinising an AFA $\mathcal{A}$, in terms of left languages of the intermediate $\cup$-NFA $\mathcal{A}_\cup$. However, it would be desirable to find a direct state-based formulation on the original AFA.

Remark 2. The two-step perspective decomposes the combinatorial complexity: there are at most 2^n distinct conjunctive clauses (and dually 2^n disjunctive clauses) when $n = |Q|$. Applying determinisation to the intermediate $\cup$-NFA or $\cap$-NFA then yields at most 2^{2^n} subset states. The overall worst-case bound for direct AFA determinisation remains $D(n)$, the n-th Dedekind number (see [2]); the point of the two-step view is structural rather than asymptotic.

Example 3. (Two-step determinisation). Consider the AFA $\mathcal{A} = (\{p,q\}, \{a\}, \delta, p \vee q, \{p\})$ with $\delta(p,a) = p \wedge q$ and $\delta(q,a) = p \vee q$.

Step 1 (DNF expansion to a $\cup$-NFA). The initial formula $p \vee q$ is already in DNF with clauses $\{p\}$ and $\{q\}$, which are the initial states of the intermediate $\cup$-NFA. The reachable clause-states are $\{p\}$, $\{q\}$, and $\{p,q\}$. The transitions are: $\{p\} \xrightarrow{a} \{p,q\}, \{q\} \xrightarrow{a} \{p\}, \{q\} \xrightarrow{a} \{q\}, \{p,q\} \xrightarrow{a} \{p,q\}$. With the final set $F = \{p\}$, only the clause $\{p\}$ is accepting, since $[\![p]\!](F) = \mathit{true}$, while $[\![q]\!](F) = \mathit{false}$ and $[\![p \wedge q]\!](F) = \mathit{false}$, so the clause-states $\{q\}$ and $\{p,q\}$ are rejecting.

Step 2 ($\cup$-Determinisation*).* Starting from the subset $\{\{p\},\{q\}\}$, which is accepting because it contains $\{p\}$, reading a yields $\{\{p,q\},\{p\},\{q\}\}$, which is also accepting. Reading a again from this subset returns the same subset. Thus, the subset construction yields two reachable accepting states. However, these two subset states represent the same Boolean formula: $p \vee q$ and $(p \wedge q) \vee p \vee q \equiv p \vee q$. Once logically equivalent formulas are identified, the two-step process constructs the same 1-state DFA as the direct determinisation. This is precisely the logical equivalence identification of Proposition 3.

The two-step framework extends naturally to BFAs in Sect. 5, where each Boolean formula is expressed by its full DNF/CNF.

5 Minterm and Maxterm Automata

The two-step approach of Sect. 4 works for AFAs because positive Boolean formulas can be put in DNF (or CNF) using only the original states as literals. For a BFA, transitions may also use negation, so a generic Boolean formula on n states is no longer captured by "conjunctions of states"—it must be expanded to a Boolean combination of all 2^n possible assignments to the state variables. The natural objects for this expansion are the *minterms* and *maxterms*: a minterm M_P is the unique conjunction asserting that the states in P are true and those in $Q \setminus P$ are false, so each minterm singles out exactly one row of the truth table of a Boolean function over Q. Dually, a maxterm V_P singles out one row by a single disjunctive clause.

The minterm $\cup$-NFA $\mathcal{M}$ defined below is then "the canonical $\cup$-NFA hidden inside a BFA": its states are the minterms of Q, the writing of φ_0 in full DNF picks out its initial states, and the transition rule simply asks, for each input symbol a, which minterms appear in the full DNF expansion of the formula obtained by combining the transitions $\tau(q_i, a)$ in the way prescribed by a given minterm M_P. The maxterm $\cap$-NFA $\mathcal{V}$ is defined dually using full CNF. The key property (Proposition 7) is that determinising any of $\mathcal{B}$, $\mathcal{M}$ or $\mathcal{V}$ produces the same DFA, so to study minimality of $\mathcal{B}^D$ we may always pass to the minterm $\cup$-NFA $\mathcal{M}$ and apply the classical $\cup$-NFA criterion.

Let $\mathcal{B} = (Q, \Sigma, \tau, \varphi_0, F)$ be a BFA for a regular language L, with the set of states $Q = \{q_1, \ldots, q_n\}$. For any subset $P \subseteq Q$ of states, we define the conjunction

$$M_P = \bigwedge_{q_i \in P} q_i \wedge \bigwedge_{q_i \in Q \setminus P} \neg q_i$$

and we call any such conjunction a *minterm* of Q. Let M be the set of minterms of Q.

Let the initial function φ_0 of $\mathcal{B}$ and the transition function $\tau(q_i, a)$, for $q_i \in Q$ and $a \in \Sigma$, be given in full DNF, consisting of a disjunction of minterms. For a Boolean function φ, we denote the set of minterms involved in its full DNF, by $Minterms(\varphi)$.

We define the *minterm $\cup$-NFA* $\mathcal{M} = (M, \Sigma, \delta, Minterms(\varphi_0), \{M_F\})$ based on the BFA $\mathcal{B}$ in the following way: we use the set M of minterms of Q as the set of states, the set of minterms of the initial function φ_0 as the set of the initial states, and the minterm $M_F = \bigwedge_{q_i \in F} q_i \wedge \bigwedge_{q_i \in Q \setminus F} \neg q_i$ as the single final state. The transition function is given by

$$\delta(M_P, a) = Minterms(\bigwedge_{q_i \in P} \tau(q_i, a) \wedge \bigwedge_{q_i \in Q \setminus P} \neg(\tau(q_i, a)))$$

for $M_P \in M$ and $a \in \Sigma$.

Similarly to the above, for a subset $P \subseteq Q$, we define the disjunction

$$V_P = \bigvee_{q_i \in P} q_i \vee \bigvee_{q_i \in Q \setminus P} \neg q_i$$

and we call such a disjunction a *maxterm* of Q. We denote the set of maxterms of Q by V.

The initial function φ_0 of $\mathcal{B}$ and the transition function $\tau(q_i, a)$, for $q_i \in Q$ and $a \in \Sigma$, can be presented in full CNF, consisting of a conjunction of maxterms. We denote the set of maxterms in the full CNF of a Boolean function φ, by $Maxterms(\varphi)$.

We define the *maxterm* $\cap$-NFA $\mathcal{V} = (V, \Sigma, \eta, Maxterms(\varphi_0), V \setminus V_{Q \setminus F})$ based on the BFA $\mathcal{B}$ as follows: we use the set V of maxterms of Q as the set of states, the set of maxterms of the initial function φ_0 as the set of initial states, and the set $V \setminus V_{Q \setminus F}$ as the set of final states. The transition function is given by

$$\eta(V_P, a) = Maxterms(\bigvee_{q_i \in P} \tau(q_i, a) \vee \bigvee_{q_i \in Q \setminus P} \neg(\tau(q_i, a)))$$

for $V_P \in V$ and $a \in \Sigma$.

We note that the maxterm $V_{Q \setminus F} = \bigvee_{q_i \in Q \setminus F} q_i \vee \bigvee_{q_i \in F} \neg q_i$ is the only maxterm that is not included in the set of final states of $\mathcal{V}$, and it is the only maxterm such that its language does not contain the empty word.

We also note that there is an obvious one-to-one correspondence $M_P \leftrightarrow V_{Q \setminus P}$ between minterms M_P and maxterms $V_{Q \setminus P}$, such that $\neg M_P = V_{Q \setminus P}$. Clearly, the following proposition holds:

Proposition 4. *For any Boolean function φ, $M_P \in Minterms(\varphi)$ holds if and only if $V_{Q \setminus P} \in Maxterms(\neg\varphi)$ holds, where $P \subseteq Q$.*

Consequently, a minterm M_P is initial if and only if the maxterm $V_{Q \setminus P}$ is not initial, and similarly, M_P is final if and only if $V_{Q \setminus P}$ is not final. Also, the transition functions of $\mathcal{M}$ and $\mathcal{V}$ are symmetric: $M_R \in \delta(M_P, a)$ holds if and only if $V_{Q \setminus R} \in \eta(V_{Q \setminus P}, a)$ holds.

One observes that the minterm $\cup$-NFA of a language L and the maxterm $\cap$-NFA of the complement language $\overline{L}$ have a similar structure. This correspondence can also be observed in the diagram of Fig. 1.

Proposition 5. *The reverse automaton of $\mathcal{M}$ is deterministic.*

The intuition is that minterms are mutually exclusive Boolean functions, so for a fixed input symbol a and a fixed target minterm M_P, the source minterm in $\mathcal{M}$ from which M_P can possibly arise, is determined uniquely by asking, for each state q_i, whether M_P appears in the full DNF of $\tau(q_i, a)$ or in that of its negation; only one of the two can hold. Combined with the fact that $\mathcal{M}$ has a single final state, this gives a deterministic reverse.

Proof. Since the minterm ∪-NFA $\mathcal{M}$ has a single final state, the reverse automaton $\mathcal{M}^R$ has one initial state. We also show that for every minterm $M_P \in M$ and every $a \in \Sigma$, there is exactly one minterm $M_S \in M$ such that $M_P \in \delta(M_S, a)$.

Let $M_P \in M$ be a minterm of Q. Clearly, for a Boolean function φ over Q, M_P is a minterm of either φ or $\neg\varphi$. However, it cannot be a minterm of both functions. Hence, precisely one of the following two conditions, either $M_P \in Minterms(\tau(q_i, a))$ or $M_P \in Minterms(\neg\tau(q_i, a))$, holds for any $q_i \in Q$ and $a \in \Sigma$. Let $S = \{q_i \in Q \mid M_P \in Minterms(\tau(q_i, a))\}$. Then $M_P \in \delta(M_S, a)$ holds, and clearly, there is no other minterm $M_{S'} \neq M_S$ such that $M_P \in \delta(M_{S'}, a)$. Hence, $\mathcal{M}^R$ is deterministic. □

We note that the left/right languages of some states of $\mathcal{M}$ may be empty.

Boolean atoms [8] are non-empty intersections of uncomplemented and complemented right languages of a BFA. That is, a Boolean atom of $\mathcal{B}$ is any non-empty intersection $B_P = \bigcap_{q_i \in P} RL(\mathcal{B}, q_i) \cap \bigcap_{q_i \in Q \setminus P} \overline{RL(\mathcal{B}, q_i)}$, where $P \subseteq Q$. Clearly, $B_P \cap B_{P'} = \varnothing$ if $P \neq P'$, and $\bigcup_{P \subseteq Q} B_P = \Sigma^*$.

Proposition 6. *If the right language of a state M_P of $\mathcal{M}$ is non-empty, then it is the Boolean atom B_P.*

The minterms of Q tile Σ^* into disjoint pieces—one piece per assignment of truth values to the states—and so do the Boolean atoms of $\mathcal{B}$. The proposition asserts that the minterm ∪-NFA realises this tiling state-by-state: the right language of M_P, when non-empty, picks out exactly the words for which the assignment "state q_i accepts iff $q_i \in P$" is correct, which is the definition of the Boolean atom B_P.

Proof. We show that for any word $w \in \Sigma^*$, there is exactly one minterm M_P, such that $w \in RL(\mathcal{M}, M_P)$, and equivalently, $w \in B_P$.

The proof is by induction on the length of w. If $|w| = 0$, then $w = \varepsilon$. By the definition of $\mathcal{M}$, we know that $\varepsilon \in RL(\mathcal{M}, M_P)$ if and only if $P = F$. Also, since $\varepsilon \in RL(\mathcal{B}, q_i)$ if and only if $q_i \in F$, we obtain that $\varepsilon \in B_F$.

Now, assume that for a word $w \in \Sigma^*$, there is a minterm M_T such that $w \in RL(\mathcal{M}, M_T)$, and $w \in B_T$. As shown in Proposition 5, for every $a \in \Sigma$, there is exactly one minterm M_S, with $S = \{q_i \in Q \mid M_T \in Minterms(\tau(q_i, a))\}$, such that $M_T \in \delta(M_S, a)$. Hence, $aw \in RL(\mathcal{M}, M_S)$ and there is no other minterm that would have aw in its right language. Also, $aw \in B_S$ holds, and since Boolean atoms are disjoint, there is no other set $S' \subseteq Q$, $S' \neq S$, such that $aw \in B_{S'}$ would hold. Hence, the right language of any non-empty state M_P of $\mathcal{M}$ is the Boolean atom B_P. □

Similarly to the concept of *partial átomaton* of an NFA [4], one could define the *Boolean átomaton* of $\mathcal{B}$ as a ∪-NFA with its states corresponding to Boolean atoms. Then, the subautomaton of $\mathcal{M}$ that involves the states with non-empty right languages, together with their transitions, is isomorphic to the Boolean átomaton.

Proposition 7. *The DFAs $\mathcal{M}^D$, $\mathcal{V}^D$ and $\mathcal{B}^D$ are isomorphic to one another.*

Proof. The proof is similar to the proof of Proposition 3. The states of $\mathcal{M}^D$, $\mathcal{V}^D$ and $\mathcal{B}^D$ are the same reachable equivalence classes of Boolean formulas over Q. □

Theorem 2. *Given a BFA $\mathcal{B}$, the DFA $\mathcal{B}^D$ is minimal if and only if the left languages of the minterm $\cup$-NFA $\mathcal{M}$ are unions of right atoms.*

Proof. The theorem follows from Propositions 1 and 7. □

6 Discussion

We now briefly discuss the criteria for obtaining a minimal DFA by applying determinisation to AFAs and BFAs, together with future challenges in this area.

For $\cup$-NFAs we have the classical criterion of Proposition 1. For $\cap$-NFAs, the singleton-evaluation from Definition 4 already fails to serve as the correct dual notion. Instead, Corollary 2 provides the dual statement, using the support-based concept of Definition 8.

For AFAs, Theorem 1 presents a characterisation in terms of left languages of the intermediate $\cup$-NFA of the two-step determinisation. However, it would be desirable to obtain a stronger and more direct statement about the original AFA states. This is not trivial, because for a general AFA, a naive occurrence-based notion of "support" depends on the chosen formula representation: logically equivalent positive formulas can mention different sets of states. Thus, the full AFA case requires additional work.

Remark 3 (Support-based minimality for AFAs). This suggests the existence of a canonical support-based notion of left language for states of an AFA that reduces to the usual left language for $\cup$-NFAs and to Definition 8 for $\cap$-NFAs, such that the determinised version of an AFA is minimal if and only if these support-based left languages are unions of right atoms.

The following examples provide useful evidence. In the $\cup$-case, they are already covered by Proposition 1, and hence are also consistent with Remark 3.

Example 4 (An AFA yielding a minimal DFA). Let us recall the $\cup$-NFA $\mathcal{N} = (\{q_0, q_1\}, \{a, b\}, \delta, q_0, \{q_1\})$ for $L = \Sigma^* a \Sigma^*$ from Example 1, with $\delta(q_0, a) = q_0 \vee q_1$, $\delta(q_0, b) = q_0$, and $\delta(q_1, a) = \delta(q_1, b) = q_1$. As shown in Example 2, the left languages are $LL(\mathcal{N}, q_0) = \Sigma^* = A^R_{\{0,1\}} \cup A^R_{\{1\}}$ and $LL(\mathcal{N}, q_1) = L = A^R_{\{0,1\}}$, both unions of right atoms. By Proposition 1, the 2-state DFA $\mathcal{N}^D$ is minimal. Viewed merely as an AFA, this is consistent with Remark 3.

Example 5 (An AFA yielding a non-minimal DFA). Consider the $\cup$-NFA $\mathcal{N}' = (\{q_0, q_a, q_b\}, \{a, b\}, \delta', q_0, \{q_a, q_b\})$ for $L' = \{a, b\}$ from Example 1, with $\delta'(q_0, a) = q_a$, $\delta'(q_0, b) = q_b$. The left languages $LL(\mathcal{N}', q_a) = \{a\}$ and $LL(\mathcal{N}', q_b) = \{b\}$ are proper subsets of the right atom $A^R_{\{1\}} = \{a, b\}$ and are not unions of right atoms. The DFA $\mathcal{N}'^D$ (with 4 states) is not minimal (the minimal DFA has 3 states) as established by Proposition 1. Again, this is consistent with Remark 3.

For BFAs, no direct atomic criterion on the original states is currently available. Determinisation of a BFA produces equivalence classes of *non-monotone* Boolean functions, and the monotone correspondence between formula classes and right atoms may fail to hold.

However, as shown in Sect. 5, every Boolean automaton $\mathcal{B}$ has an associated minterm $\cup$-NFA $\mathcal{M}$ and maxterm $\cap$-NFA $\mathcal{V}$, with the DFAs $\mathcal{B}^D$, $\mathcal{M}^D$ and $\mathcal{V}^D$ being isomorphic to one another (Proposition 7). Theorem 2 provides the characterisation for obtaining a minimal DFA in terms of left languages of the minterm $\cup$-NFA $\mathcal{M}$. This reduces the question of minimality after determinisation to the properties of the associated minterm $\cup$-NFA—a classical $\cup$-NFA, for which the known theory applies.

Example 6 (A BFA producing a non-minimal DFA). Consider the BFA $\mathcal{B} = (\{p,q\}, \{a\}, \tau, p, \{p\})$ with $\tau(p,a) = \neg q, \tau(q,a) = \neg p$. Reading a from the initial formula p: $\delta^*(p,a) = \neg q$, and $[\![\neg q]\!](\{p\}) = \neg\,\mathtt{false} = \mathtt{true}$, so a is accepted. Reading aa: $\delta^*(p,aa) = \neg(\neg p) \equiv p$, so aa is accepted. By induction, every string is accepted, and $L = a^*$. The minimal DFA has one state, but the determinisation of $\mathcal{B}$ produces DFA states $[p]$ and $[\neg q]$, which are not logically equivalent (they are different Boolean functions of p,q). Hence, $\mathcal{B}^D$ has 2 states and is not minimal.

By Theorem 2, we can check minimality via the minterm $\cup$-NFA $\mathcal{M}$ of $\mathcal{B}$. The minterms are $M_{\{p,q\}}, M_{\{p\}}, M_{\{q\}}, M_{\varnothing}$. The initial formula p in full DNF is $M_{\{p,q\}} \vee M_{\{p\}}$. One can verify that $\mathcal{M}^D$ is also not minimal for a^*, confirming that the left languages of $\mathcal{M}$ are not all unions of right atoms.

7 Conclusions

We considered the problem of what conditions are required so that determinising an AFA or a BFA yields a minimal DFA. For $\cup$-NFAs, Proposition 1 recovers the classical atomic characterisation. For $\cap$-NFAs, we used the support-based left languages induced via the negation duality of Fig. 1. For general AFAs, this support-based viewpoint suggests a natural extension, but we leave this to future work. Instead, we presented a characterisation via left languages of the intermediate $\cup$-NFA of the two-step determinisation for AFAs. Also, for BFAs, the criterion passes through the associated minterm $\cup$-NFA.

We established a duality between $\cup$-NFAs and $\cap$-NFAs via negation of Boolean formulas, and showed that this duality commutes with determinisation. We presented the two-step determinisation for AFAs, using an intermediate $\cup$-NFA on conjunctive clauses of the Boolean formulas in DNF. When using CNF, the same approach leads dually to an intermediate $\cap$-NFA on disjunctive clauses. This viewpoint was extended to the minterm and maxterm constructions for BFAs.

The minterm and maxterm automata provide canonical $\cup$-NFA and $\cap$-NFA representations of any BFA, bridging the Boolean and classical automata-theoretic worlds. The connection to Boolean atoms [8] offers further structural insight.

Acknowledgements. Thank you to Lynette van Zijl for contributing to our earlier discussions on this topic.

References

1. Bell, J.P., Smertnig, D., Tamm, H.: Duality of lattices associated to left and right quotients. In: Gazdag, Z., Iván, S., Kovásznai, G. (eds.) Proceedings of the 16th International Conference on Automata and Formal Languages, AFL 2023, Eger, Hungary, September 5-7, 2023. EPTCS, vol. 386, pp. 35–50 (2023). https://doi.org/10.4204/EPTCS.386.5
2. Berglund, M., van der Merwe, B., van Litsenborgh, S.: Regular expressions with lookahead. J. Univers. Comput. Sci. **27**(4), 324–340 (2021). https://doi.org/10.3897/JUCS.66330
3. Brzozowski, J.A., Leiss, E.L.: On equations for regular languages, finite automata, and sequential networks. Theor. Comput. Sci. **10**, 19–35 (1980). https://doi.org/10.1016/0304-3975(80)90069-9
4. Brzozowski, J.A., Tamm, H.: Theory of átomata. Theor. Comput. Sci. **539**, 13–27 (2014). https://doi.org/10.1016/J.TCS.2014.04.016
5. Chandra, A.K., Kozen, D., Stockmeyer, L.J.: Alternation. J. ACM **28**(1), 114–133 (1981). https://doi.org/10.1145/322234.322243
6. Fellah, A., Jürgensen, H., Yu, S.: Constructions for alternating finite automata. Int. J. Comput. Math. **35**(1–4), 117–132 (1990). https://doi.org/10.1080/00207169008803893
7. van der Merwe, B., Tamm, H., van Zijl, L.: Minimal DFA for symmetric difference NFA. In: Kutrib, M., Moreira, N., Reis, R. (eds.) Descriptional Complexity of Formal Systems - 14th International Workshop, DCFS 2012, Braga, Portugal, 23–25 July 2012. Proceedings, pp. 307–318. Lecture Notes in Computer Science, Springer (2012). https://doi.org/10.1007/978-3-642-31623-4_24
8. Tamm, H.: Boolean automata and atoms of regular languages. In: Bonchi, F., Puglisi, S.J. (eds.) 46th International Symposium on Mathematical Foundations of Computer Science, MFCS 2021, 23–27 August 2021, Tallinn, Estonia. LIPIcs, vol. 202, pp. 86:1–86:13. Schloss Dagstuhl - Leibniz-Zentrum für Informatik (2021). https://doi.org/10.4230/LIPICS.MFCS.2021.86

Author Index

T. J. Smith (Ed.): CIAA 2026, LNCS 16695, p. 217, 2026.
https://doi.org/10.1007/978-3-032-31176-4

MIX
Papier aus verantwortungsvollen Quellen
Paper from responsible sources
FSC® C105338

FSC
www.fsc.org

If you have any concerns about our products,
you can contact us on
ProductSafety@springernature.com

In case Publisher is established outside the EU,
the EU authorized representative is:
Springer Nature Customer Service Center GmbH
Europaplatz 3, 69115 Heidelberg, Germany

Printed by Libri Plureos GmbH
in Hamburg, Germany